INVERSE HEAT CONDUCTION

INVERSE HEAT CONDUCTION

Ill-posed Problems

JAMES V. BECK
Department of Mechanical Engineering
Michigan State University

BEN BLACKWELL
Sandia National Laboratories
Albuquerque, New Mexico

CHARLES R. ST. CLAIR, JR.
Department of Mechanical Engineering
Michigan State University

A Wiley-Interscience Publication
New York • Chichester • Brisbane • Toronto • Singapore

Copyright © 1985 by John Wiley & Sons, Inc.

All rights reserved. Published simultaneously in Canada.

Reproduction or translation of any part of this work
beyond that permitted by Section 107 or 108 of the
1976 United States Copyright Act without the permission
of the copyright owner is unlawful. Requests for
permission or further information should be addressed to
the Permissions Department, John Wiley & Sons, Inc.

Library of Congress Cataloging in Publication Data:
Beck, J. V. (James Vere), 1930–
  Inverse heat conduction.

  "A Wiley-Interscience publication."
  Includes bibliographies and index.
   1. Heat—Conduction.   2. Numerical analysis—
Improperly posed problems.   I. Blackwell, Ben.
II. St. Clair, Charles R.   III. Title.
QC320.B4   1985   536',23   85-5391
ISBN 0-471-08319-4

Printed in the United States of America

10 9 8 7 6 5 4 3 2 1

To my wife, Barbara; children, Sharon and Douglas; and father and mother, Peter and Louise Beck

J.V.B.

To my wife, Betty; and children, Jeffrey and Gregory Blackwell

B.B.

To the greatest engineers in the family: Charles R. St. Clair, Sr. and Deborah S. Short (my daughter). And to those for whom the engineers live and labor: my mother Erla, my wife Jeanette, and our children and grandchildren from oldest to youngest with the greatest in doubt: Charles III, Scott, Judy, Timothy B. Short, Gregory, and Kevin. And to those who are as family: John and Ann Polomsky

C.R.S.

PREFACE

This book presents a study of the *I*nverse *H*eat *C*onduction *P*roblem (IHCP), which is the estimation of the surface heat flux history of a heat conducting body. Transient temperature measurements *inside* the body are utilized in the calculational procedure. The presence of errors in the measurements as well as the ill-posed nature of the problem lead to "estimates" rather than the "true" surface heat flux and/or temperature.

This book was written because of the importance and practical nature of the IHCP; furthermore, at the time of writing there is no available book on the subject written in English. The specific problem treated is only one of many ill-posed problems but the techniques discussed herein can be applied to many others. The basic objective is to estimate a function given measurements that are "remote" in some sense. Other applications include remote sensing, oil exploration, nondestructive evaluation of materials, and determination of the Earth's interior structure.

The authors became interested in the IHCP over two decades ago while employed in the aerospace industry. One of the applications was the determination of the surface heat flux histories of reentering heat shields.

This book is written as a textbook in engineering with numerical examples and exercises for students. These examples will be useful to practicing engineers who use the book to become acquainted with the problem and methods of solution. A companion book, *Parameter Estimation in Engineering and Science* by J. V. Beck and K. J. Arnold (Wiley, 1977), discusses estimation of certain constants or parameters rather than functions as in the IHCP. Though many of the ideas relating to least squares and sensitivity coefficients are present in both books, the present book does not require a mastery of parameter estimation.

The book is written at the advanced B.S. or the M.S. level. A course in heat conduction at the M.S. level or courses in partial differential equations and numerical methods are recommended as prerequisite materials.

Our philosophy in writing this book was to emphasize general techniques rather than specialized procedures unique to the IHCP. For example, basic techniques developed in Chapter 4 can be applied either to integral equation representations of the heat diffusion phenomena or to finite difference (or element) approximations of the heat conduction equation. The basic procedures in Chapter 4 can treat nonlinear cases, multiple sensors, nonhomogeneous media, multidimensional bodies, and many equations, in addition to the transient heat conduction equation.

The two general procedures that are used are called (a) function specification and (b) regularization. A method of combining these (the trial function method) is also suggested. One of the important contributions of this book is the demonstration that all of these methods can be implemented in a sequential manner. The sequential method in some case gives nearly the same result as whole domain estimation and yet is much more computationally efficient.

One of our goals was to provide the reader with an insight into the basic procedures that provide analytical tools to compare various procedures. We do this by using the concepts of sensitivity coefficients, basic test cases, and the mean squared error. The reader is also shown that optimal estimation involves the compromise between minimum sensitivity to random measurement errors and the minimum bias.

Preliminary notes have been used for an ASME short course and for a graduate course at Michigan State University.

There are many people who have helped in the preparation of this text and to whom we express our appreciation. These include D. Murio, M. Raynaud, and other colleagues and students who have read and commented on the notes. Thanks are also due to Judy Duncan, Phyllis Murph, Terese Stuckman, Alice Montoya, and Jeana Pineau, who have aided in typing the manuscript.

James V. Beck wishes to express appreciation for the contributions to his education made by Kenneth Astill of Tufts University, Warren Rohsenow of Massachusetts Institute of Technology, and A. M. Dhanak of Michigan State University.

Ben Blackwell would like to acknowledge the contributions that several people made to his heat transfer education: H. Wolf of the University of Arkansas, M. W. Wildin of the University of New Mexico, and W. M. Kays of Stanford University.

A special and deep appreciation is extended to George A. Hawkins for the education and philosophy that he imparted to Charles R. St. Clair, Jr. as his graduate student.

JAMES V. BECK
BEN BLACKWELL
CHARLES R. ST. CLAIR, JR.

East Lansing, Michigan
Albuquerque, New Mexico
East Lansing, Michigan
August 1985

CONTENTS

Nomenclature — xv

1. DESCRIPTION OF THE INVERSE HEAT CONDUCTION PROBLEM — 1

- 1.1 Introduction, 1
- 1.2 Examples of Inverse Problems, 3
  - 1.2.1 Inverse Heat Conduction Problem Examples, 3
  - 1.2.2 Other Inverse Function Estimation Problems, 7
- 1.3 Function Estimation Versus Parameter Estimation, 9
- 1.4 Measurements, 9
  - 1.4.1 Description of Measurement Errors, 9
  - 1.4.2 Statistical Description of Errors, 10
- 1.5 Why is the IHCP Difficult?, 13
  - 1.5.1 Sensitivity to Errors, 13
  - 1.5.2 Examples of Damping and Lagging; Exact Solutions, 13
- 1.6 Sensitivity Coefficients, 19
  - 1.6.1 Definition of Sensitivity Coefficients and Linearity, 19
  - 1.6.2 One-Dimensional Sensitivity Coefficient Examples, 22
    - 1.6.2.1 Lumped Body Case, 22
    - 1.6.2.2 Semi-Infinite Body, 25
    - 1.6.2.3 Plate Insulated on One Side, 30
  - 1.6.3 Two-Dimensional Sensitivity Coefficient Example, 31
- 1.7 Classification of Methods, 36
- 1.8 Criteria for Evaluation of IHCP Methods, 38
- 1.9 Scope of Book, 39

References, 40

Problems, 43

2. EXACT SOLUTIONS OF THE INVERSE HEAT CONDUCTION PROBLEM   51

  2.1   Introduction, 51
  2.2   Steady-State Solution, 52
  2.3   Transient Analysis of Bodies with Small Internal Thermal Resistance, 54
        2.3.1   Exact Solution, 54
        2.3.2   Approximate Solutions, 54
        2.3.3   Temperature Errors and Approximate Solutions, 55
  2.4   Heat Flux From Measured Surface Temperature History, 59
        2.4.1   Exact Results for Continuous Surface Temperature History, 59
        2.4.2   Approximate Results for Semi-Infinite Body with Surface Temperature Measured at Discrete Times, 61
        2.4.3   Temperature Error Propagation in Eq. (2.4.8), 63
  2.5   Exact Solutions of Inverse Heat Conduction Problems, 67
        2.5.1   Literature Review, 67
        2.5.2   Derivation of Exact Solution for Planar Geometry, 67
        2.5.3   Expressions for Cylinders and Spheres, 71
        2.5.4   Example Results for Planar Geometry, 72
  References, 75
  Problems, 76

3. APPROXIMATE METHODS FOR DIRECT HEAT CONDUCTION PROBLEMS   78

  3.1   Introduction, 78
        3.1.1   Various Numerical Approaches, 78
        3.1.2   Scope of Chapter, 79
  3.2   Duhamel's Theorem, 80
        3.2.1   Derivation of Duhamel's Theorem, 80
        3.2.2   Numerical Approximation of Duhamel's Theorem, 82
        3.2.3   Matrix Form of Duhamel's Theorem, 83
  3.3   Difference Methods, 87
        3.3.1   Finite Control Volume Procedure for Constant Property Planar Geometries, 87
        3.3.2   Other Boundary Conditions and Material Interfaces, 93
        3.3.3   Numerical Techniques for Solving Systems of First-Order Ordinary Differential Equations, 94
        3.3.4   General Form of Difference Equations for Heat Conduction in Planar Body, 96
        3.3.5   Standard Form for Temperature Equation for IHCP, 99
  References, 102
  Problems, 103

4. INVERSE HEAT CONDUCTION ESTIMATION PROCEDURES   108

- 4.1 Introduction, 108
- 4.2 Ill-Posed Problems, 110
  - 4.2.1 Partial Differential Equation Perspective, 110
  - 4.2.2 Integral Equation Perspective, 112
  - 4.2.3 Difference Equation Perspective, 113
- 4.3 Single Future Time Step Method, 115
  - 4.3.1 Introduction, 115
  - 4.3.2 Exact Matching of Measured Temperatures (Single Sensor), 115
  - 4.3.3 Multiple Temperature Sensors, 118
- 4.4 Function Specification Method, 119
  - 4.4.1 Introduction, 119
  - 4.4.2 Whole Domain Estimation, 119
    - 4.4.2.1 Smoothly Changing Heat Flux, 120
    - 4.4.2.2 Abruptly Changing Heat Flux Histories, 122
  - 4.4.3 Sequential Estimation, 125
    - 4.4.3.1 Constant Heat Flux Functional Form, 125
    - 4.4.3.2 Linear Heat Flux Functional Form, 131
    - 4.4.3.3 Alternative Interpretation, 133
- 4.5 Regularization Method, 134
  - 4.5.1 Introduction, 134
  - 4.5.2 Physical Significance of Regularization Terms, 135
  - 4.5.3 Whole Domain Regularization Method, 137
    - 4.5.3.1 Algebraic Formulation, 137
    - 4.5.3.2 Matrix Formulation, 138
    - 4.5.3.3 Selection of Regularization Parameter, 140
  - 4.5.4 Sequential Regularization Method, 141
- 4.6 Trial Function Method, 145
  - 4.6.1 Introduction, 145
  - 4.6.2 Matrix Analysis, 145
  - 4.6.3 Zeroth-Order Regularization Method, 147
  - 4.6.4 Generalized Sequential Function Specification Method, 147
- 4.7 Filter Form of Linear IHCP, 148
  - 4.7.1 Introduction, 148
  - 4.7.2 Sequential Filter Algorithm, 148
  - 4.7.3 Prefiltering Temperature Measurements, 153
- 4.8 Two Conflicting Objectives, 153
  - 4.8.1 Minimum Deterministic Bias, 153
  - 4.8.2 Minimum Sensitivity to Random Errors, 154
  - 4.8.3 Mean Squared Error, 154
  - 4.8.4 Variance of Estimated Heat Flux Component, 156
  - 4.8.5 Estimate of Deterministic Error in Surface Heat Flux, 157

References, 159
Problems, 161

5. INVERSE CONVOLUTION PROCEDURES FOR A SINGLE SURFACE HEAT FLUX   165

- 5.1 Introduction, 165
- 5.2 Test Cases, 167
  - 5.2.1 Introduction, 167
  - 5.2.2 Step Change in Surface Heat Flux, 168
  - 5.2.3 Triangular Heat Flux, 169
  - 5.2.4 Random Errors, 170
  - 5.2.5 Heat Flux Impulse Test Case ($\delta \hat{q}_M/\delta q_r$), 173
  - 5.2.6 Temperature Impulse Test Case ($\delta \hat{q}_M/\delta Y_r$), 174
  - 5.2.7 Test Cases with Units, 174
- 5.3 Function Specification Algorithms, 176
  - 5.3.1 Introduction, 176
  - 5.3.2 Single Future Temperature Algorithm (Stolz Method), 176
    - 5.3.2.1 Step Heat Flux Test Case, 177
    - 5.3.2.2 Triangular Heat Flux Test Case, 177
    - 5.3.2.3 Heat Flux Impulse Test Case ($\delta \hat{q}_M/\delta q_r$), 178
    - 5.3.2.4 Temperature Impulse Test Case ($\delta \hat{q}_M/\delta Y_r$), 179
  - 5.3.3 Multiple Future Temperatures Algorithm, 181
    - 5.3.3.1 Step Heat Flux Test Case, 181
    - 5.3.3.2 Triangular Heat Flux Test Case, 182
    - 5.3.3.3 Heat Flux Impulse Test Case ($\delta \hat{q}_M/\delta q_r$), 184
    - 5.3.3.4 Temperature Impulse Test Case ($\delta \hat{q}_M/\delta Y_r$), 184
- 5.4 Regularization Algorithms, 186
  - 5.4.1 Introduction, 186
  - 5.4.2 Whole Domain Regularization Method, 187
    - 5.4.2.1 Triangular Heat Flux Test Case, 189
    - 5.4.2.2 Heat Flux Impulse Test Case, 190
    - 5.4.2.3 Temperature Impulse Test Case ($\delta \hat{q}_M/\delta Y_r$), 191
  - 5.4.3 Sequential Regularization Method, 191
    - 5.4.3.1 Triangular Heat Flux Test Case, 193
    - 5.4.3.2 Heat Flux Impulse Test Case, 194
    - 5.4.3.3 Temperature Impulse Test Case, 194
    - 5.4.3.4 Comparison of Whole Domain and Sequential Regularization Methods, 196
    - 5.4.3.5 Comparison of Sequential Regularization and Function Specification Methods, 196

CONTENTS   xiii

    5.5   Digital Filter Algorithm, 196
           5.5.1   Introduction, 196
           5.5.2   Function Specification-Based Filter, 197
                  5.5.2.1   Finite Plate Case, 197
                  5.5.2.2   Semi-Infinite Body, 200
           5.5.3   Whole Domain Regularization Filter, 201
    5.6   Optimal Considerations, 203
           5.6.1   Optimal Function Specification Algorithm, 204
           5.6.2   Optimal Whole Domain Regularization Method, 210
References, 212
Problems, 213

6. DIFFERENCE METHODS FOR THE SOLUTION OF THE ONE-DIMENSIONAL INVERSE HEAT CONDUCTION PROBLEM   218

    6.1   Introduction, 218
    6.2   Sensitivity Coefficients and Their Calculation by Difference Methods, 219
    6.3   Single Temperature Sensor, Function Specification ($q=C$), Single Future Time Step (Exact Matching of Data), 222
           6.3.1   Modification of Difference Equations of the Direct Heat Conduction Problem for the Solution of the IHCP, 222
           6.3.2   Sensitivity Coefficient Approach for Exactly Matching Data from a Single Sensor, 223
    6.4   Multiple Temperature Sensors, Function Specification ($q=C$), Single Future Time Step, 230
    6.5   Whole Domain Estimation With Difference Methods, 233
    6.6   Single Temperature Sensor, Function Specification ($q=C$), r Future Time Steps, 237
    6.7   Multiple Temperature Sensors; Function Specification ($q=C$), Arbitrary Future Time Steps, 241
    6.8   Single Temperature Sensor, Function Specification, Linear Heat Flux (Connected Segments), 242
    6.9   Second Order Sequential Regularization Methods, 243
    6.10 Space Marching Techniques for One-Dimensional Problems, 247
           6.10.1   Analytical Solution, 248
           6.10.2   Method of D'Souza, 249
           6.10.3   Method of Weber, 252
           6.10.4   Method of Raynaud and Bransier, 253
           6.10.5   Method of Hills and Hensel, 254
           6.10.6   Comparison with Prior Methods, 256
    6.11 Numerical Calculations, 256
    6.12 Computer Programs, 262

References, 264
Problems, 265

7. MULTIPLE HEAT FLUX ESTIMATION   267

7.1 Introduction, 267
7.2 Two Independent Heat Fluxes Case, 268
    7.2.1 Sequential Function Specification Method, 271
    7.2.2 Sequential Regularization Method, 273
7.3 Multiple Heat Flux Case, 275
    7.3.1 Sequential Function Specification Method for Multiple Heat Flux Components, 276
    7.3.2 Sequential Regularization Method for Multiple Heat Fluxes, 277
References, 279
Problems, 279

8. HEAT TRANSFER COEFFICIENT ESTIMATION   281

8.1 Introduction, 281
8.2 Sensitivity Coefficients, 283
    8.2.1 Lumped Body Case, 283
    8.2.2 Semi-Infinite Body, 287
8.3 Lumped Body Analyses, 290
    8.3.1 Exact Matching of the Measured Temperatures, 290
    8.3.2 Regression Method, 292
    8.3.3 Function Specification Procedure with q=Constant, 293
    8.3.4 Function Specification Procedure with h=Constant, 294
8.4 Bodies With Internal Temperature Gradients, 297
    8.4.1 Analysis for r Future Temperatures Using $q=C$ Function Specification Method, 297
    8.4.2 Examples, 299
8.5 Estimation of Contact Conductance, 301
References, 301
Problems, 301

Author Index   303

Subject Index   305

NOMENCLATURE

a, b, c, d, e, f	Coefficients in tridiagonal matrix algorithm; see (6.3.9–10)
a	Radius of cylinder or sphere
c	Heat capacity
cov (., .)	Covariance operator; see (1.4.8)
e_i	Residual temperature error; see (1.4.7)
erf	Error function
erfc	Complementary error function
E	Sensor depth below heated surface
$E(\cdot)$	Expected value operator
f_j	Filter coefficient; see (4.7.2)
f_j^+	Dimensionless filter coefficient
G	Green's function
h	Convective heat transfer coefficient
h_c	Contact conductance
$\mathbf{H}_0, \mathbf{H}_1, \ldots$	Regularization matrices; see (4.5.16)
ierfc	Integral error function
$\mathbf{1}$	A vector of ones
$\mathbf{I}$	Identity matrix
J	Number of temperature sensors
$J_0, J_1, \ldots$	Bessel functions of the first kind, order 0, 1, ...
k	Thermal conductivity
K_i	Gain coefficient at time t_i; see (4.4.25)
K_i^+	$=(q_c L/k)K_i$, $q_c=1$ in consistent units, dimensionless gain coefficient

K_{ji}	Gain coefficient for sensor j at time t_i
L	Slab thickness
M	General time index
p	$\alpha\Delta t/\Delta x^2$, grid scale Fourier number
q	Heat flux
q_c	Constant value of heat flux
q_i	Heat flux at time t_i
$\hat{q}_i$	Estimated value of q_i
$\tilde{q}_i$	Heat flux at time t_i that exactly matches the temperature data Y_i; see (4.4.3.3)
$\hat{q}_M$	Estimated heat flux for interval t_{M-1} to t_M
$\tilde{\mathbf{q}}$	Heat flux vector; see (4.6.4)
$\tilde{\mathbf{q}}^f$	Heat flux vector; see (4.6.4)
q^*	Trial value of q
r	Number of future time steps
r	Radial coordinate
r_a^+	r/a
S	Least square function
t	Time
t^+	Dimensionless time, $\alpha t/L^2$
t_a^+	$\alpha t/a^2$
T	Temperature
T_0	Initial temperature
T_∞	Ambient temperature at which convection or radiation is taking place
T^+	$(T-T_0)/(qL/k)$
$\hat{T}$	Estimated value of T
$\hat{\mathbf{T}}\|\mathbf{q}=\mathbf{0}$	Vector of estimated temperatures for $\mathbf{q}=\mathbf{0}$
$\overset{*}{T}$	Temperature corresponding to $q=q^*$; see (6.3.8)
$u(x,t)$	Temperature response function for a body at zero initial temperature and subjected to a unit step in surface temperature; see (2.4.1)
$V(\cdot)$	Variance operator
w_i	Heat flux weighting factor; see (4.4.40c)
$W_0, W_1, \ldots$	Regularization constants; see (4.5.1)
x	Spatial coordinate
x^+	x/L
X	Sensitivity coefficient for heat flux pulse
X^+	$=(k/x)\partial T/\partial q_c$

NOMENCLATURE

X	Matrix of sensitivity coefficients
Y_i	Measured value of temperature at time t_i
Z	Sensitivity coefficient for heat flux step

GREEK SYMBOLS

α	Thermal diffusivity, also regularization parameter
α^+	See (5.6.14)
$\beta_1, \beta_2, \ldots$	Parameters; see (4.4.1–5)
γ	Thermal wave speed
δq_r	Heat flux impulse
δY_r	Temperature impulse or error
$\dfrac{\delta q_M}{\delta Y_j}$	Heat flux error for a unit error in temperature
Δt	Time step
Δt_M	$= t_{M+1} - t_M$
Δx	Spatial grid size for difference methods
$\Delta \phi(\mathbf{r}, t_n)$	$= \phi(\mathbf{r}, t_{n+1}) - \phi(\mathbf{r}, t_n)$
$\Delta \tau_E$	$\alpha \Delta t / E^2$
ε_i	Temperature error; see (1.4.1)
θ	Time difference weighting parameter
λ	Dummy time variable
ρ	Density
σ	Standard deviation
$\phi(x, t)$	Temperature response to a unit step in heat flux
ω	Tridiagonal matrix coefficient; see (6.3.10)

CHAPTER 1

DESCRIPTION OF THE INVERSE HEAT CONDUCTION PROBLEM

1.1 INTRODUCTION

If the heat flux or temperature histories at the surface of a solid are known as functions of time, then the temperature distribution can be found. This is termed a direct problem. In many dynamic heat transfer situations, the surface heat flux and temperature histories of a solid must be determined from transient temperature measurements at one or more interior locations; this is an inverse problem. In particular, during the past two decades the special case of estimating a surface condition from interior measurements has come to be known as *the inverse heat conduction problem*. There are numerous other inverse problems in transient conduction and diffusion, but this particular problem has been so named and is the main subject of this book.

The inverse heat conduction problem is much more difficult to solve analytically than the direct problem. But in the direct problem many experimental impediments may arise in measuring or producing given boundary conditions. The physical situation at the surface may be unsuitable for attaching a sensor, or the accuracy of a surface measurement may be seriously impaired by the presence of the sensor. Although it is often difficult to measure the temperature history of the heated surface of a solid, it is easier to measure accurately the temperature history at an interior location or at an insulated surface of the body. Thus there is a choice between relatively inaccurate measurements or a difficult analytical problem. An accurate and tractable inverse problem solution would thus minimize both disadvantages at once.

The problems of determining the surface temperature and the surface heat flux histories are equivalent in the sense that if one is known the other can be found in a straightforward fashion. They cannot be independently found since in direct heat conduction problems only one boundary condition can be imposed at a given time and boundary. Even though this is true, the following seemingly contradictory statement can be made: the heat flux is more difficult to calculate accurately than the surface temperature. For this reason the emphasis in this book is on the calculation of the surface heat flux history. (The surface temperature is a by-product of the heat flux calculations for difference procedures.)

For the purposes of this book the *inverse heat conduction problem* (IHCP) is defined as follows: The IHCP is the estimation of the surface heat flux history given one or more measured temperature histories inside a heat-conducting body. The word "estimation" is used because in measuring the internal temperatures, errors are always present to some extent and they affect the accuracy of the heat flux calculation. Furthermore even if discrete data accurate to a large but finite number of significant figures are used, the heat flux cannot be exactly determined.

One of the earliest papers on the IHCP was published by Stolz[1] in 1960; it addressed calculation of heat transfer rates during quenching of bodies of simple finite shapes. Stolz[1] claimed use of his method as early as June 1957. For semi-infinite geometries Mirsepassi[2] maintained that he had used the same technique both numerically[2,3] and graphically[3] for several years prior to 1960. A Russian paper by Shumakov[4] on the IHCP was translated in 1957. The space program, starting about 1956, gave considerable impetus to the study of the inverse heat conduction problem. The applications therein were related to nose cones of missiles and probes, to rocket nozzles, and other devices. Beck also initiated his work on the IHCP about that time and developed the basic concepts[5-11] that permitted much smaller time steps than the Stolz method.[1] Others whose work had application to the space program included Blackwell,[12,13] Imber,[16-23] Mulholland,[24-27] and Williams and Curry.[31] Another research area that extensively required solutions of the IHCP was the testing of nuclear reactor components.[32-38] Many of the computer programs in current use in the United States[35-38] appear to be based on the method described in a 1970 paper.[9] Other applications reported for the IHCP included (1) periodic heating in combustion chambers of internal combustion engines,[39] (2) solidification of glass,[40] (3) indirect calorimetry for laboratory use,[41] and (4) transient boiling curve studies.[42] Over 300 papers have been written to date on the IHCP or closely related problems.

There have been extremely varied approaches to the inverse heat conduction problem. These have included the use of Duhamel's theorem (or convolution integral) which is restricted to linear problems.[1-3,5-7,10,28,58] Numerical procedures such as finite differences[8,9,11-13,29-31,35-38] and finite elements[35,36] have also been employed due to their inherent ability to treat nonlinear problems. Exact solution techniques were proposed by Burggraf,[43] Imber and Khan,[23] Langford,[14] and others; such techniques have limited use for realistic problems (as discussed in Chapter 2) but they can give considerable insight into the IHCP. Some techniques used Laplace transforms and were also limited to linear cases.[41,44]

The IHCP is one of many mathematically "ill-posed" problems. Such

SEC. 1.2 EXAMPLES OF INVERSE PROBLEMS

problems are typically inverse problems and are extremely sensitive to measurement errors. (See Section 4.2 for further discussion of ill-posed problems.) There are a number of procedures that have been advanced for the solution of ill-posed problems in general. One of these was developed by Tikhonov and Arsenin in 1963.[45] Tikhonov introduced what he called the regularization method to reduce the sensitivity of ill-posed problems to measurement errors. A modification of this method is presented herein for more efficient solution of the IHCP (see Section 4.5). Numerous other general procedures for ill-posed problems have been proposed including a technique, well known to geophysicists, called the Backus-Gilbert technique.[28,46,47] The mathematical techniques for solving sets of ill-conditioned algebraic equations called single-value decomposition techniques can also be used for the IHCP.[48,57]

The purposes of this chapter are to introduce the inverse heat conduction and related problems and to provide a general description of various aspects of the IHCP. The contents of the remainder of this chapter are as follows: Section 1.2 provides some explicit examples of the IHCP and related problems. The IHCP is related to function and parameter estimation in Section 1.3. Section 1.4 gives a description of the nature of the measurement errors. In Section 1.5 an answer is given why the IHCP is difficult. The important subject of sensitivity coefficients is discussed in Section 1.6. A brief classification of solution methods of the IHCP is given in Section 1.7. Criteria for evaluating various IHCP methods are suggested in Section 1.8. The final section, 1.9, gives the scope of the book.

1.2 EXAMPLES OF INVERSE PROBLEMS

1.2.1 Inverse Heat Conduction Problem Examples

One example of the IHCP is the estimation of the heating history experienced by a shuttle or missile reentering the earth's atmosphere from space. The heat flux at the heated surface is needed. Figure 1.1 depicts a reentering body and an enlarged section of its skin. Though the heat flux, denoted q, may be in general a function of both position y and time t, it is assumed at present that lateral conduction can be neglected compared to the heat flow normal to the surface. Thus the net surface heat flux as a function of time is estimated from measurements obtained from an interior temperature sensor at position x_1 as shown in Figure 1.1b. The measurements are made at discrete times, $t_1, t_2, \ldots$ or in general at time t_i at which the temperature measurement is denoted Y_i. (The word "discrete" means at several particular times, such as 1 second, 2 seconds, 3 seconds, etc., but not continuously.) Figure 1.2 is an illustration of postulated values.

An estimated surface heat flux, denoted $\hat{q}_i$, is associated with the time t_i at which the corresponding temperature measurement, Y_i, is made. The *true* value of the surface heat flux is simply denoted q_i. The surface heat flux history,

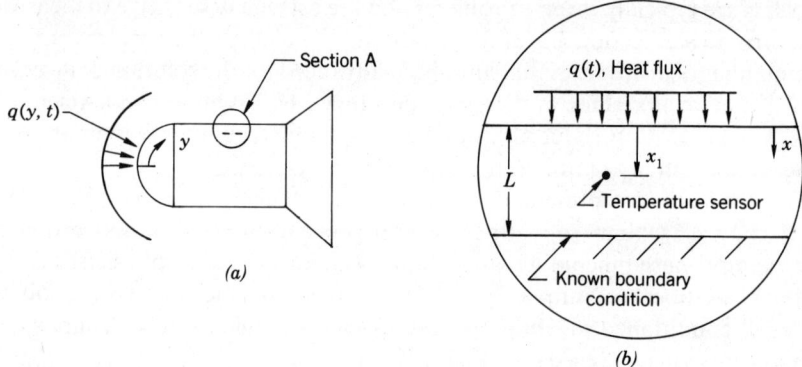

FIGURE 1.1 Example of reentering vehicle for which the surface heat flux is needed. (*a*), Reentering vehicle schematic; (*b*), section A.

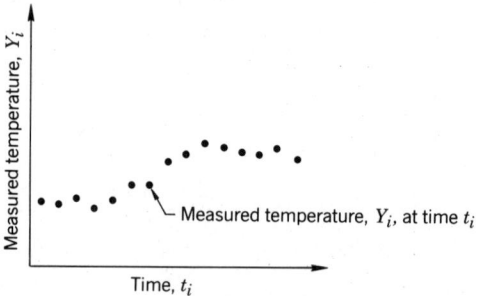

FIGURE 1.2 Measured temperatures at discrete times.

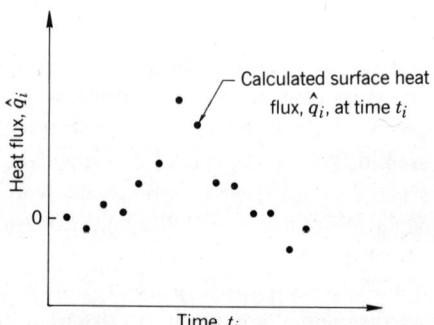

FIGURE 1.3 Representations of calculated surface heat fluxes.

$q(t)$, can be an arbitrary single-valued time function; see Figure 1.3, which is a representation of estimated values of $q(t)$ at times t_i. In general the heat flux can rise and fall abruptly and can be both positive and negative where negative values indicate heat losses from the surface.

The source of heating is immaterial to the IHCP procedures. Convective

SEC. 1.2 EXAMPLES OF INVERSE PROBLEMS

sources include high-temperature fluids that flow in reactor heat exchangers, over reentry vehicle surfaces, or across turbine blades. The heating can also be by radiation from any source or by conduction from an adjacent solid that is in thermal contact with the boundary in question.

To estimate the surface heat flux history it is necessary to have a mathematical model of the heat transfer process. For example, in the reentry vehicle case shown in Figure 1.1b, it is assumed that the section of the skin is of a single material, homogeneous and isotropic, and that it closely approximates a flat plate. (A radial segment of a cylinder or other one dimensional coordinate can be treated in a similar manner.) Then a possible mathematical model for the temperature T in the plate is:

$$\frac{\partial}{\partial x}\left(k\frac{\partial T}{\partial x}\right) = \rho c \frac{\partial T}{\partial t} \qquad (1.2.1)$$

$$T(x, 0) = T_0(x) \qquad (1.2.2)$$

$$\frac{\partial T}{\partial x} = 0 \quad \text{at } x = L \qquad (1.2.3a)$$

$$T(x_1, t_i) = Y_i \qquad (1.2.3b)$$

The objective is to estimate the surface heat flux at discrete times, t_i, from

$$q(t_i) = -k\left.\frac{\partial T(x, t_i)}{\partial x}\right|_{x=0} \qquad (1.2.4)$$

This problem is quite different from the direct problem in that the boundary condition is not specified at $x=0$ but instead a measured temperature history is given at one or more *internal* locations. Further complications are that the measured temperatures are obtained only at discrete times and they inherently have errors in them. Clearly interior measurements contain much less information than given for classical direct problems where the surface conditions are continuous, errorless relations.

The thermal conductivity, k, density, ρ, and specific heat, c, are postulated to be known functions of temperature. If any one of these thermal properties varies with temperature, the IHCP becomes nonlinear. The initial temperature distribution, $T_0(x)$, is also taken as known. The location, x_1, of the sensor is assumed to be measured and to have negligible error. The thickness of the plate, L, is also known and considered errorless.

The known boundary condition of perfect insulation given by Eq. (1.2.3a) is only one of many that can be prescribed at $x=L$. There can be a convective and/or a radiation condition at $x=L$. For an IHCP with a single unknown heat flux, it is only necessary that the boundary condition at $x=L$ be known. For a temperature-dependent heat transfer coefficient h and also for a radiation condition, the inverse heat conduction problem again becomes nonlinear.

For the case of a single interior temperature history, the problem can be subdivided into two separate problems, one of which is a direct problem as shown in Figure 1.4. The portion of the body from $x=x_1$ to L, body 2, can be analyzed as a direct problem because there are known boundary conditions at both boundaries [$T(t)=Y(t)$ at $x=x_1$, $\partial T/\partial x=0$ at $x=L$]. From this direct problem the heat flux at x_1 can be found from the solution for the temperature distribution in $x_1 \leqslant x \leqslant L$ by using

$$\hat{q}_{x_1}(t) = -k \frac{\partial T}{\partial x}\bigg|_{x=x_1} \quad (1.2.5)$$

This same heat flux must *leave* body 1 ($0 \leqslant x < x_1$). Consequently, *two* conditions are specified at $x=x_1$ in body 1 and none at $x=0$. Such a set of boundary conditions for the transient heat conduction equation, Eq. (1.2.1), is related to the mathematical problem being ill-posed.

The IHCP for a single unknown surface heat flux can be complicated in many ways, some of which are illustrated in Figure 1.5. There are four temperature sensors shown which preclude the simple subdivision shown in Figure 1.4 because the heat flux calculated to leave one subdivision would not in general equal the calculated heat flux entering the next subdivision. Another complication is the composite body of three different materials which may be joined together with either perfect or imperfect contacts. In addition the plates might

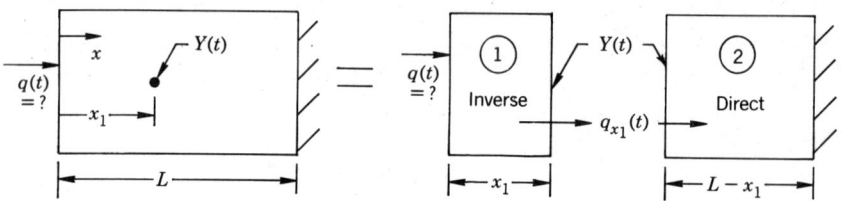

FIGURE 1.4 Subdivision of a single interior sensor IHCP into inverse and direct problems.

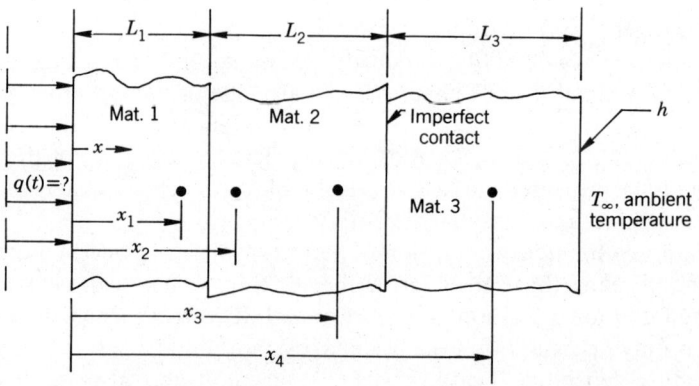

FIGURE 1.5 Composite plate with multiple temperature sensors.

SEC. 1.2 EXAMPLES OF INVERSE PROBLEMS

not be flat but rather be parts of a cylindrical wall. A satisfactory solution of the inverse heat conduction problem should permit treatment of each of these complicating factors.

1.2.2 Other Inverse Function Estimation Problems

There are several other problems related to the inverse heat conduction problem. The IHCP involves estimation of the surface heat-flux time-function utilizing measured interior temperature histories. It is a linear problem (see Section 1.6.1) if the thermal properties are independent of temperature and the boundary condition at the "known" boundary is linear. A closely related problem involves the convective boundary condition,

$$-k \frac{\partial T}{\partial x}\bigg|_{x=0} = h[T_\infty(t) - T(0, t)] \qquad (1.2.6)$$

If the heat transfer coefficient, h, is known either as a constant or as a function of time, the estimation of the ambient temperature $T_\infty(t)$ from given internal temperature measurements is a linear, inverse function estimation problem. If h is a known function of T, then the inverse problem becomes nonlinear.[49]

Another important function estimation problem in connection with Eq. (1.2.6) is the determination of h as a function of time. This is a *nonlinear* problem even if the differential equation is linear. See Section 1.6 for further discussion of nonlinearity. The determination of the transient heat transfer coefficient is an important technique, for example, for investigating the complete boiling curve.[42] The estimation of $h(t)$ is discussed in Chapter 8.

An interface contact conductance, $h_c(t)$, is often used to model imperfect contact. For the interface in Figure 1.5 the heat flux is related to h_c by

$$-k \frac{\partial T}{\partial x}\bigg|_{x=(L_1+L_2)^-}$$

$$= h_c[T|_{x=(L_1+L_2)^-} - T|_{x=(L_1+L_2)^+}]$$

$$= -k \frac{\partial T}{\partial x}\bigg|_{x=(L_1+L_2)^+} \qquad (1.2.7)$$

where the sign $+$ means the material 3 side of the interface and the sign $-$ means the material 2 side. The problem of estimating $h_c(t)$ is very similar to that for the convective heat transfer coefficient.

Endothermic or exothermic chemical reactions can occur inside materials. These can be of unknown magnitudes. Also there can be an energy source due to electric heating, a nuclear source, or frictional heating. In these cases an appropriate describing equation for one-dimensional plane geometries is

$$\frac{\partial}{\partial x}\left(k \frac{\partial T}{\partial x}\right) + g(x, t) = \rho c \frac{\partial T}{\partial t} \qquad (1.2.8)$$

where $g(x, t)$ is a volume energy source term. If g is a function of time only, then estimation of $g(t)$ from transient interior temperature measurements is quite similar to the one-dimensional IHCP. If Eq. (1.2.8) is linear and the boundary conditions are linear, estimation of $g(t)$ is a linear problem; however, if $k = k(T)$, the problem of estimating $g(t)$ becomes nonlinear. When g is a function of both x and t, the estimation of $g(x, t)$ is similar to that of a two-dimensional IHCP.

Two inverse function estimation problems that have received a great deal of attention from mathematicians are called the (improperly posed) Cauchy problem for the two-dimensional Laplace's equation[45,53-55] and the initial-boundary value problem for the backward heat equation.[53,56]

One form of the Cauchy problem for the equation,

$$\frac{\partial^2 T}{\partial x^2} + \frac{\partial^2 T}{\partial y^2} = 0 \tag{1.2.9}$$

is for incomplete specification of the boundary conditions but some interior *measurements* of temperature are given. The objective is to obtain an estimate of $T(x, y)$ for the complete domain including the boundaries.

One example of a backward heat equation problem is the determination of the initial temperature distribution, $T_0(x)$, in a finite body given the boundary conditions and some internal transient measurements of temperature. (See Problem 1.26.)

The inverse problems mentioned become more complex as more functions are determined simultaneously. For example, one might attempt to simultaneously estimate for Figure 1.5 the heat flux $q(t)$ on the left boundary of the body and $T_\infty(t)$ on the right. This would involve simultaneous estimation of two time functions.

If the surface heat flux is a function of position across the surface as shown in Figure 1.6, a number of heat flux components would be simultaneously estimated; this is the two-dimensional IHCP and is discussed further in Chapter 7.

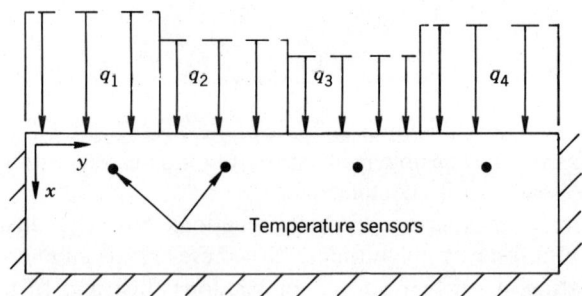

FIGURE 1.6 Surface heat flux as a function of position for a flat plate.

1.3 FUNCTION ESTIMATION VERSUS PARAMETER ESTIMATION

The words "function estimation" were used in the previous section in connection with the IHCP. In the IHCP, the heat flux is found as an arbitrary, single-valued function of time. The heat flux can be positive or negative, constant or abruptly changing, periodic or nonperiodic, and so on. It may be influenced by human decisions. For example, the pilot of a shuttle can change the reentry trajectory. In the IHCP problem the surface heat flux is a function of time and may require hundreds of individually estimated heat flux components, $\hat{q}_i$, to define it adequately.

Related estimation problems are those called "parameter estimation" problems which are also inverse problems but with the emphasis on the estimation of certain "parameters" or constants or physical properties. In the context of heat conduction one might be interested in determining the thermal conductivity of a solid given some internal temperature histories *and* the surface heat flux and other boundary conditions.[49] The thermal conductivity of ARMCO iron near room temperature, for example, could be a parameter; it is not a function and does not require hundreds of values of k_i to describe it. The parameter estimation and function estimation problems start to merge if estimates are made of the thermal conductivity, k, as a function of temperature, T. However, the $k(T)$ function is not arbitrary and is not adjustable by humans.

Parameter estimation is a companion subject; a book by Beck and Arnold has been written on the subject.[49] A background in parameter estimation is not required to understand this book. The subject of parameter estimation has been built on a statistical base but that is not as true for function estimation problems. This book stresses the numerical and mathematical aspects of function estimation rather than the statistical aspects. Certain statistical aspects are included, however, as in Section 1.4 in which measurement errors are discussed.

1.4 MEASUREMENTS

1.4.1 Description of Measurement Errors

In the inverse heat conduction problem there are a number of measured quantities in addition to temperature; such as time, sensor location, and specimen thickness. Each is assumed to be accurately known except the temperature. If this is not true, then it may be necessary, for example, to simultaneously estimate sensor location and the surface heat flux. The latter problem would involve both the inverse heat conduction and parameter estimation problems and is beyond the scope of this book. If the thermal properties are not accurately known, they should be determined as accurately as possible using parameter estimation techniques.

The temperature measurements are assumed to contain the major sources of error or uncertainty. Any known systematic effects due to calibration errors, presence of the sensor, conduction and convection losses or whatever are assumed to be removed to the extent that the remaining errors may be considered to be *random*. These random errors can then be statistically described.

The information provided by the sensors inside the heat-conducting body is incomplete in several respects. First, these measurements are at discrete locations. There is only a finite number of sensors, sometimes only one. Hence the spatial variation of temperature is quite incompletely known. Moreover, the measurements obtained from any sensor are available only at discrete times, rather than continuously. Due to the nature of the measurement errors, a *continuous* temperature record might contribute little more information than the discrete values, however.

1.4.2 Statistical Description of Errors

A set of eight standard statistical assumptions regarding the temperature measurements is given in this section. These are *standard* assumptions and may not be valid for a particular case. These eight assumptions[49] do provide a yardstick with which to compare the actual conditions. The random errors in the temperature measurements cause random errors in the surface heat flux values. The standard assumptions permit simplifications in the analysis of random errors. The eight standard assumptions discussed in Beck and Arnold[49] are:

1. The first standard assumption is that the errors are additive or

$$Y_i = T_i + \varepsilon_i \quad \text{(additive errors)} \tag{1.4.1}$$

where Y_i is the temperature measurement at time t_i, T_i is the "true" temperature at time t_i, and ε_i is the random error at time t_i.

2. The second standard assumption is that the temperature errors, ε_i, have a zero mean (a theoretical quantity),

$$E(\varepsilon_i) = 0 \quad \text{(zero mean errors)} \tag{1.4.2}$$

where $E(\cdot)$ is the "expected value operator."[49] A random error is one that varies as the measurement is repeated but the theoretical mean does not have to be equal to zero. There can be a *bias*; that is, the error might tend to be positive. It is frequently possible to calculate and remove the bias.

A *sample* mean is the average that is based on actual measurements. The true average of ε_i cannot be determined because ε_i is unknown; a typical equation for finding the sample mean of a random variable such as Y_i is

$$\bar{Y}_i = \frac{1}{J} \sum_{j=1}^{J} Y_{ji} \tag{1.4.3}$$

SEC. 1.4 MEASUREMENTS

where Y_{ji} is the jth measurement at time t_i and there are J measurements at time t_i. If Eq. (1.4.2) is true, the expected value of $\bar{Y}_i$ is T_i. The expression given by Eq. (1.4.3) is called the *sample* mean and can sometimes be used to check the assumption given by Eq. (1.4.2).

3. The third standard assumption is that of a constant variance,

$$V(Y_i) = \sigma^2 \quad \text{(constant variance error)} \qquad (1.4.4)$$

where $V(\cdot)$ is the "variance operator" and is related to the expected value operator by

$$V(Y_i) = E\{[Y_i - E(Y_i)]^2\} \qquad (1.4.5)$$

The symbol σ^2 does not contain an i subscript, thus Eq. (1.4.4) means that the variance of Y_i is independent of time t_i and is a constant. If the constant variance assumption embodied in Eq. (1.4.4) is valid and there is only a single sensor, an estimate of the "variance of Y_i", denoted s^2, is

$$s^2 = \frac{1}{n-p} \sum_{i=1}^{n} (Y_i - \hat{Y}_i)^2 = \frac{1}{n-p} \sum_{i=1}^{n} e_i^2 \qquad (1.4.6)$$

for n measurement times; p is the number of parameters being used to estimate T_i, the estimate of which is denoted $\hat{Y}_i$, and e_i is the residual defined by

$$e_i = Y_i - \hat{Y}_i \qquad (1.4.7)$$

Expressions of the type given by Eq. (1.4.6) can be employed to investigate the validity of the constant variance assumption given by Eq. (1.4.4).

4. The fourth standard assumption relates to the correlations among measurements. For two measurement errors ε_i and ε_j where $i \neq j$, the two errors are uncorrelated if the covariance of ε_i and ε_j is zero or

$$\operatorname{cov}(\varepsilon_i, \varepsilon_j) \equiv E\{[\varepsilon_i - E(\varepsilon_i)][\varepsilon_j - E(\varepsilon_j)]\} = 0 \quad \text{for } i \neq j \quad \text{(uncorrelated errors)}$$

$$(1.4.8)$$

The different errors ε_i and ε_j are uncorrelated if each has no effect on or relationship to the other. An example of correlated errors is $\varepsilon_i = \rho \varepsilon_{i-1} + u_i$ where u_i is uncorrelated to the ε_i's and ρ is a constant. As the sampling rate of an automatic data acquisition system increases, the errors tend to become more correlated. High correlation between succeeding temperature measurements indicates that each new measurement is contributing much less information than if the correlation were zero. Very high sampling rates (which approach continuous measurements) may contribute little more information than considerably lower rates; that is, larger time steps, Δt, between the measurements.

A measure of the correlation between the two succeeding data points Y_i and

Y_{i+1} is the sample correlation coefficient, $\hat{\rho}$, defined as:

$$\hat{\rho} = \frac{\sum_{i=1}^{n-1} e_i e_{i+1}}{\sum_{i=1}^{n-1} e_i^2} \quad (1.4.9)$$

(This is an appropriate estimator for ρ if $\varepsilon_i = \rho \varepsilon_{i-1} + u_i$ which was mentioned previously.) A low correlation is near zero and a high correlation is near ± 1. (For further discussion of correlated errors, see Reference 49, pp. 301–326.)

5. The fifth standard assumption is that the temperature measurement errors have a normal (that is, gaussian) distribution,

ε_i has a normal distribution (1.4.10)

If the second, third, and fourth standard assumptions are valid, the probability density of ε_i is given by

$$f(\varepsilon_i) = \frac{1}{\sigma \sqrt{2\pi}} \exp\left(\frac{-\varepsilon_i^2}{2\sigma^2}\right) \quad (1.4.11)$$

The assumption of normality is frequently valid even if standard assumptions 2, 3, and 4 are not; in that case a joint probability density for the errors is needed (Reference 49, p. 230).

6. The sixth standard assumption is that the statistical parameters such as σ^2 and ρ are known,

Known statistical parameters (1.4.12)

7. The seventh standard assumption is that the times $t_1, t_2, \ldots, t_n$, positions $x_1, x_2, \ldots, x_J$, specimen dimensions, and thermal properties are accurately known.

Errorless time, dimensions, and configuration of object
in question and thermal property values (1.4.13)

In other words, the only source of error is in the measured temperatures. In statistical terms, the variances of time, and so on, are zero.

8. The last standard assumption is that there is no prior information regarding the shape of the surface heat flux,

No prior information regarding the surface heat flux (1.4.14)

"Prior information" means information known before any temperature measurements are made for a particular case. If prior information exists, then it can be utilized to obtain better estimates. If, for example, from experience with previous similar tests the heat flux is constant over some time period or is periodic, this information can be used to improve upon the estimators given in this book. It is assumed herein that little is known about the surface heat flux except that it can vary abruptly with time. High-frequency fluctuations of $\hat{q}_i$,

SEC. 1.5 WHY IS THE IHCP DIFFICULT?

that is, those that vary significantly between successive values, are not permitted, however.

1.5 WHY IS THE IHCP DIFFICULT?

1.5.1 Sensitivity to Errors

The inverse heat conduction problem is difficult because it is extremely sensitive to measurement errors. The difficulties are particularly pronounced as one tries to obtain the maximum amount of information from the data. For the one-dimensional IHCP when discrete values of the q curve are estimated, maximizing the amount of information implies small time steps between q_i values (see Figure 1.3). However, the use of small time steps frequently introduces instabilities in the solution of the IHCP unless restrictions are employed that will be discussed in later chapters. Notice the condition of *small* time steps has the opposite effect in the IHCP compared to that in the numerical solution of the heat conduction equation. In the latter, stability problems often can be corrected by reducing the size of the time steps.

1.5.2 Examples of Damping and Lagging; Exact Solutions

The transient temperature response of an internal point in an opaque, heat-conducting body is quite different from that of a point at the surface. The internal temperature excursions are much diminished internally compared to the surface temperature changes. This is a damping effect. A large time delay or lag in the internal response can also be noted. These damping and lagging effects for the direct problem are important to study because they provide engineering insight into the difficulties encountered in the inverse problem.

One interesting case is that of a semi-infinite body heated by a sinusoidal surface heat flux of frequency ω,

$$q = q'_0 \cos(\omega t) \quad (1.5.1)$$

where q'_0 is the maximum value of the surface heat flux. After a sufficiently long time, the temperature solution also becomes periodic and is given by

$$T = T_0 + \frac{q'_0}{k}\left(\frac{\alpha}{\omega}\right)^{1/2} \exp\left[-x\left(\frac{\omega}{2\alpha}\right)^{1/2}\right] \cos\left[\omega t - x\left(\frac{\omega}{2\alpha}\right)^{1/2} - \frac{\pi}{4}\right] \quad (1.5.2)$$

where α is the thermal diffusivity, k is thermal conductivity, and T_0, a constant, is the initial temperature distribution. The envelope of Eq. (1.5.2) is

$$(T - T_0)_{\text{env}} = q'_0 k^{-1} \left(\frac{\alpha}{\omega}\right)^{1/2} \exp\left[-x\left(\frac{\omega}{2\alpha}\right)^{1/2}\right] \quad (1.5.3)$$

As the frequency ω increases, the envelope decreases. The maximum temperature rise occurs at $x=0$ and is proportional to $\omega^{-1/2}$. For an interior location

$$\frac{(T-T_0)_{env}}{(T-T_0)_{env, x=0}} = \exp\left[-x\left(\frac{\omega}{2\alpha}\right)^{1/2}\right] \qquad (1.5.4)$$

which shows that the envelope of interior temperatures sharply decreases for increased x values. The exponential in Eq. (1.5.4) also indicates a large effect as ω is increased. To obtain some insight from this equation, the case is considered wherein the right-hand side is less than 0.01 or

$$x\left(\frac{\omega}{2\alpha}\right)^{1/2} > 4.6$$

For steel with $\alpha = 10^{-5}$ m^2/s and $\omega = 2\pi$ rad/s $= 1$ Hz, there is negligible response for $x \geq 0.82$ cm ($=0.32$ in.). This is large damping but if ω were further increased, say by a factor of 100, there would be negligible response for $x > 0.08$ cm ($=0.03$ in.).

The lagging effect can also be investigated through an examination of Eq. (1.5.2). The surface temperature lags $\pi/4$ radians or 45° behind the surface heat flux, and any interior location lags even more. For example, for the values of $\alpha = 10^{-5}$ m^2/s, $\omega = 2\pi$ rad/s, and $x = 0.82$ cm, there is a lag of 4.6 rad or 264° which corresponds to a 0.73-s lag of the internal temperature compared to the surface T.

Returning now to the IHCP, consider a transient interior temperature with small fluctuations imposed on its changing value. These fluctuations can be the result of high-frequency sinusoidal surface heat flux components or random measurement errors. For a given sensitivity in the temperature sensor, it is possible to specify many different heat flux curves (each having high-frequency components) that will produce interior temperatures indistinguishable from one another. This implies that the inverse problem does not have a unique solution. However, the heat flux history that caused a thermocouple response can usually be determined to acceptable accuracy for properly designed experiments using the methods to be given.

Similar effects to those just described can be found through an examination of the problem of a flat plate exposed to a constant heat flux q_c at $x=0$ and insulated at $x=L$. See Figure 1.7 and Table 1.1. The solution for the temperature distribution for this problem is given by

$$T^+(x^+, t^+) = t^+ + \frac{1}{3} - x^+ + \frac{1}{2}(x^+)^2 - \frac{2}{\pi^2} \sum_{n=1}^{\infty} \frac{1}{n^2} e^{-n^2\pi^2 t^+} \cos(n\pi x^+) \qquad (1.5.5)$$

where

$$T^+ \equiv \frac{T-T_0}{q_c L/k}, \quad t^+ \equiv \frac{\alpha t}{L^2}, \quad x^+ \equiv \frac{x}{L} \qquad (1.5.6a,b,c)$$

The dimensionless time defined by Eq. (1.5.6b) is sometimes called the Fourier number. For $x^+ = 1$, the insulated surface, the time $t^+ = 0.05$ can be considered as small and above 0.5 as large since little temperature response occurs before

TABLE 1.1 Dimensionless Temperature Values, $T^+(x^+, t^+)$, for Various Dimensionless Time and Distances for a Finite Plate Heated at $x=0$ and Insulated at $x=L$, $T^+(x^+, t^+) \equiv [T(x, t) - T_o]/q_c L/k$; $x^+ \equiv x/L$; $t^+ \equiv \alpha t/L^2$

t^+	$x^+ = 0.0$	$x^+ = 0.25$	$x^+ = 0.50$	$x^+ = 0.75$	$x^+ = 1.0$
0.01	0.112838	0.004377	0.000014	0.000000	0.000000
0.02	0.159577	0.020235	0.000802	0.000008	0.000000
0.03	0.195441	0.039238	0.003722	0.000150	0.000005
0.04	0.225676	0.058510	0.008754	0.000702	0.000057
0.05	0.252313	0.077297	0.015366	0.001879	0.000269
0.06	0.276395	0.095405	0.023074	0.003764	0.000786
0.07	0.298541	0.112807	0.031528	0.006360	0.001735
0.08	0.319154	0.129537	0.040486	0.009630	0.003207
0.09	0.338514	0.145644	0.049784	0.013523	0.005251
0.10	0.356826	0.161180	0.059311	0.017986	0.007885
0.11	0.374245	0.176198	0.068992	0.022969	0.011104
0.12	0.390892	0.190745	0.078777	0.028422	0.014887
0.13	0.406863	0.204865	0.088632	0.034302	0.019205
0.14	0.422240	0.218598	0.098535	0.040569	0.024024
0.15	0.437089	0.231980	0.108469	0.047187	0.029306
0.16	0.451466	0.245044	0.118425	0.054123	0.035017
0.17	0.465422	0.257820	0.128395	0.061347	0.041121
0.18	0.479000	0.270335	0.138375	0.068831	0.047584
0.19	0.492236	0.282614	0.148361	0.076553	0.054375
0.20	0.505165	0.294679	0.158352	0.084488	0.061464
0.25	0.566146	0.352432	0.208336	0.126735	0.100516
0.30	0.622842	0.407165	0.258334	0.172002	0.143824
0.35	0.676928	0.460054	0.308333	0.219112	0.189738
0.40	0.729423	0.511818	0.358333	0.267348	0.237244
0.45	0.780946	0.562895	0.408333	0.316271	0.285721
0.50	0.831876	0.613553	0.458333	0.365614	0.334791
0.55	0.882444	0.663954	0.508333	0.415212	0.384223
0.60	0.932790	0.714199	0.558333	0.464967	0.433877
0.65	0.983002	0.764349	0.608333	0.514818	0.483665
0.70	1.033131	0.814440	0.658333	0.564726	0.533536
0.75	1.083210	0.864496	0.708333	0.614671	0.583457
0.80	1.133258	0.914530	0.758333	0.664637	0.633409
0.85	1.183287	0.964551	0.808333	0.714616	0.683379
0.90	1.233305	1.014563	0.858333	0.764603	0.733361
0.95	1.283316	1.064571	0.908333	0.814595	0.783351
1.00	1.333323	1.114576	0.958333	0.864591	0.833344

SEC. 1.5 WHY IS THE IHCP DIFFICULT?

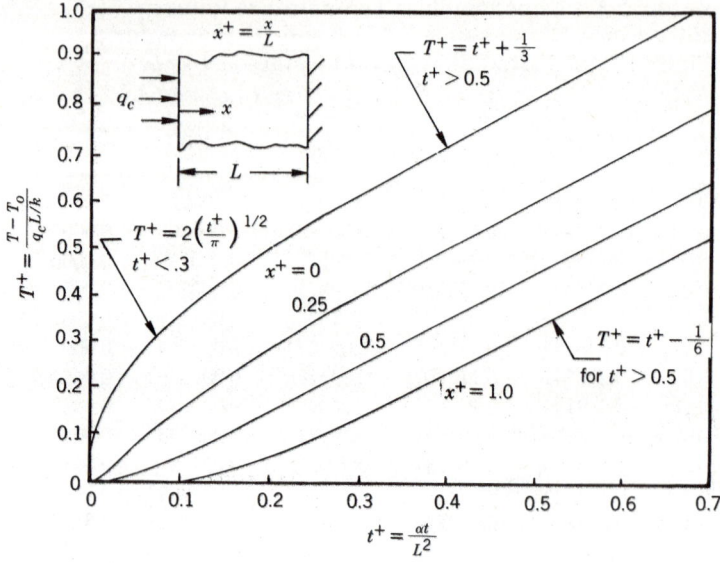

FIGURE 1.7 Temperatures inside a plate with a constant heat flux at $x=0$ and insulated at

$t^+ = 0.05$ and a "fully-developed" linear-with-time response occurs after $t^+ =$ For small times the response at $x^+ = 0$ is quite rapid; in fact, the dimensio temperature is expressed by

$$T^+(0, t^+) = 2\left(\frac{t^+}{\pi}\right)^{1/2} \quad \text{for } t^+ < 0.3$$

As $t^+ \to 0$ the time derivative of Eq. (1.5.7) goes to infinity, indicating a stantaneous change in the surface temperature when the surface heat fl applied. For an interior point the response is slow, being both lagged damped. As an example, for $x^+ = 1$ and for small times the T^+ expressic (see Reference 51, p. 484)

$$T^+(1, t^+) \approx 4(t^+)^{1/2} \text{ ierfc}[(4t^+)^{-1/2}]$$
$$\approx 8\left[\frac{(t^+)^3}{\pi}\right]^{1/2} \exp\left(-\frac{1}{4t^+}\right)$$

These expressions yield very small temperatures at early times and the derivative is zero as $t^+ \to 0$. See also Figure 1.7 for $x^+ = 1$ and small t^+ val

Some numerical values for T^+ are provided by Table 1.1. For $t^+ = 0.05$ example, $T^+(0, t^+) = 0.2523$, whereas $T^+(1, t^+) = 0.000269$, a factor of aln 1000 smaller. This factor increases as t^+ becomes smaller. On the other hand sufficiently large times, the factor approaches unity. This can be demonstra as follows: The summation in Eq. (1.5.5) can be dropped for large times to

$$T^+(x^+, t^+) \approx t^+ + \tfrac{1}{3} - x^+ + \tfrac{1}{2}(x^+)^2, \quad t^+ > 0.5$$

SEC. 1.6 WHY IS THE IHCP DIFFICULT?

so that

$$T^+(0, t^+) \approx t^+ + \tfrac{1}{3} \quad (1.5.10a)$$

$$T^+(1, t^+) \approx t^+ - \tfrac{1}{6} \quad (1.5.10b)$$

Hence, the temperature ratio is

$$\frac{T^+(0, t^+)}{T^+(1, t^+)} \approx \frac{t^+ + \tfrac{1}{3}}{t^+ - \tfrac{1}{6}} \approx 1 + \frac{1}{2t^+} \quad \text{for } t^+ \gg 1 \quad (1.5.11)$$

This result is appropriate for a "thin" plate which is defined as one with a negligible temperature difference across it compared to the temperature rise.

The detailed values for Eq. (1.5.5) given in Table 1.1 are provided for examples and problems in subsequent chapters.

A geometry related to the finite plate is the semi-infinite body. This case with a constant heat flux at $x=0$ is discussed further in Section 1.6.2.2 and numerical values are given in Table 1.2. Two other cases of interest are the solid cylinder and solid sphere, both subjected to a constant heat flux. The center location for each has the greatest lagging and damping; for that reason only

TABLE 1.2 Dimensionless Temperature Values, for $T^+(t_x^+)$, Various Dimensionless Times for a Semi-Infinite Body, $T^+(t_x^+) \equiv [T(x, t) - T_o]/(q_c x/k)$; $t_x^+ \equiv \alpha t/x^2$

t_x^+	$T^+(t_x^+)$	t_x^+	$T^+(t_x^+)$	t_x^+	$T^+(t_x^+)$
0.05	0.000135	1.25	0.50579	80.0	9.1241
0.06	0.000393	1.50	0.60612	90.0	9.7345
0.10	0.003943	1.75	0.70101	100.0	10.3120
0.12	0.007444	2.00	0.79119	200.0	14.9776
0.15	0.014653	2.50	0.95962	300.0	18.5604
0.18	0.023792	3.00	1.11505	400.0	21.5817
0.20	0.030732	3.50	1.26002	500.0	24.2439
0.24	0.046147	4.00	1.39635	600.0	26.6510
0.25	0.050254	5.00	1.64825	700.0	28.8648
0.30	0.071893	6.00	1.87832	800.0	30.9254
0.35	0.094800	7.00	2.09140	900.0	32.8608
0.40	0.118437	8.00	2.29076	1000.0	34.6914
0.45	0.142456	9.00	2.47874		
0.50	0.166631	10.00	2.65708		
0.55	0.190810	12.00	2.98997		
0.60	0.214891	15.00	3.44283		
0.65	0.238808	20.00	4.10921		
0.70	0.262515	25.00	4.69822		
0.75	0.285982	30.00	5.23182		
0.80	0.309190	35.00	5.72321		
0.85	0.332128	40.00	6.18105		
0.90	0.354791	50.00	7.01871		
0.95	0.377175	60.00	7.77678		
1.00	0.399282	70.00	8.47439		

CHAP.1 DESCRIPTION OF THE INVERSE HEAT CONDUCTION PROBLEM

the center temperatures are tabulated in Table 1.3 for the solid cylinder and in Table 1.4 for the solid sphere. The equation for the temperature distribution in the cylinder is (see Reference 51, p. 203)

$$T_a^+(r^+, t_a^+) = 2t_a^+ + \frac{1}{2}(r^+)^2 - \frac{1}{4} - 2\sum_{n=1}^{\infty} e^{-\beta_n^2 t_a^+} \frac{J_0(r^+\beta_n)}{\beta_n^2 J_0(\beta_n)} \quad (1.5.12)$$

where β_n, $n = 1, 2, \ldots$, are the positive roots of the Bessel function,

$$J_1(\beta_n) = 0 \quad (1.5.13)$$

and where a is the cylinder radius and

$$T_a^+(r^+, t_a^+) \equiv \frac{[T(r, t) - T_0]k}{q_c a}, \quad t_a^+ \equiv \frac{\alpha t}{a^2}, \quad r^+ \equiv \frac{r}{a} \quad (1.5.14)$$

The equation for the temperature distribution in the sphere is (see Reference 51, p. 242)

$$T_a^+(r^+, t_a^+) \equiv 3t_a^+ + \frac{1}{2}(r^+)^2 - \frac{3}{10} - \frac{2}{r^+}\sum_{n=1}^{\infty} \frac{\sin(r^+\beta_n)}{\beta_n^2 \sin \beta_n} e^{-\beta_n^2 t_a^+} \quad (1.5.15)$$

where β_n, $n = 1, 2, \ldots$, are the positive roots of

$$\tan \beta_n = \beta_n \quad (1.5.16)$$

TABLE 1.3 Dimensionless Temperatures at the Center of a Solid Cylinder, $T^+(0, t_a^+) \equiv [T(0, t) - T_0]/(q_c a/k)$, $t_a^+ \equiv \alpha t/a^2$

t_a^+	$T^+(0, t_a^+)$	t_a^+	$T^+(0, t_a^+)$
0.010	0.00000	0.100	0.02692
0.020	0.00000	0.150	0.08731
0.030	0.00003	0.200	0.16794
0.040	0.00028	0.250	0.25861
0.050	0.00120	0.300	0.35413
0.060	0.00325	0.350	0.45198
0.070	0.00676	0.400	0.55095
0.080	0.01187	0.450	0.65046
0.090	0.01862	0.500	0.75022
0.100	0.02692	0.550	0.85011
0.110	0.03667	0.600	0.95005
0.120	0.04771	0.650	1.05002
0.130	0.05993	0.700	1.15001
0.140	0.07317	0.750	1.25001
0.150	0.08731	0.800	1.35000
0.160	0.10223	0.850	1.45000
0.170	0.11784	0.900	1.55000
0.180	0.13405	0.950	1.65000
0.190	0.15077	1.000	1.75000
0.200	0.16794		

SEC. 1.6 SENSITIVITY COEFFICIENTS

TABLE 1.4 Dimensionless Temperatures at the Center of a Solid Sphere

t_a^+	$T^+(0, t_a^+)$	t_a^+	$T^+(0, t_a^+)$
0.010	0.00000	0.100	0.05988
0.020	0.00000	0.150	0.17203
0.030	0.00009	0.200	0.30804
0.040	0.00088	0.250	0.45293
0.050	0.00342	0.300	0.60107
0.060	0.00865	0.350	0.75039
0.070	0.01699	0.400	0.90014
0.080	0.02848	0.450	1.05005
0.090	0.04288	0.500	1.20002
0.100	0.05988	0.550	1.35001
0.110	0.07911	0.600	1.50000
0.120	0.10023	0.650	1.65000
0.130	0.12293	0.700	1.80000
0.140	0.14694	0.750	1.95000
0.150	0.17203	0.800	2.10000
0.160	0.19801	0.850	2.25000
0.170	0.22472	0.900	2.40000
0.180	0.25203	0.950	2.55000
0.190	0.27983	1.000	2.70000
0.200	0.30804		

the first few of which are $\beta_1 = 4.4934$, $\beta_2 = 7.7253$ and $\beta_3 = 10.9041$. Both Tables 1.3 and 1.4 clearly exhibit the same lagging behavior as the $x^+ = 1$ location in Table 1.1.

1.6 SENSITIVITY COEFFICIENTS

1.6.1 Definitions of Sensitivity Coefficients and Linearity

In function estimation as in parameter estimation a detailed examination of the sensitivity coefficients can provide considerable insight into the estimation problem. These coefficients can show possible areas of difficulty and also lead to improved experimental design. The sensitivity coefficient is defined as the first derivative of a dependent variable, such as temperature, with respect to an unknown parameter, such as a heat flux component. If the sensitivity coefficients are either small or correlated with one another, the estimation problem is difficult and very sensitive to measurement errors.

For the inverse heat conduction problem, the sensitivity coefficients of interest are those of the first derivatives of temperature T at location x_j and time t_i with respect to a heat flux component, q_M, and are defined by

$$X_{jM}(x_j, t_i) \equiv \frac{\partial T(x_j, t_i)}{\partial q_M} \qquad (1.6.1a)$$

for $j = 1, \ldots, J$, $i = 1, 2, \ldots, n$, and $M = 1, 2, \ldots, n$. Note that the number of times t_i equals the number of heat flux components. The heat flux component, q_M, is shown in Figures 1.8 and 1.9; q_M is the constant heat flux between times t_{M-1} and t_M. If there is only one interior location, that is, $J = 1$, the sensitivity coefficient is simply given by

$$X_M(t_i) \equiv \frac{\partial T_i}{\partial q_M} \quad (1.6.1b)$$

For the transient problems considered in the IHCP, the sensitivity coefficients are zero for $M > i$. In other words, the temperature at time t_i is independent of a yet-to-occur future heat flux component of q_M, $M > i$.

One of the important characteristics of the IHCP is that it is a linear problem if the heat conduction equation is linear and the boundary conditions are linear. The thermal properties (k, ρ, and c), can be functions of position and not affect the linearity. They cannot, however, be functions of temperature without causing the IHCP to be nonlinear. Linearity, if it exists, is an important property because it allows superposition in various ways and it generally eliminates the need for iteration in the solution. If the *linear* IHCP is treated as if it were nonlinear, excessive computer time would be used due to unnecessary iterations.

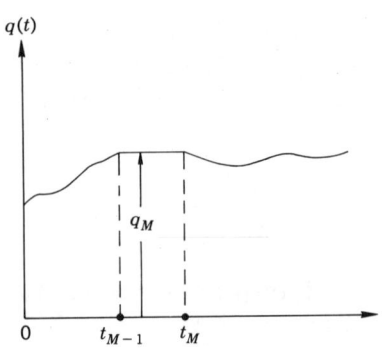

FIGURE 1.8 Heat flux history with a constant heat flux q_M and arbitrary elsewhere.

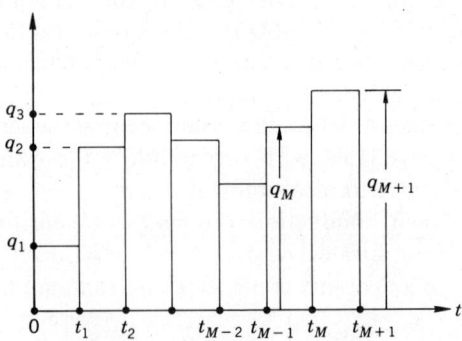

FIGURE 1.9 Heat flux components.

SEC. 1.6 SENSITIVITY COEFFICIENTS

One way to determine the linearity of an estimation problem is to inspect the sensitivity coefficients. If the sensitivity coefficients are *not* functions of the parameters, then the estimation problem is linear. If they are, then the problem is nonlinear. This can be illustrated using Eq. (1.5.7) and differentiating T with respect to q_c. From Eq. (1.5.7):

$$\frac{\partial T(0, t)}{\partial q_c} = \frac{L}{k} 2 \left(\frac{\alpha t}{\pi L^2} \right)^{1/2}, \quad t^+ < 0.3 \tag{1.6.2}$$

which is independent of q_c. More generally, Eq. (1.5.5) can be used to arrive at the same conclusion.

An example of a nonlinear estimation problem is that of estimating α. Taking the derivative of T in Eq. (1.5.7) with respect to α yields

$$\frac{\partial T(0, t)}{\partial \alpha} = \frac{q_c L}{k} \left(\frac{t}{\pi \alpha L^2} \right)^{1/2}, \quad t^+ < 0.3 \tag{1.6.3}$$

The right side of Eq. (1.6.3) is a function of α; thus the estimation of α from transient temperature measurements is a nonlinear problem. (If the more lengthy Eq. (1.5.5) is used, the same conclusion is reached.)

This principle of the sensitivity coefficients being independent of the parameter to be estimated can be employed in cases when the solution is not explicitly known. The equations for a flat plate with temperature-independent properties are given as an example:

$$\frac{\partial}{\partial x} \left[k(x) \frac{\partial T}{\partial x} \right] = \rho c(x) \frac{\partial T}{\partial t} \tag{1.6.4}$$

$$-k \frac{\partial T}{\partial x} \bigg|_{x=0} = \begin{cases} q_M = \text{constant}, & t_{M-1} < t < t_M \\ q(t), & t > t_M \end{cases} \tag{1.6.5}$$

$$-k \frac{\partial T}{\partial x} \bigg|_{x=L} = q_{\text{loss}} \tag{1.6.6}$$

$$T(x, t_{M-1}) = T_{M-1}(x) \tag{1.6.7}$$

where $T_{M-1}(x)$ denotes the temperature distribution at time t_{M-1} and q_{loss} is a heat flux history due to losses that are independent of q_M. The heat flux, $q(t)$, for $t > t_M$ is an arbitrary function of time. The thermal properties can be functions of x. And the parameter of interest is the heat flux, q_M, which is a constant between times t_{M-1} and t_M. See Figure 1.9. The temperature distribution at time t_{M-1} is known and is given by Eq. (1.6.7).

It is desired to find the differential equation and boundary conditions for the q_M sensitivity coefficient defined by

$$X_M(x, t) \equiv \frac{\partial T(x, t)}{\partial q_M} \tag{1.6.8}$$

For $t < t_{M-1}$, the solution is $X_M(x, t) = 0$; that is, the body has not yet been exposed to q_M. For times greater than t_{M-1}, Eqs. (1.6.4)–(1.6.7) are differentiated with respect to q_M to obtain,

$$\frac{\partial}{\partial x}\left[k(x)\frac{\partial X_M}{\partial x}\right] = \rho c(x)\frac{\partial X_M}{\partial t} \tag{1.6.9}$$

$$-k\frac{\partial X_M}{\partial x}\bigg|_{x=0} = \begin{cases} 1, & t_{M-1} < t < t_M \\ 0, & t > t_M \end{cases} \tag{1.6.10}$$

$$\frac{\partial X_M}{\partial x}\bigg|_{x=L} = 0 \tag{1.6.11}$$

$$X_M(x, t_{M-1}) = 0 \tag{1.6.12}$$

Equations (1.6.9)–(1.6.12) describe the mathematical problem for the sensitivity coefficient X_M which can be explicitly found if the functions $k(x)$ and $\rho c(x)$ are known. Notice that it is not necessary to know q_M, $q(t)$, or even $T_{M-1}(x)$ to obtain a solution for the sensitivity coefficient $X_M(x, t)$ because Eqs. (1.6.9)–(1.6.12) are not functions of q_M, $q(t)$, or $T_{M-1}(x)$.

An important conclusion that can be drawn from Eqs. (1.6.9)–(1.6.12) is that the estimation problem for q_M is linear, a consequence of $X_M(x, t)$ not being a function of q_M. This means that the (unknown) value of q_M is not needed to find its sensitivity coefficient. It is also significant that the same differential equation is given for $X_M(x, t)$, Eq. (1.6.9), as for $T(x, t)$, Eq. (1.6.4). Also the boundary conditions are of the same type; that is, the gradient conditions given by Eqs. (1.6.5) and (1.6.6) for $T(x, t)$ are still gradient conditions, Eqs. (1.6.10) and (1.6.11) for $X_M(x, t)$. The main differences are that the $X_M(x, t)$ boundary conditions are simpler and the initial condition is zero. Due to this similarity between the $T(x, t)$ and $X_M(x, t)$ problems, the same solution procedure or computer program can be used for the $X_M(x, t)$ solution as for the $T(x, t)$, which can result in considerable programming and computational efficiency.

1.6.2 One-Dimensional Sensitivity Coefficient Examples

1.6.2.1 Lumped Body Case The simplest transient heat conduction sensitivity coefficient is for a lumped body described by the differential equation and initial condition,

$$q(t)A = \rho c V \frac{dT}{dt}, \quad T(0) = T_0 \tag{1.6.13a,b}$$

where ρc, A, and V are all constants. The $q(t)$ function can be such as those given in Figures 1.8 or 1.9. The sensitivity coefficient for q_M is obtained by differentiating Eq. (1.6.13a) with respect to q_M to get ($X_M \equiv \partial T/\partial q_M$),

$$\rho c \frac{V}{A}\frac{dX_M}{dt} = \begin{cases} 1, & t_{M-1} < t < t_M \\ 0, & \text{otherwise} \end{cases} \tag{1.6.14}$$

SEC. 1.6 SENSITIVITY COEFFICIENTS

and thus $X_M(t)=0$, for $t \leq t_{M-1}$. The initial condition for $X_M(t)$ for $t > t_{M-1}$ is then

$$X_M(t_{M-1}) = 0 \qquad (1.6.15)$$

The complete solution for X_M is

$$X_M(t) = 0, \quad t \leq t_{M-1} \qquad (1.6.16a)$$

$$X_M(t) = \frac{A(t - t_{M-1})}{\rho c V}, \quad t_{M-1} < t < t_M \qquad (1.6.16b)$$

$$X_M(t) = \frac{A(t_M - t_{M-1})}{\rho c V}, \quad t > t_M \qquad (1.6.16c)$$

which is shown in Figure 1.10a. The sensitivity coefficients $X_1, \ldots, X_4$ are shown in Figure 1.10b.

In general, well-designed experiments for estimating the surface heat flux component q_M would have large values of the sensitivity coefficient $X_M(t)$. Hence, Eq. (1.6.16b) indicates that the ratio of the surface area A to the thermal capacity ($\rho c V$) should be as large as possible. That is, a thin foil is desired.

Several observations can be made regarding the IHCP through the examination of Figure 1.10. First, for the heat flux component q_M, measurements before time t_{M-1} yield no information regarding q_M because Eq. (1.6.16a) is true. Consequently measurements after time t_{M-1} are needed to estimate q_M. Because X_M remains greater than zero for $t > t_{M-1}$, there is an "infinite memory" of q_M. In other words, the temperature at any time subsequent to t_{M-1} is affected by q_M.

A second observation concerns the case of estimating more than one component of q_M, such as q_1, q_2, and q_3. To estimate several parameters simultaneously, it is necessary that the sensitivity coefficients be linearly independent.[49] This means that at least one $C_i \neq 0$ and that constants C_1, C_2, and C_3

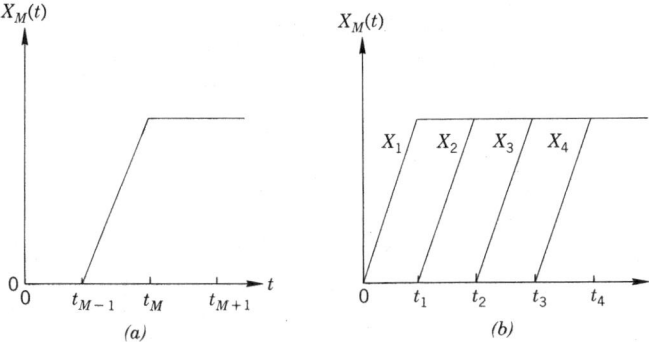

FIGURE 1.10 Sensitivity coefficients for the lumped body case.

cannot be found such that

$$C_1 X_1(t) + C_2 X_2(t) + C_3 X_3(t) = 0 \qquad (1.6.17a)$$

over the domain of the measurements. Referring to Figure 1.10b, if measurements of T are made at t_1, t_2, and t_3, there would be no set of constants that would make Eq. (1.6.17a) true. Hence, q_1, q_2, and q_3 can be estimated if Y_1, Y_2, and Y_3 are known along with the temperature at time zero.

Another way to look at the linear dependence of the sensitivity coefficients X_1, X_2, and X_3 is to write Eq. (1.6.17a) in matrix form for the three different times, t_i, t_j, and t_k,

$$\begin{bmatrix} X_1(t_i) & X_2(t_i) & X_3(t_i) \\ X_1(t_j) & X_2(t_j) & X_3(t_j) \\ X_1(t_k) & X_2(t_k) & X_3(t_k) \end{bmatrix} \begin{bmatrix} C_1 \\ C_2 \\ C_3 \end{bmatrix} = \begin{bmatrix} 0 \\ 0 \\ 0 \end{bmatrix} \qquad (1.6.17b)$$

For linear dependence this equation must be true for arbitrary C_i values, but not all C_i can equal zero. Equation (1.6.17b) is true if and only if the determinant of the square matrix on the left is equal to zero. One such example is for the matrix associated with t_2, $t_{2.5}$, and t_3,

$$\begin{bmatrix} 1 & 1 & 0 \\ 1 & 1 & 0.5 \\ 1 & 1 & 1 \end{bmatrix} \qquad (1.6.17c)$$

which has a determinant of zero. Note that the first two columns of Eq. (1.6.17c) are equal and thus linear dependence is seen by simple inspection. Equation (1.6.17a) is satisfied by setting $C_1 = -C_2$ and $C_3 = 0$. The heat flux components q_1, q_2, and q_3 cannot be simultaneously and uniquely estimated by using Y_i, Y_j, and Y_k if Eq. (1.6.17b) is valid.

The final observation regarding the sensitivities shown in Figure 1.10 is that, though there is an infinite memory of q_M, there is a perfect correlation between q_M and q_{M+1} for times greater than t_{M+1}. This is because the sensitivity coefficients for X_M and X_{M+1} are equal for $t > t_{M+1}$ as shown in Figure 1.10b. Expressed another way, the measurement Y_{100} contains information regarding q_1 but it may not be efficient to use this information; Y_{100} also contains information about $q_2, \ldots, q_{100}$ and is affected in *exactly* the same manner by changes in each of these q components. On the other hand Y_1 is not affected by $q_2, \ldots, q_{100}$ and neither is Y_{100} by $q_{101}, \ldots$. Consequently an estimation scheme that *simultaneously* estimates $q_1, \ldots, q_{100}$ by using $Y_1, \ldots, Y_{100}$ may not be much better than one that estimates the q_i values in a sequential manner that depends mainly on previous times. (By sequential, it is meant that q_{M-1} is estimated before and is completely independent of q_M.) Even though a temperature at a large time contains information about a heat flux component at an early time, it may not be advisable to use that information to estimate the early q component. This conclusion is a result of the high correlation between the interven-

SEC. 1.6 SENSITIVITY COEFFICIENTS

ing q_i's as revealed by the proportionality of the sensitivity coefficients to each other over large time periods as shown in Figure 1.10b.

1.6.2.2 Semi-Infinite Body. The solution for the temperature in a semi-infinite planar body (defined by $x \geq 0$) subjected to a constant surface heat flux, q_c, is

$$T(x,t) = T_0 + 2 \frac{q_c}{k} (\alpha t)^{1/2} \text{ ierfc}\left[\left(\frac{4\alpha t}{x^2}\right)^{-1/2}\right] \quad (1.6.18\text{a})$$

which can be written in dimensionless form as

$$T^+(t_x^+) = 2(t_x^+)^{1/2} \text{ ierfc}[(4t_x^+)^{-1/2}] \quad (1.6.18\text{b})$$

$$T^+ \equiv \frac{k[T(x,t) - T_0]}{q_c x}, \quad t_x^+ \equiv \frac{\alpha t}{x^2} \quad (1.6.18\text{c,d})$$

This dimensionless temperature is tabulated for some values of t_x^+ in Table 1.2. The function, ierfc(z), is

$$\text{ierfc}(z) = \pi^{-1/2} \exp(-z^2) - z \, \text{erfc}(z) \quad (1.6.19)$$

At the heated surface the temperature is

$$T(0,t) = T_0 + 2\left(\frac{q_c}{k}\right)\left(\frac{\alpha t}{\pi}\right)^{1/2} \quad (1.6.20)$$

which is similar to Eq. (1.5.7) for early times in a finite plate. Notice that for Eqs. (1.6.19) and (1.6.20) the q_c sensitivity coefficient,

$$X(x,t) \equiv \frac{\partial T(x,t)}{\partial q_c} \quad (1.6.21)$$

is equal to the temperature rise for $q_c = 1$. It is convenient to examine the case of $x = 0$ and $x \neq 0$ separately. For $x = 0$,

$$X(0,t) = \frac{2}{k}\left(\frac{\alpha t}{\pi}\right)^{1/2} = 2\left(\frac{t}{\pi k \rho c}\right)^{1/2} \quad (1.6.22\text{a})$$

and for $x \neq 0$

$$X(x, t_x^+) \equiv 2\frac{x}{k}(t_x^+)^{1/2} \text{ ierfc}[(4t_x^+)^{-1/2}] \quad (1.6.22\text{b})$$

The dimensionless sensitivity coefficient,

$$X^+ \equiv \frac{k}{x}\frac{\partial T}{\partial q_c} = T^+ \quad (1.6.23)$$

is tabulated in Table 1.2 and plotted in Figure 1.11. The plot for $x = 0$ is included for comparison; for the $x = 0$ case, the x in Eq. (1.6.18d) and in Eq. (1.6.23) should be interpreted as the x of the interior location. [This is confusing but it permits the $x = 0$ and $x \neq 0$ cases to be plotted in the same graph. Note that

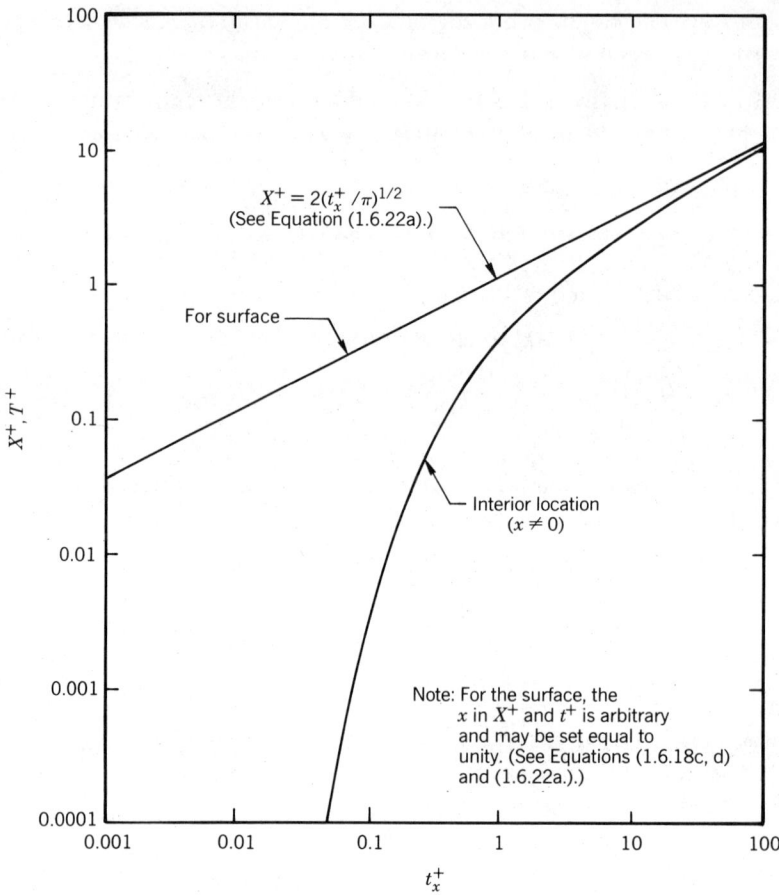

FIGURE 1.11 Sensitivity coefficients for semi-infinite body heated with a constant heat flux.

Eq. (1.6.22a) can be normalized by multiplying by k/x only if $x \neq 0$.] For a single constant heat flux, the sensitivity coefficient behaves in the same manner as the temperature. The temperature rise for a semi-infinite body is similar to that of a finite body at early times. The surface temperatures (and sensitivity coefficients) are numerically equal for the finite and semi-infinite bodies at small times and differ by only 1% at $t^+ = t_x^+ = 0.3$ (see Problem 1.20). Because of similarities between the temperature rise and the sensitivity coefficient and also between the finite plate and semi-infinite body, many of the same comments made in Section 1.5 regarding the lagging and damping effects can be applied to a semi-infinite body as well as to a finite plate.

In the IHCP the complete heat flux history is sought. Hence, it is necessary to examine the sensitivity coefficients for a general heat flux component, q_M, which is depicted in Figure 1.9. As can be seen from Figure 1.11, the behavior of the surface temperature is quite different from that of an interior temperature.

SEC. 1.6 SENSITIVITY COEFFICIENTS

In order to demonstrate this difference, the $x=0$ and $x \neq 0$ cases are treated separately. The $x=0$ case is considered first. For $t<t_{M-1}$, the sensitivity coefficient at $x=0$,

$$X_M(0, t) \equiv \frac{\partial T(0, t)}{\partial q_M} \tag{1.6.24}$$

is zero. For $t>t_{M-1}$ there are two nonzero expressions,

$$\begin{aligned}X_M(0, t) &= \frac{2}{k}\left[\frac{\alpha(t-t_{M-1})}{\pi}\right]^{1/2}, \quad t_{M-1}<t<t_M \\ &= \frac{2}{k}\left\{\left[\frac{\alpha(t-t_{M-1})}{\pi}\right]^{1/2} - \left[\frac{\alpha(t-t_M)}{\pi}\right]^{1/2}\right\}, \quad t>t_M\end{aligned} \tag{1.6.25}$$

The second form of Eq. (1.6.25) is constructed using the principle of superposition. See Problems 1.21 and 1.22. These expressions given by Eq. (1.6.25) can be plotted on a single curve if one lets $\Delta t = t_M - t_{M-1}$ and rewrites Eq. (1.6.25) as

$$\begin{aligned}kX_M(0, t)\left(\frac{\Delta t}{\alpha}\right)^{1/2} &= \frac{2}{\pi^{1/2}}\left(\frac{t-t_{M-1}}{\Delta t}\right)^{1/2}, \quad t_{M-1}<t<t_M \\ &= \frac{2}{\pi^{1/2}}\left[\left(\frac{t-t_{M-1}}{\Delta t}\right)^{1/2} - \left(\frac{t-t_{M-1}}{\Delta t}-1\right)^{1/2}\right], \quad t>t_M\end{aligned} \tag{1.6.26}$$

This expression is plotted in Figure 1.12 for $M=1, 2, 3$, and 4. The graph is valid for any choice of Δt provided Δt is a positive constant.

The sensitivity coefficients plotted in Figure 1.12 are for the surface of a semi-infinite body and are similar in some respects to those shown in Figure 1.10 which are for a lumped body. The similarities include the instantaneous response to changes in the surface heat flux. A difference is that the semi-infinite body responds more rapidly and the effect of heating over a time interval gradually

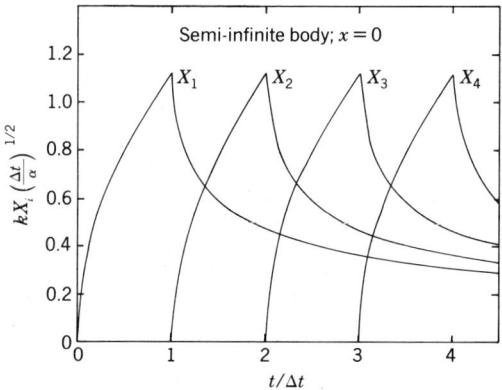

FIGURE 1.12 Sensitivity coefficients for heat flux components for surface of a semi-infinite body.

diminishes. There is an "infinite memory" in the sense that the sensitivity coefficient given for $t > t_M$ by Eq. (1.6.26) only approaches zero for $t - t_M \to \infty$.

The decision regarding the times at which to make measurements to estimate certain parameters is called experimental design. As noted previously, because $X_M(0, t) = 0$ for $t < t_{M-1}$ temperature measurements prior to t_{M-1} cannot be used to estimate q_M. On the other hand, if only measurement times much larger than t_M are used to estimate q_M, difficulties are encountered for two reasons. First, the X_M value is relatively small, indicating little sensitivity regarding q_M and hence little information. Second, the $X_M, X_{M+1}, \ldots$ functions tend to become correlated, that is, have nearly the same shape and thus approach linear dependence. Consequently the best choice of measurement times for estimating q_M from surface temperature measurements of a semi-infinite body must include t_M, t_{M+1}, and slightly larger t values.

The case of an interior temperature measurement history has substantially different characteristics than that for $x = 0$ just discussed. The dimensionless sensitivities for q_1, q_2, and q_3 are shown in Figure 1.13. Unlike the $x = 0$ curve, the results for $x \neq 0$ cannot be combined into a single curve. The dimensionless sensitivity curves for q_1, q_2, and q_3 are denoted X_1^+, X_2^+, and X_3^+; the curves for $X_4^+, X_5^+, \ldots$ have exactly the same shape but they are shifted to the right Δt_x^+ from X_M^+ to X_{M+1}^+.

There are several important observations that can be made in regard to Figure 1.13. First, the X_M^+'s display a lag which is most clearly shown in Figure 1.13a by X_1^+. Even though q_1 starts at $t = 0$ as shown by Figure 1.9, X_1^+ in Figure 1.13a does not appreciably rise until $t_x^+ > 0.05$. This is quite unlike the $x = 0$ behavior for a semi-infinite body for which Figure 1.12 shows the greatest gradient of X_1^+ as $t \to 0$.

The second observation is that the magnitudes of the X_i^+'s increase with Δt_x^+. Small values of Δt_x^+ give small values of X_i^+ causing the estimation of the surface heat flux to become inaccurate.

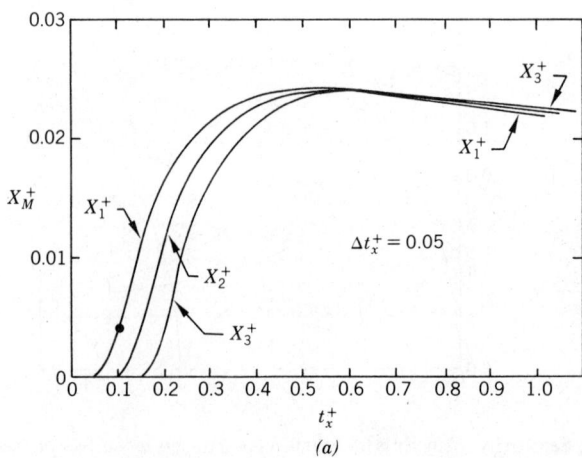

(a)

SEC. 1.6 SENSITIVITY COEFFICIENTS

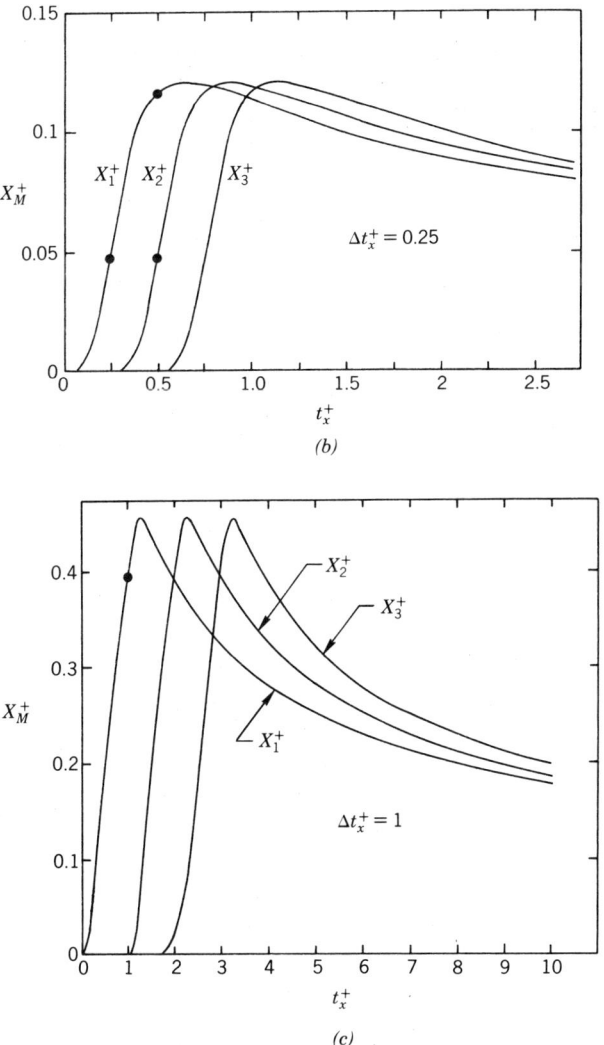

FIGURE 1.13 Dimensionless sensitivity coefficients for q_1, q_2 and q_3 for an interior point in a semi-infinite body. $\Delta t_x^+ = 0.05$, 0.25, and 1.

Third, each X_i^+ curve has a maximum which occurs farther from the time of the end of the heating pulse as Δt_x^+ becomes small. For example, for $\Delta t_x^+ = 1$, X_1^+ has a maximum at $t_x^+ = 1.3$. Also the value of X_1^+ at t_x^+ equal to one Δt_x^+ is about 87% of the maximum X_1^+ for $\Delta t_x^+ = 1$. This is in contrast to the $\Delta t_x^+ = 0.05$ case where the maximum X_1^+ is at $t_x^+ = 0.45$ which is 8 Δt_x^+ values after the heating ends. Moreover, X_1^+ at $t_x^+ = \Delta t_x^+$ is only 0.7% of the maximum X_1^+.

The fourth and final observation is that all the X_M^+'s tend to become correlated as t_x^+ increases.

30 CHAP. 1 DESCRIPTION OF THE INVERSE HEAT CONDUCTION PROBLEM

These characteristics of sensitivity coefficients provide some insight into the design of an estimator for the IHCP utilizing an interior sensor. One of the most obvious points is that temperature measurements at times greater than t_M are needed to estimate q_M, particularly as Δt_x^+ becomes small. Because X_1^+ for $\Delta t_x^+ = 0.05$ is nearly zero at time $t_x^+ = 0.05$ (the end of nonzero q_1), Y_1 would yield very little information regarding q_1. The Y_2 measurement, however, is significantly affected by q_1, but insignificantly by q_2. Hence, a more effective strategy for estimating q_M for the case of $\Delta t_x^+ = 0.05$ would use $Y_0, \ldots, Y_M$ and Y_{M+1} rather than omitting Y_{M+1}. Even a better procedure might be to utilize more *future* (relative to t_M) temperature measurements than Y_{M+1}. If, however, a large number of future temperature measurements is used, for example, $Y_2, \ldots, Y_{100}$ to estimate q_1 then the procedure might not be efficient due to the high correlation among the sensitivity coefficients.

When r future temperatures such as $Y_M, Y_{M+1}, \ldots, Y_{M+r-1}$ are used to estimate q_M, there is extra information for estimating q_M which manifests itself as extra algebraic equations involving q_M. This information is needed, but due to measurement errors, it is not completely consistent. One way to use all this information is to employ the method of least squares which is discussed in Chapter 4. Also see Problems 1.9–1.14.

1.6.2.3 Plate Insulated on One Side

The heat flux sensitivity coefficients for a flat plate heated on one side and insulated on the other are discussed in this section. This is the same geometry as shown in the inset of Figure 1.7. For a heat flux of infinite duration the temperature distribution is given in equation form by Eq. (1.5.5), in tabular form by Table 1.2, and in graphical form by Figure 1.7. These results can also be interpreted as being equal to the dimensionless heat flux sensitivity coefficient of

$$X^+(x^+, t^+) \equiv \frac{k}{L} \frac{\partial T}{\partial q_c} \tag{1.6.27}$$

The sensitivity coefficient for q_1 where

$$q = \begin{cases} q_1, & 0 < t < t_1 = \Delta t \\ 0, & t > t_1 \end{cases} \tag{1.6.28}$$

can be readily found using Eq. (1.5.5). Because the other coefficients for $q_2, \ldots$ have exactly the same shape but are displaced Δt apart, it is only necessary to examine X_1^+ which is given by

$$X_1^+ \equiv \frac{k}{L} \frac{\partial T}{\partial q_c} = T^+(x^+, t^+), \quad 0 < t^+ < \Delta t^+$$

$$= T^+(x^+, t^+) - T^+(x^+, t^+ - \Delta t^+), \quad t^+ > \Delta t^+ \tag{1.6.29}$$

where

$$\Delta t^+ \equiv \frac{\alpha \Delta t}{L^2} \tag{1.6.30}$$

SEC. 1.6 SENSITIVITY COEFFICIENTS

and Δt is the duration of heating. This result is even more complicated than that for a semi-infinite body because X_1^+ depends on t^+ and Δt^+ and also x^+.

For the large dimensionless times of

$$t^+ - \Delta t^+ > 0.5 \tag{1.6.31}$$

the second expression of Eq. (1.6.29) goes to the time- and space-independent value of

$$X_{1,\max}^+ = \Delta t^+ \tag{1.6.32}$$

This expression shows the dependence of X_1^+ on the duration of heating; hence, as more components of q are estimated over a fixed time period, the sensitivity coefficients become smaller and thus the $q(t)$ curve becomes more difficult to estimate. Notice that as t^+ increases, the temperatures increase as shown by Figure 1.7 but the sensitivity coefficients do not.

The sensitivity coefficient X_1^+ normalized with respect to $X_{1,\max}^+$ is plotted in Figure 1.14. In order to present results compactly, the time has been made dimensionless by normalization with respect to the heating duration, Δt.

1.6.3 Two-Dimensional Sensitivity Coefficient Example

In this section the estimation of two-dimensional heat flux histories is investigated. The heat flux is a function of time as previously discussed but also is a function of position over the surface. Further, the heat flux is subdivided into a

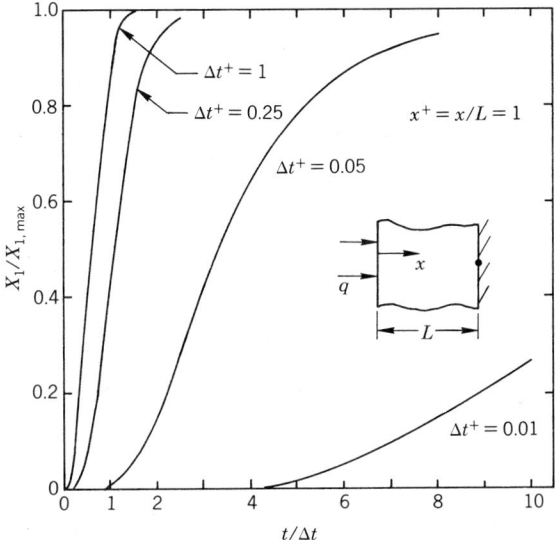

FIGURE 1.14 Heat flux sensitivity coefficients at insulated surface of flat plate.

number of simple building blocks of short duration over small regions. As before, each heat flux component is assumed constant over position as well as over time. See Figure 1.6 for the spatial variation and Figure 1.9 for the time variation. Other functional variations such as linear, parabolic, sinusoidal, and so on, are possible but the basic ideas can be more easily presented for the constant approximation.

The two-dimensional and time-dependent heat flux, $q(y, t)$ is approximated by (see Problem 1.23)

$$q(y, t) = \sum_i \sum_j f_{ji}(y, t) \qquad (1.6.33)$$

where i refers to time and j to position. The function $f_{ji}(y, t)$ is given by

$$f_{ji}(y, t) = \begin{cases} q_{ji}, & t_{i-1} \leqslant t < t_i \text{ and } y_{j-1/2} \leqslant y < y_{j+1/2} \\ 0, & \text{otherwise} \end{cases} \qquad (1.6.34)$$

for constant heat flux "building blocks." The sensitivity coefficient for q_{ji} is then

$$X_{ji}(x, y, t) \equiv \frac{\partial T(x, y, t)}{\partial q_{ji}} \qquad (1.6.35)$$

An example of a two-dimensional body is a semi-infinite body heated with a space-variable flux as shown in Figure 1.15. The coordinate x goes into the body starting at the surface; y is parallel to the surface but there is no natural starting point. The heat flux $q(y, t)$ is usually a continuous function in both time and position but it is approximated by a series of uniform pulses over short distances and time periods, as discussed previously. The sensitivity coefficients for this problem can be developed from the $T(x, y, t)$ solution for a semi-infinite body heated continuously over negative y values; this problem can be described

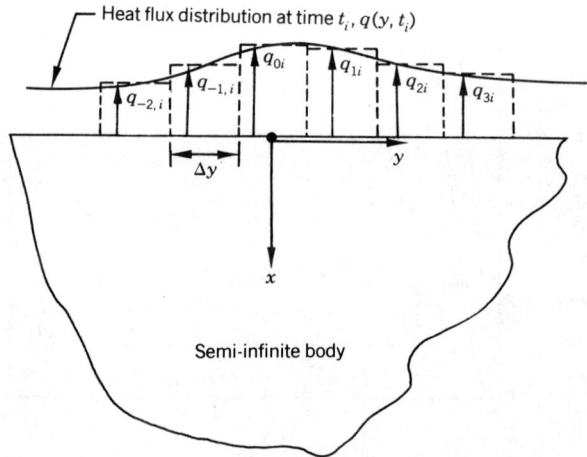

FIGURE 1.15 Semi-infinite body with a space- and time-variable heat flux that is approximated by constant elements.

SEC. 1.6 SENSITIVITY COEFFICIENTS

mathematically by

$$\alpha \left(\frac{\partial^2 T}{\partial x^2} + \frac{\partial^2 T}{\partial y^2} \right) = \frac{\partial T}{\partial t} \tag{1.6.36}$$

At

$$x=0 \begin{cases} -k \dfrac{\partial T}{\partial x} = q_0, & y<0 \\ \dfrac{\partial T}{\partial x} = 0, & y>0 \end{cases} \tag{1.6.37}$$

$$\frac{\partial T}{\partial y} \to 0 \quad \text{as } y \to \pm \infty \tag{1.6.38a}$$

$$T \to T_0 \quad \text{for } x \to \infty \tag{1.6.38b}$$

$$T(x, y, 0) = T_0 \tag{1.6.38c}$$

The solution for all x's is given in Reference 50, but to simplify the treatment only the solution for $x=0$ is used, which is given in Carslaw and Jaeger [Reference 51, p. 264].

The solution[51] for Eqs. (1.6.36)–(1.6.38) for $x=0$ and $y \neq 0$ is

$$T_y^+(y^+) \equiv \frac{T(0, y, t) - T_0}{q_0 (\alpha t/\pi)^{1/2}/k} = \operatorname{erfc}\left(\frac{y^+}{2}\right) - \frac{y^+}{2\pi^{1/2}} E_1\left[\frac{(y^+)^2}{4}\right], \quad y>0 \tag{1.6.39a}$$

$$T_y^+(y^+) = 2 - \operatorname{erfc}\left(\frac{|y^+|}{2}\right) + \frac{|y^+|}{2\pi^{1/2}} E_1\left[\frac{(y^+)^2}{4}\right], \quad y<0 \tag{1.6.39b}$$

where y^+ and $E_1(z)$ (see Reference 52, p. 227) are defined by

$$y^+ \equiv \frac{y}{(\alpha t)^{1/2}}, \quad E_1(z) \equiv \int_z^\infty u^{-1} e^{-u} du \tag{1.6.40a,b}$$

The expression given by Eq. (1.6.39) can be plotted as a function of y^+ with attention to whether y is positive or negative. Figure 1.16 gives a plot of T_y^+ as a function of $|y^+|/2$ for positive and negative y values. For the special locations corresponding to $y \to -\infty$, $y=0$, and $y \to \infty$, the surface temperatures are:

$$T = T_0 + 2q_0 \left(\frac{t}{\pi k \rho c}\right)^{1/2}, \quad y \to -\infty \tag{1.6.41a}$$

$$T = T_0 + q_0 \left(\frac{t}{\pi k \rho c}\right)^{1/2}, \quad y = 0 \tag{1.6.41b}$$

$$T = T_0, \quad y \to \infty \tag{1.6.41c}$$

The sensitivity coefficients for the various terms q_{ji} can be constructed using the principle of superposition with Eq. (1.6.39). To be more specific, the surface

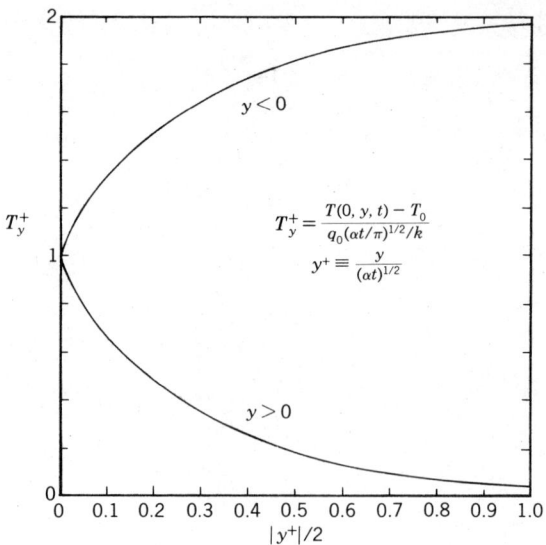

FIGURE 1.16 Dimensionless surface temperature for a semi-infinite body heated uniformly over one-half the surface.

shown in Figure 1.15 which is exposed to a number of equally spaced heat flux components for time t_i is considered. Let the heated regions be equal to $\Delta y = 2a$. A prototype sensitivity coefficient is the one for q_{01} which is for heating over the region $-a < y < a$ and for a heating duration of $0 < t < \Delta t$. The dimensionless sensitivity coefficient for the location, $x = 0$, $y = 2sa$, $(s = 0, 1, \ldots)$ and time $t = n\,\Delta t$ $(n = 1, 2, \ldots)$ is needed. The notation is

$$X^+_{01,sn}$$

where the first two subscripts refer to the location of the application of the heat flux component (i.e., $j = 0$) and to the time of the applied heat flux component $(i = 1)$, and the last two subscripts are for the location and time of evaluation. The first two subscripts are for the cause (i.e., the impulse) and the last two are for the effect (i.e., response). The sensitivity coefficient at time $t = \Delta t$, $(n = 1)$, at location $y_s = 2sa$ is obtained by superimposing two solutions taken from Eq. (1.6.39),

$$X^+_{01,s1} \equiv \frac{k}{a}\frac{\partial T(0, y_s, \Delta t)}{\partial q_{01}} = 2\left(\frac{t_a^+}{\pi}\right)^{1/2}\left\{T_y^+\left[\frac{2s-1}{(t_a^+)^{1/2}}\right] - T_y^+\left[\frac{2s+1}{(t_a^+)^{1/2}}\right]\right\} \quad (1.6.42)$$

where

$$t_a^+ \equiv \frac{\alpha\,\Delta t}{a^2} \quad (1.6.43)$$

SEC. 1.6 SENSITIVITY COEFFICIENTS

The notation $T_y^+ [(2s-1)/(t_a^+)^{1/2}]$ means that y^+ in Eq. (1.6.39) is to be replaced by $(2s-1)/(t_a^+)^{1/2}$. For positive values of $2s-1$ (or $2s+1$), Eq. (1.6.39a) is used, whereas Eq. (1.6.39b) is used for negative values. For the time of $n \Delta t$ and $n = 2, 3, \ldots$, the sensitivity coefficient for q_{01} is

$$X_{01,sn}^+ = 2 \left(\frac{nt_a^+}{\pi} \right)^{1/2} \left\{ T_y^+ \left[\frac{2s-1}{(nt_a^+)^{1/2}} \right] - T_y^+ \left[\frac{2s+1}{(nt_a^+)^{1/2}} \right] \right\}$$
$$- 2 \left(\frac{(n-1)t_a^+}{\pi} \right)^{1/2} \left\{ T_y^+ \left[\frac{2s-1}{((n-1)t_a^+)^{1/2}} \right] - T_y^+ \left[\frac{2s+1}{((n-1)t_a^+)^{1/2}} \right] \right\} \quad (1.6.44)$$

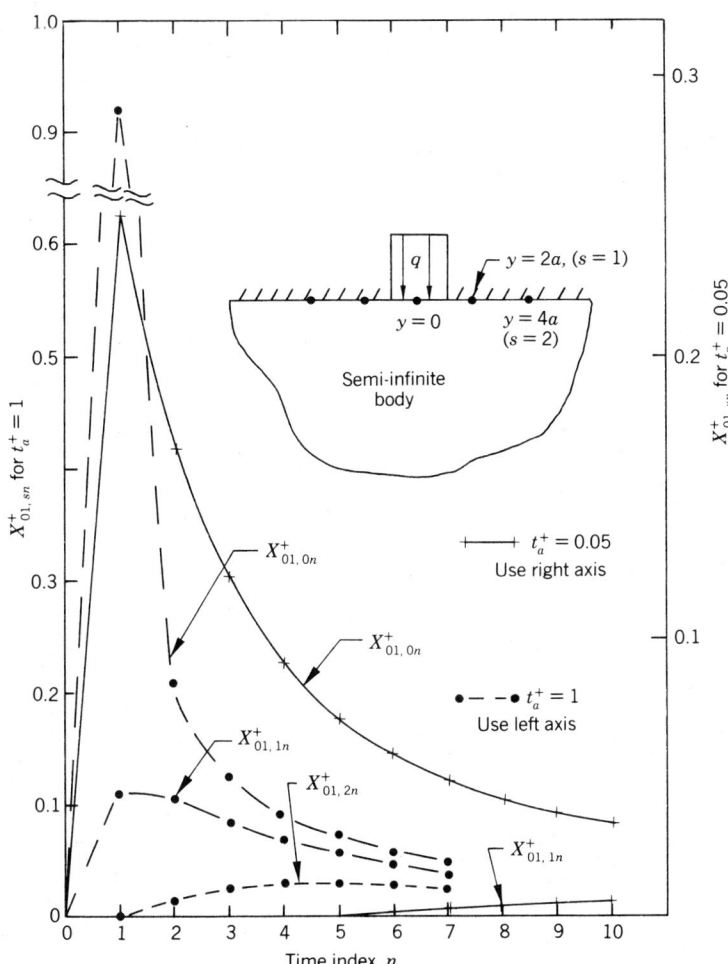

FIGURE 1.17 Sensitivity coefficients for heat flux components along surface of a semi-infinite body.

The general sensitivity coefficient for q_{ji} at y_s and t_n is related to the one for q_{01} by

$$X^+_{ji,sn} = \begin{cases} 0, & n < i \\ X^+_{01, ls-jl, n-i+1}, & n \geq i \end{cases} \qquad (1.6.45)$$

Hence, if the $X^+_{01,sn}$ sensitivity coefficients are known for $s = 0, 1, 2, \ldots$ and $n = 1, 2, \ldots$, at the surface of a semi-infinite body, the sensitivity coefficients at the surface can be found for all other $X^+_{ji,sn}$.

Figure 1.17 gives plots of $X^+_{01,0n}$, $X^+_{01,1n}$, and $X^+_{01,2n}$ for $t_a^+ = 1$. These curves are those with dots and dashed lines. Use the left axis for these curves. The $X^+_{01,0n}$ curve is the largest since it is for the response directly below the heat pulse; it is a maximum at $n = 1$, the end of the heat pulse, and decays thereafter. The $X^+_{01,2n}$ curve is much smaller in magnitude than the $X^+_{01,0n}$ curve and its maximum is displaced in time. Also shown in Figure 1.17 are two curves for $t_a^+ = 0.05$, which is a relatively small value. These are the continuous curves that have crosses for which the right axis is used. In this case the $X^+_{01,0n}$ curve is similar to that for $t_a^+ = 1$ but the $X^+_{01,1n}$ curve is very small in magnitude and there is a very large lag in response. This means that the heat impulse for the small time steps of $t_a^+ = 0.05$ affects mainly the temperature response at the location of the pulse. There is little correlation (or cross-coupling) of the individual heat pulses.

The uncorrelated nature of the spatial variation of the sensitivity coefficients for small time steps and with surface temperature measurements has experimental design implications. If, for example, the heat transfer coefficients (or equivalently heat fluxes if the ambient temperatures are known) are needed around a jet engine turbine blade, a transient experiment with surface thermocouples could be used. The calculational time steps should be small to reduce spatial correlation between the components of the heat transfer coefficient.

1.7 CLASSIFICATION OF METHODS

The methods for solving the inverse heat conduction problem can be classified in several ways, some of which are discussed in this section.

One classification relates to the ability of a method to treat nonlinear as well as linear IHCP's. This book emphasizes basic algorithms that can be employed for both linear and nonlinear problems. The two basic procedures given herein are the function specification and regularization methods. Both of these can be used for nonlinear problems provided the nonlinear heat conduction equation is solved. Some methods of solution of the IHCP are inherently linear such as those based on the Laplace transform; such methods are not considered because the nonlinear case is more important for industrial applications.

SEC. 1.7 CLASSIFICATION OF METHODS

The method of solution of the heat conduction equation is another way to classify the IHCP. Methods of solution include the use of Duhamel's theorem, finite differences, finite elements and finite control volumes. The use of Duhamel's theorem restricts the IHCP algorithm to the linear case, whereas the other procedures can treat the nonlinear problem. Duhamel's theorem is used frequently in this book because the basic IHCP algorithms are easier to use and program for simple calculations than the finite difference and other methods. Moreover for linear problems, the answers for the surface heat flux are nearly identical for all of the methods mentioned provided the heat conduction equation is solved accurately. Consequently, experience acquired using Duhamel's theorem incorporated in a basic IHCP algorithm is also relevant to the other methods when used for linear IHCP's.

The time domain utilized in the IHCP can also be used to classify the method of solution. Three time domains have been proposed: (1) only to the present time, (2) to the present time plus a few time stops, and (3) the complete time domain. The use of measurements only to the present time with a single temperature sensor allows the calculated temperatures to match the corresponding measured temperatures in an exact manner; that is, the calculated temperatures equal the measured values. This is called the Stolz method.[1] Such exact matching is intuitively appealing but the algorithms based on it frequently are extremely sensitive to measurement errors. In the second method, a few future temperatures (associated with future times) are used; the associated algorithms are called "sequential." Greatly improved algorithms are obtained compared with exact matching. The improvements are noted in the considerably reduced sensitivity to measurement errors and in the much smaller time steps that are possible. Small time steps permit more detailed information regarding the time variation of the surface heat flux to be found. The whole domain estimation procedure is also very powerful because very small time steps can be taken but it is not as computationally efficient as the use of only a few future temperatures. Both the function specification and regularization methods can be employed in sequential and whole domain estimation forms.

The last classification to be mentioned is relative to the dimensionality of the IHCP. If a single heat flux history is to be determined, the IHCP can be considered as one-dimensional. In the use of Duhamel's theorem, the physical dimensions of the problem are not of concern; that is, the same procedure is used for physically one-, two-, or three-dimension bodies provided a single heat flux history is to be estimated. If two or more heat flux histories are estimated and Duhamel's theorem is used, the problem is multidimensional. When the finite difference or the other methods for nonlinear problems are employed, the dimensionality of the problem depends on the number of space coordinates needed to describe the heat-conducting body; one coordinate would give a one-dimensional problem, two-coordinates a two-dimensional problem, and so on.

1.8 CRITERIA FOR EVALUATION OF IHCP METHODS

In order to evaluate the several IHCP procedures, various criteria are needed. The criteria proposed in Reference 10 are given in this section and are as follows:

1. The predicted temperatures and heat fluxes should be accurate if the measured data are of high accuracy.
2. The method should be insensitive to measurement errors.
3. The method should be stable for small time steps or intervals. This permits the extraction of more information regarding the time variation of surface conditions than is permitted by large time steps.
4. Temperature measurements from one or more sensors should be permitted.
5. The method should not require continuous first-time derivatives of the surface heat flux. Furthermore, step changes or even more abrupt changes in the surface heat fluxes should be permitted.
6. Knowledge of the precise starting time of the application of the surface heat flux should not be required. The start of heating is frequently not synchronized with the discrete times that temperatures are measured. Reasons for this might be that the starting time is not accurately known or is difficult to measure. Precise times at which abrupt changes in the heat flux occur may also be unknown.
7. The method should not be restricted to any fixed number of observations.
8. Composite solids should be permitted.
9. Temperature-variable properties should be permitted.
10. Contact conductances should not be excluded.
11. The method should be easy to program.
12. The computer cost should be moderate.
13. The user should not have to be highly skilled in mathematics in order to use the method or to adapt it to other geometries.
14. The method should be capable of treating various one-dimensional coordinate systems.
15. The method should permit extension to more than one heating surface.
16. The method should have a statistical basis and permit various statistical assumptions for the measurement errors.

The function specification and regularization methods are capable of satisfying all these criteria provided the nonlinear heat conduction equation is approximated using methods such as the finite difference, finite element, or finite control volume methods.

If the heat conduction equation is solved by Duhamel's theorem and the function specification or regularization method is used, all the criteria can be satisfied except that of treating the nonlinear problem (criterion number 9).

1.9 SCOPE OF BOOK

The scope of the book is mainly limited to the inverse heat conduction problem. It is one of a class of ill-posed problems. However, many of the techniques given herein apply to a wide variety of ill-posed problems. A number of solution methods are presented and the emphasis is on general methods that can meet the criteria in Section 1.8.

Chapter 1 has given an introduction to the subject.

Chapter 2 presents some exact solutions, one of which is due to Burggraf.[43] Though this exact solution is restricted in its application, the insight gained from it is very important.

Chapter 3 presents two different basic procedures for solving direct transient heat conduction problems. The first is based on a numerical form of Duhamel's integral. The second method approximates the partial differential equation for transient heat conduction by a set of algebraic equations, that is, difference equations. The second method is more powerful in that it can treat the nonlinear problems.

Chapter 4 opens with a discussion of ill-posed problems and then presents a number of methods for the IHCP. Many researchers have contributed to these methods. The two basic classes are the function specification and regularization methods. A procedure called the trial function method unifies these approaches. Two ways that the function specification and regularization methods can be used are called sequential and whole domain. Various modifications of them are discussed. Each of these methods can utilize the Duhamel and difference equation procedures and thus is applicable for both linear and nonlinear problems. An important part of this chapter is a discussion of a digital filter form of the IHCP. Such a form can be used for the very efficient implementation of the linear IHCP algorithms. The final section of Chapter 4 discusses some criteria for comparing estimation procedures.

Chapter 5 presents a number of test cases. Utilizing a numerical approximation of Duhamel's integral for the solution of the heat conduction equation, these test cases are used to investigate a number of the IHCP algorithms. The study of IHCP algorithms is facilitated by using a numerical form for Duhamel's integral because only a single equation is needed at each time rather than a set of difference equations. Chapter 5 ends with a discussion of optimal choices of parameters in the function specification and regularization methods.

Chapter 6 uses the finite control volume method to approximate the heat conduction equation for the one-dimensional inverse heat conduction problem. The sequential function specification and regularization methods developed in Chapter 4 are used. In addition, some space marching techniques are discussed; these methods are unique to the difference equation approach since analogous equations based on Duhamel's integral are not available. The chapter concludes with a list of computer programs available for the IHCP.

Chapter 7 is for multiple heat flux estimation. Two or more heat flux histories can be estimated at the same time. This case is commonly encountered in two- or three-dimensional inverse heat conduction problems.

Chapter 8, the last chapter, discusses methods for estimating the heat transfer coefficient. One way is to use the IHCP methods to calculate the surface heat flux history. An alternate procedure calculates the heat transfer coefficient directly.

REFERENCES

1. Stolz, G., Jr., Numerical Solutions to an Inverse Problem of Heat Conduction for Simple Shapes, *J. Heat Transfer* **82**, 20–26 (1960).
2. Mirsepassi, T. J., Heat-Transfer Charts for Time-Variable Boundary Conditions, *Br. Chem. Eng.* **4**, 130–136 (1959).
3. Mirsepassi, T. J., Graphical Evaluation of a Convolution Integral, *Mathematical Tables and Other Aides to Computation* **13**, 202–212 (1959).
4. Shumakov, N. V., A Method for the Experimental Study of the Process of Heating a Solid Body, *Soviet Physics–Technical Physics* (Translated by American Institute of Physics) **2**, 771 (1957).
5. Beck, J. V., Correction of Transient Thermocouple Temperature Measurements in Heat-Conducting Solids, Part II, The Calculation of Transient Heat Fluxes Using The Inverse Convolution, AVCO Corp., Res. and Adv. Dev. Div., Wilmington, MA., Tech. Report RAD-TR-7-60-38 (Part II), March 30, 1961.
6. Beck, J. V., Calculation of Surface Heat Flux From an Internal Temperature History, ASME Paper 62-HT-46 (1962).
7. Beck, J. V., "Surface Heat Flux Determination Using an Integral Method," *Nucl. Eng. Des.* **7**, 170–178 (1968).
8. Beck, J. V. and Wolf, H., The Nonlinear Inverse Heat Conduction Problem, ASME Paper, 65-HT-40 (1965).
9. Beck, J. V., Nonlinear Estimation Applied to the Nonlinear Heat Conduction Problem, *Int. J. Heat Mass Transfer* **13**, 703–716 (1970).
10. Beck, J. V., Criteria for Comparison of Methods of Solution of the Inverse Heat Conduction Problem, *Nucl. Eng. Des.* **53**, 11–22 (1979).
11. Beck, J. V., Litkouhi, B., and St. Clair, C. R., Jr., Efficient Sequential Solution of the Nonlinear Inverse Heat Conduction Problem, *Numer. Heat Transfer* **5**, 275–286 (1982).
12. Blackwell, B. F., A New Iterative Technique for Solving the Implicit Finite-Difference Equations for the Inverse Problem of Heat Conduction, unpublished technical report, Sandia Laboratories, Albuquerque, NM (1968).
13. Blackwell, B. F., An Efficient Technique for the Numerical Solution of the One-Dimensional Inverse Problem of Heat Conduction, *Numer. Heat Transfer* **4**, 229–239 (1981).
14. Langford, D., New Analytic Solutions of the One-Dimensional Heat Equation for Temperature and Heat Flow Rate Both Prescribed at the Same Fixed Boundary (with Applications to the Phase Change Problem), *Q. Appl. Math.* **24**, 315–322 (1967).
15. Woo, K. C. and Chow, L. C., Inverse Heat Conduction by Direct Inverse Laplace Transform, *Numer. Heat Transfer* **4**, 499–504 (1981).

REFERENCES

16. Imber, M., A Temperature Extrapolation Method for Hollow Cylinders, *AIAA J.* **11**, 117–118 (1973).
17. Imber, M., Comments on "On Transient Cylindrical Surface Heat Flux Predicted from Interior Temperature Responses," *AIAA J.* **14**, 542–543 (1975).
18. Imber, M., Inverse Problem for a Solid Cylinder, *AIAA J.* **17**, 91–94 (1979).
19. Imber, M., Nonlinear Heat Transfer in Planar Solids: Direct and Inverse Applications, *AIAA J.* **17**, 204–212 (1979).
20. Imber, M., A Temperature Extrapolation Mechanism for Two-Dimensional Heat Flow, *AIAA J.* **12**, 1087–1093 (1974).
21. Imber, M., The Two Dimensional Inverse Problem in Heat Conduction, Fifth International Heat Transfer Conference, Tokyo, Japan (1974).
22. Imber, M., Two-Dimensional Inverse Conduction Problem—Further Observations," *AIAA J.* **13**, 114–115 (1975).
23. Imber, M. and Khan, J., Prediction of Transient Temperature Distributions with Embedded Thermocouples, *AIAA J.* **10**, 784–789 (1972).
24. Mulholland, G. P., Gupta, B. P., and San Martin, R. L., Inverse Problem of Heat Conduction in Composite Media, ASME Paper, 75-WA/HT-83 (1975).
25. Mulholland, G. P. and San Martin, R. L., Inverse Problem of Heat Conduction in Composite Media, Third Canadian Congress of Applied Mechanics, Calgary, Alberta, Canada (May 1971).
26. Mulholland, G. P. and Cobble, M. H., Diffusion Through Composite Media, *Int. J. Heat Mass Transfer* **15**, 147–152 (1972).
27. Mulholland, G. P. and San Martin, R. L., Indirect Thermal Sensing In Composite Media, *Int. J. Heat Mass Transfer* **16**, 1056–1060 (1973).
28. Hills, R. G. and Mulholland, G. P., Accuracy and Resolving Power of One-Dimensional Transient Inverse Heat Conduction Theory as Applied to Discrete and Inaccurate Measurements, *Int. J. Heat Mass Transfer* **22**, 1221–1229 (1979).
29. Randall, J. D., Embedding Multidimensional Ablation Problems in Inverse Heat Conduction Problems Using Finite Differences, 6th Int. Heat Transfer Conf., Toronto, Ont., Aug. 7–11, 1978. Publ. by Natl. Res. Council of Can., Toronto, Ont., 1978. Available from Hemisphere Publ. Corp, Washington, D.C., Vol. 3, 129–134.
30. Randall, J. D., Finite Difference Solution of the Inverse Heat Conduction Problem and Ablation, Technical Report, Johns Hopkins University, Laurel, Maryland (1976), Proceedings of the 25th Heat Transfer and Fluid Mechanics Institute, Univ. of California, Davis (1976).
31. Williams, S. D. and Curry, D. M., An Analytical Experimental Study for Surface Heat Flux Determination, *J. Spacecraft* **14**, 632–637 (1977).
32. France, D. M., Chiang, T., Carlson, R. D., and Minkowycz, W. J., Measurements and Correlation of Critical Heat Flux in a Sodium Heated Steam Generator Tube, Technical Memorandum, ANL-CT-78-15, Argonne National Laboratory, Argonne, IL (1978).
33. France, D. M., Carlson, R. D., Chiang, T., and Priemer, R., CHF-Induced Thermal Oscillations Measured in an LMFBR Steam Generated Tube Wall, Technical Report, ANL-CT-78-1, Argonne National Laboratory Argonne, IL (1977).
34. France, D. M. and Chiang, T., Analytical Solution to Inverse Heat Conduction Problems with Periodicity, *J. Heat Transfer* **102**, 579–581 (1980).
35. Bass, B. R., Applications of the Finite Element to the Inverse Heat Conduction Problem Using Beck's Second Method, *J. Eng. Ind.* **102**, 168–176 (1980).
36. Bass, B. R., Incap: A Finite Element Program for One-Dimensional Nonlinear Inverse Heat Conduction Analysis, Technical Report NRC/NUREG/CSD/TM-8, Oak Ridge National Laboratory (1979).
37. Muzzy, R. J., Avila, J. H. and Root, R. E., Topical Report: Determination of Transient Heat Transfer Coefficients and the Resultant Surface Heat Flux from Internal Temperature Measurements, General Electric, GEAP-20731 (1975).

38. Snider, D. M., INVERT 1.0—A Program for Solving the Nonlinear Inverse Heat Conduction Problem for One-Dimensional Solids, EG&G Idaho, Inc., Idaho Falls, Idaho, EGG-2068 (1981).
39. Alkidas, A. L., Heat Transfer Characteristics of a Spark-Ignition Engine," *J. Heat Transfer* **102**, 189–193 (1980).
40. Howse, T. K. J., Kent, R., and Rawson, H., The Determination of Glass-Mould Heat Fluxes from Mould Temperature Measurements, *Glass Technol.* **12**, 91–93 (1971).
41. Sparrow, E. M., Haji-Sheikh, A., and Lundgren, T. S., The Inverse Problem in Transient Heat Conduction, *J. Appl. Mech., Trans. ASME, Series E*, **86**, 369–375 (1964).
42. Lin, D. Y. T., and Westwater, J. W., Effect of Metal Thermal Properties on Boiling Curves Obtained by the Quenching Method, *Heat Transfer 1982—Munchen Conference Proceedings*, Hemisphere Publ. Corp., New York, 1982, pp. 155–160, Vol. 4.
43. Burggraf, O. R., An Exact Solution of the Inverse Problem in Heat Conduction Theory and Applications, *J. Heat Transfer* **86C**, 373–382 (1964).
44. Grysa, K., Cialkowski, M. J. and Kaminski, H., An Inverse Temperature Field Problem of the Theory of Thermal Stresses, *Nucl. Eng. Des.* **64**, 169–184 (1981).
45. Tikhonov, A. N. and Arsenin, V. Y., *Solutions of Ill-Posed Problems*, V. H. Winston & Sons, Washington, D.C., 1977.
46. Backus, G. and Gilbert, F., Uniqueness in the Inversion of Inaccurate Gross Earth Data, *Phil. Trans. R. Soc. London Ser. A* **266**, 123–192 (1970).
47. Nolet, G., Simultaneous Inversion of Seismic Data, *Geophys. J. R. Astr. Soc.* **55**, 679–691 (1978).
48. Mandrel, J., Use of the Singular Value Decomposition in Regression Analysis, *Am. Stat.* **36**, 15–24 (1982).
49. Beck, J. V. and Arnold, K. J., *Parameter Estimation in Engineering and Science*, Wiley, New York, 1977.
50. Litkouhi, B. and Beck, J. V., Temperatures in Semi-Infinite Body Heated by Constant Heat Flux Over Half Space, *Heat Transfer 1982—Munchen Conference Proceedings*, Hemisphere Publ. Corp., New York, 1982, pp. 21–27, Vol. 2.
51. Carslaw, H. S. and Jaeger, J. C., *Conduction of Heat in Solids*, 2nd ed., Oxford Univ. Press, London, 1959.
52. Abramowitz, M. and Stegun, I. A., *Handbook of Mathematical Functions with Formulas, Graphs and Mathematical Tables*, National Bureau of Standards, Applied Mathematics Series, Vol. 55, 1964.
53. Payne, L. E., *Improperly Posed Problems in Partial Differential Equations*, SIAM, Philadelphia, PA., 1975.
54. Hadamard, J., *Lectures on Cauchy's Problem in Linear Partial Differential Equations*, Yale University Press, New Haven, CT, 1923.
55. Cannon, J. R. and Douglas, J., The Cauchy Problem for the Heat Equation, *SIAM J. Numer. Anal.* **3**, 317–336 (1967).
56. John, F., Numerical Solution of the Heat Equation for Preceeding Times, *Ann. Mat. Para. Appl.* **40**, 129–142 (1955).
57. Lawson, C. L. and Hanson, R. J., *Solving Least Squares Problems*, Prentice-Hall, Englewood Cliffs, NJ, 1974.
58. Murio, D. A., The Mollification Method and the Numerical Solution of an Inverse Heat Conduction Problem, *SIAM J. Sci. Stat. Comput.* **2**, 17–34 (1981).

PROBLEMS

1.1. Give a mathematical description of the IHCP for a solid, homogeneous sphere when the temperature is measured at the center point. The sphere radius is denoted a. Make a sketch of a sphere showing the various quantities. Form a graph that shows typical discrete temperatures measured at time steps of Δt with an initial temperature of $T_0 =$ constant. Also show an associated surface heat flux history.

1.2. A certain automobile brake is composed of a cast iron drum and two brake shoes. The rotating drum has an inner radius of 10 cm and an outer radius 10.7 cm. The heat transfer in the drum is assumed to be only in the radial direction and there is a convective boundary condition at the outer drum radius. The brake shoes are 0.5 cm thick, are inside the drum, and are stationary. They cover only 75% of the angular area of the drum. The inner surface of the brake shoes (the surface not in contact with the drum) can also be considered to have a convective boundary condition. Describe the inverse heat conduction problem(s) for determining the surface heat flux based on the drum surface area. Describe the problem mathematically and through the use of any needed sketches.

1.3. The temperature distribution in a semi-infinite body ($x \geq 0$) is given by

$$T - T_0 = (T_s - T_0)\,\mathrm{erfc}\left[\frac{x}{(4\alpha t)^{1/2}}\right]$$

where T_s is the surface temperature and T_0 is the initial temperature.

a. Derive an expression for the heat flux for any x. Also give the expression for $x = 0$.

Answer: $k(T_s - T_0)(\pi\alpha t)^{-1/2}$ for $x = 0$.

b. Plot the heat flux versus $\alpha t/E^2$ for $x = E$ and also show on the same plot the $x = 0$ curve.

1.4. For a semi-infinite body with a surface temperature given by

$$T(0, t) = 0, \quad t < 0$$
$$T(0, t) = Ct^{n/2}, \quad n = 1, 2, 3, \ldots, \quad t \geq 0$$

the temperature distribution is given by (Reference 51, p. 63)

$$T = C\Gamma\left(\frac{n+2}{2}\right)(4t)^{n/2} i^n\,\mathrm{erfc}\left[\frac{x}{(4\alpha t)^{1/2}}\right]$$

where (Reference 51, p. 483)

$$i^n\,\mathrm{erfc}(z) = \int_z^\infty i^{n-1}\,\mathrm{erfc}(u)\,du, \quad n = 1, 2, \ldots, \quad i^0\,\mathrm{erfc}(z) = \mathrm{erfc}(z)$$

44 CHAP.1 DESCRIPTION OF THE INVERSE HEAT CONDUCTION PROBLEM

Derive the expression for the surface heat flux,

$$q(0, t) = k\alpha^{-1/2} C t^{(n-1)/2} \frac{\Gamma\left(\frac{n}{2}+1\right)}{\Gamma\left(\frac{n}{2}+\frac{1}{2}\right)}$$

Find an expression for $T(0, t)/q(0, t)$ and comment on the result. Does this provide a relation that can be used for the IHCP? Give reasons.

1.5. Use the relation in Problem 1.3 for T at x not equal to zero to evaluate

$$q(0, t) \approx k \frac{T_s - T(x, t)}{x}$$

for the $\alpha t/x^2$ values of 0.01, 0.1, 1, 10, and 100. Compare with the exact result of Problem 1.3a and give some conclusions.

1.6. A solid copper billet 1.82 in. long and 1.00 in. in diameter is heated in a furnace and then removed. The density-specific heat product of the copper is 51 Btu/ft^3-F. Some of the temperature measurements are given below. The heat transfer model is

$$\rho c V \frac{dT}{dt} = qA$$

where V is the volume and A is the heated surface area (the cylindrical sides and the flat ends).

a. Using the forward difference approximation,

$$\left.\frac{dT}{dt}\right|_{t_i} = \frac{Y_{i+1} - Y_i}{\Delta t}$$

calculate the heat flux for $\Delta t = 96$ s using all the given data. Plot the results versus time.

b. Plot results on the same figure for the backward difference approximation.

c. Repeat part (a) using the central difference approximation,

$$\left.\frac{dT}{dt}\right|_{t_i} = \frac{Y_{i+1} - Y_{i-1}}{2\Delta t}$$

d. Which approximation (forward, backward, or central) gives the best results?

i	Time(s)	Y_i(°F)	i	Time(s)	Y_i(°F)
1	1632	142.93	8	2304	122.46
2	1728	139.34	9	2400	120.18

PROBLEMS

i	Time(s)	$Y_i(°F)$	i	Time(s)	$Y_i(°F)$
3	1824	136.04	10	2496	118.04
4	1920	132.94	11	2592	115.97
5	2016	130.07	12	2688	114.13
6	2112	127.39	13	2784	112.35
7	2208	124.88			

1.7. Use the forward difference method to approximate the first derivative of e^{-t} at $t=2$. Use $\Delta t = 0.001, 0.01, 0.1$, and 1. Use the exact values of e^{-t} first and then repeat the solution by adding the random errors of 0.000464, 0.000137, 0.002455, -0.000323 and -0.000068 to e^{-t} for $t=2, 2.001, 2.01, 2.1$, and 3, respectively. For example, $Y_1 = e^{-2} + 0.000464 = 0.135799$. (These random values have a standard deviation of 0.001 and are taken from the first row of Table 5.2.) What conclusions can be drawn from the approximations as $\Delta t \to 0$ for the exact values and from those with the random errors?

1.8. Find the average of the temperatures given in Problem 1.6. Calculate and plot the residuals defined by

$$e_i = Y_i - \hat{Y}_i$$

where $\hat{Y}_i$ is the average temperature. What is the main reason for the lack of randomness in the residuals?

1.9. The method of ordinary least squares involves minimizing the sum of squares function S,

$$S = \sum_{i=1}^{n} (Y_i - T_i)^2$$

with respect to the parameters. For the model

$$T_i = \beta_1 + \beta_2(t_i - \bar{t})$$

where

$$\bar{t} = \frac{1}{n} \sum_{i=1}^{n} t_i$$

derive the estimators for β_1 and β_2 which are respectively denoted $\hat{\beta}_1$ and $\hat{\beta}_2$,

$$\hat{\beta}_1 = \bar{Y} = \frac{1}{n} \sum_{i=1}^{n} Y_i$$

$$\hat{\beta}_2 = \frac{\sum (t_i - \bar{t}) Y_i}{\sum (t_i - \bar{t})^2}$$

1.10. Using the linear model in Problem 1.9, estimate $\hat{\beta}_1$ and $\hat{\beta}_2$ for the data of Problem 1.6. Also calculate and plot the residuals. Comment on the time variation of the residuals.

1.11. For measurements equally spaced in time, orthogonal polynomials can be used in least squares estimation. (See Reference 49, p. 248.) A polynomial model is

$$T_i = \beta_0 + \beta_1 t_i + \beta_2 t_i^2 + \cdots + \beta_r t_i^r \qquad \text{(a)}$$

where $t_{i+1} = t_i + \Delta t$. This model can be rewritten as

$$T_i = \alpha_0 p_0(t_i) + \alpha_1 p_1(t_i) + \alpha_2 p_2(t_i) + \cdots + \alpha_r p_r(t_i) \qquad \text{(b)}$$

where $p_0(t_i)$, $p_1(t_i)$, and $p_2(t_i)$ are

$$p_0(t_i) = 1$$

$$p_1(t_i) = \frac{t_i - \bar{t}}{\Delta t} = i - \frac{n+1}{2}, \quad \bar{t} = \frac{1}{n}\sum_{i=1}^{n} t_i$$

$$p_2(t_i) = \left(\frac{t_i - \bar{t}}{\Delta t}\right)^2 - \frac{n^2 - 1}{12} \qquad \text{(c)}$$

Verify that these $p_j(t_i)$ functions for $r=2$ are orthogonal; that is, show that

$$\sum_{j=1}^{n} p_j(t_i) p_k(t_i) = 0 \quad \text{for } j \neq k$$

$$\neq 0 \quad \text{for } j = k$$

and $j, k = 0, 1, 2$.
    Derive

$$\hat{\alpha}_M = \sum_{i=1}^{n} Y_i p_M(t_i) \left[\sum_{j=1}^{n} p_M^2(t_j)\right]^{-1} \qquad \text{(d)}$$

for $m = 0, 1, \ldots, r$.

1.12. For the polynomial and orthogonal polynomial models of Problem 1.11, prove for $r=1$ that

$$\hat{\beta}_0 = \hat{\alpha}_0 - \hat{\alpha}_1 \frac{\bar{t}}{\Delta t}, \quad \hat{\beta}_1 = \frac{\hat{\alpha}_1}{\Delta t}$$

and for $r=2$ that

$$\hat{\beta}_0 = \hat{\alpha}_0 - \hat{\alpha}_1 \frac{\bar{t}}{\Delta t} + \hat{\alpha}_2 \left[\left(\frac{\bar{t}}{\Delta t}\right)^2 - \frac{n^2 - 1}{12}\right]$$

$$\hat{\beta}_1 = \left(\hat{\alpha}_1 - \hat{\alpha}_2 \frac{2\bar{t}}{\Delta t}\right)\frac{1}{\Delta t}, \quad \hat{\beta}_2 = \frac{\hat{\alpha}_2}{(\Delta t)^2}$$

1.13. Using the $r=2$ orthogonal polynomial model given in Problem 1.11, estimate the parameters using ordinary least squares in Eq. (b) of Problem

PROBLEMS

1.11 for the data of Problem 1.6. Calculate the average estimated standard deviation and sample correlation coefficient. Discuss your results.

1.14. Derive estimators for β_1 and β_2 in the model
$$T_i = \beta_1 X_{i1} + \beta_2 X_{i2}$$
using the ordinary least square method which requires that
$$S = \sum (Y_i - T_i)^2$$
be minimized with respect to β_1 and β_2.

Answer:
$$\hat{\beta}_1 = \frac{d_1 c_{22} - d_2 c_{12}}{\Delta}, \quad \hat{\beta}_2 = \frac{d_2 c_{11} - d_1 c_{12}}{\Delta}$$

$$c_{kl} = \sum_{i=1}^{n} X_{ik} X_{il}, \quad d_k = \sum_{i=1}^{n} Y_i X_{ik}$$

$$\Delta = c_{11} c_{22} - c_{12}^2$$

1.15. With ordinary least squares, show that the estimator, $\hat{\beta}$, of β in the model $T = \beta$ is
$$\hat{\beta} = \frac{1}{n} \sum_{i=1}^{n} Y_i$$
and for the standard statistical assumptions being valid, show that the variance of $\hat{\beta}$ is
$$V(\hat{\beta}) = \frac{\sigma^2}{n}$$

1.16. Show that for the random variables Y_i and Y_j,
a. $V(Y_j) = E(Y_j^2) - E^2(Y_j)$
b. $\text{cov}(Y_i, Y_j) = E(Y_i Y_j) - E(Y_i) E(Y_j)$

1.17. What is the expected value for the random variable ε that has the probability density of
$$f_\varepsilon(y) = \begin{cases} 1 & \text{for } 0 < y < 1 \\ 0 & \text{otherwise} \end{cases}$$
What is the variance of ε?

1.18. What are the mean and the variance of a random variable ε that has the probability density
$$f_\varepsilon(y) = a \exp[-(y-5)^2], \quad -\infty < y < \infty$$
What is the value of a?

1.19. Using the first row of the random numbers of Table 5.2, generate a set of 10 correlated random numbers using

$$\varepsilon_{i+1} = 0.5\varepsilon_i + u_{i+1}, \quad i=1, 2, \ldots, 9$$

where the u_i are the entries in Table 5.2, with $\varepsilon_1 = u_1$ the first entry. Calculate the sample average and the associated residuals. Next calculate the sample correlation coefficient. Repeat for the next two rows of Table 5.2. Give some conclusions.

1.20. a. Plot the ratio of the temperature rise at $x=0$ of a finite plate with a constant heat flux at $x=0$ and insulated at $x=L$ to the temperature rise at $x=0$ of a semi-infinite body heated at $x=0$ by the same value of constant heat flux. Evaluate the ratio (at least) for $\alpha t/L^2 = 0.05$, 0.1, 0.15, 0.5, 1, and 5. Plot on a semi-logarithmic scale.
  b. Repeat part (a) for the temperature rise at $x=L$ in both bodies.
  c. Repeat part (a) for the temperature rise at $x=L/2$ in both bodies.
  d. Compare the results for the different cases particularly for small $\alpha t/L^2$ values.
  e. For the same additive measurement errors in both geometries, over what regions of x and t would an IHCP procedure give the same results? Different results?

1.21. For the problem

$$\alpha \frac{\partial^2 T}{\partial x^2} = \frac{\partial T}{\partial t}, \quad \alpha = \text{constant}$$

$$-k \left. \frac{\partial T}{\partial x} \right|_{x=0} = q_1(t) + q_2(t), \quad k = \text{constant}$$

$$\left. \frac{\partial T}{\partial x} \right|_{x=L} = 0$$

$$T(x, 0) = F(x)$$

show that the solution $T(x, t)$ is equal to the solution of the three problems,

$$T(x, t) = T_0(x, t) + T_1(x, t) + T_2(x, t)$$

where the three problems are described by

$$\alpha \frac{\partial^2 T_i}{\partial x^2} = \frac{\partial T_i}{\partial t}, \quad i = 0, 1, 2$$

problem 0:

$$\frac{\partial T_0}{\partial x} = 0 \quad \text{at} \quad x = 0 \quad \text{and} \quad L$$

$$T_0(x, 0) = F(x)$$

PROBLEMS

problems 1 and 2:

$$-k\frac{\partial T_i}{\partial x}\bigg|_{x=0} = q_i(t)$$

$$\frac{\partial T_i}{\partial x}\bigg|_{x=L} = 0$$

$$T_i(x, 0) = 0$$

for $i=1$ for problem 1 and $i=2$ for problem 2.

1.22. Use the summation relation in Problem 1.21 to verify Eq. (1.6.25). What are $q_1(t)$ and $q_2(t)$ for Eq. (1.6.25)?

1.23. Make a three-dimensional plot of $q(y, t)$ versus y (horizontal axis) and t (drawn as an axis 30° to the y-axis) for

$$q_{11}=3, \quad q_{12}=5, \quad q_{13}=7$$
$$q_{21}=2, \quad q_{22}=3, \quad q_{23}=4$$
$$q_{31}=1, \quad q_{32}=2, \quad q_{33}=3$$

1.24. Derive Eq. (1.6.44).

1.25. Derive Eq. (1.6.45).

1.26. The solution for the temperature in an infinite body is

$$T = T_0 + \frac{T_1 - T_0}{2}\left[\text{erfc}\frac{a-x}{2(\alpha t)^{1/2}} + \text{erfc}\frac{a+x}{2(\alpha t)^{1/2}}\right]$$

where $-\infty < x < \infty$ and the initial temperature distribution is (Reference 51, pp. 54, 55)

$$T = T_0 \quad |x| > a$$
$$T = T_1 \quad |x| < a$$

Using the principle of superposition give the temperature distribution in a semi-infinite body for the initial temperature distribution of

$$T = T_1 \quad |x| < a$$
$$T = T_2 \quad a < x < 3a$$
$$T = T_0 \quad \text{otherwise}$$

Plot the sensitivity coefficients,

$$X_1 = \frac{\partial T(0, t)}{\partial T_1} \quad \text{and} \quad X_2 = \frac{\partial T(0, t)}{\partial T_2}$$

versus t for $\alpha t/a^2 = 0$ to 5. (See Reference 51, p. 55.) The backward heat conduction problem is the estimation of the initial temperature distribution from temperatures measured at later times. Comment on the relative difficulty of recovering T_1 and T_2 from measurements of T at $x=0$ and $2a$ for the two cases of measurements of $\alpha t/a^2$ equal to (a) 0.2, 0.4, 0.6, 0.8, and 1; and (b) 2, 4, 6, 8, and 10.

1.27. a. For Eqs. (1.6.4), (1.6.5), and (1.6.7) and the convective boundary condition at $x=L$ of

$$-k\frac{\partial T}{\partial x}\bigg|_L = h[T(L, t) - T_\infty(t)]$$

where h is constant, derive the differential equation and boundary conditions for the sensitivity coefficient defined by Eq. (1.6.8).

b. Relate in words the results of part (a) to the problem of a unit step increase of surface heat flux.

c. For the case of $x=L$, $t_{M-1}=0$, $t<t_M$, $k(x)=$ constant, $c(x)=$ constant, and $hL/k=1$, find numerical values for kX_M/L for $t^+=0.25, 0.5$, and 1. Compare the values with those for x/L in Table 1.1.

CHAPTER 2
EXACT SOLUTIONS OF THE INVERSE HEAT CONDUCTION PROBLEM

2.1 INTRODUCTION

Exact solutions of the inverse heat conduction problem are very important because (1) they provide closed form expressions for the heat flux in terms of temperature measurements, (2) they give considerable insight into the characteristics of inverse problems, and (3) they provide standards of comparison for approximate methods. Inverse heat conduction problems can be divided into steady-state and transient problems. The steady-state inverse problem is simpler in that the only necessary thermal property is the thermal conductivity k, and a temperature history is not required. The transient inverse heat conduction problem can be divided into two categories: lumped thermal capacitance and distributed thermal capacitance. The transient case requires many discrete temperature measurements. For lumped thermal capacitance, the important thermal property is the volumetric heat capacity ρc. If the thermal capacitance is distributed, then the thermal conductivity k must be known in addition to the volumetric heat capacity. Throughout Chapter 2, the thermal properties are assumed to be independent of temperature; this assumption is one of the weaknesses of exact solutions.

Section 2.2 considers one-dimensional steady-state problems in which the temperature is known at two or more locations. Section 2.3 examines the lumped thermal capacitance case and some numerical approximations to the exact solution. Section 2.4 considers a planar semi-infinite body for which the surface temperature history is known; an approximate technique for numerically evaluating the resulting integral and an example problem are presented. Section 2.5 presents the development of an exact solution for a one-dimensional planar body with a temperature sensor at an arbitrary depth E below the heated surface; this solution requires the existence of all order derivatives of both the experimental temperature data $Y(t)$ and the heat flux q_E passing through depth E. Results for solid spheres and cylinders are also presented.

2.2 STEADY-STATE SOLUTION

One of the simpler inverse solutions is for steady-state heat flow through a planar one-dimensional body with constant thermal properties. For this situation, Fourier's Law,

$$q = -k\frac{dT}{dx} \tag{2.2.1}$$

is a differential equation and can be integrated directly. The entering heat flux, q, is the same as at any location x. Suppose the steady-state temperature is known at two depths (x_1, x_2) below the surface, as shown in Figure 2.1. Integrating Eq. (2.2.1) between locations x_1 and x_2 and solving for q gives,

$$\hat{q} = k\frac{(Y_1 - Y_2)}{x_2 - x_1} \tag{2.2.2}$$

where Y_i represents the experimental value of temperature at depth x_i. Equation (2.2.2) demonstrates that a minimum of two experimental temperature measurements and their corresponding locations along with the thermal conductivity must be known to determine the heat flux. Note that the heat flux is linear in the experimental temperature measurements Y_i; this linearity occurs repeatedly for constant-property IHCP's.

As indicated previously, a minimum of two temperature measurements is necessary for the steady-state determination of heat flux. Suppose there are J temperature sensors located in a planar body for which the heat flow is one-dimensional. While there are J sensors, the number of distinct sensor locations may be less than J due to multiple sensors at the same depth. The steady-state heat flux can be calculated by minimizing the least squares error between the computed and experimental temperatures. In order to generalize the analysis, assume that some of the sensors are more accurate than others, as indicated

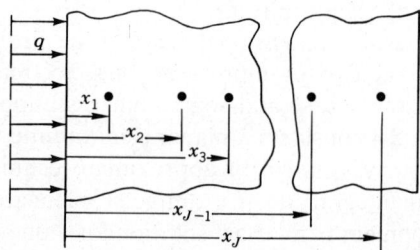

FIGURE 2.1 Steady-state temperature measurements at J locations.

SEC. 2.2 STEADY-STATE SOLUTION

by the weighting factors w_i. A weighted least squares criterion is defined as

$$S = \sum_{j=1}^{J} w_j^2 (Y_j - T_j)^2 \qquad (2.2.3)$$

Fourier's law indicates the temperature profile must be linear,

$$T(x) = ax + b = -q\frac{x}{k} + T_0 \qquad (2.2.4)$$

There are two parameters (q, T_0) in Eq. (2.2.4) that are determined such that the weighted least squares error is a minimum. (T_0 is the temperature at $x=0$.) Differentiating Eq. (2.2.3) with respect to the two unknown parameters gives

$$\frac{\partial S}{\partial T_0} = -2 \sum_{i=1}^{J} w_j^2 (Y_j - T_j) \frac{\partial T_j}{\partial T_0} = 0 \qquad (2.2.5)$$

$$\frac{\partial S}{\partial q} = -2 \sum_{j=1}^{J} w_j^2 (Y_j - T_j) \frac{\partial T_j}{\partial q} = 0 \qquad (2.2.6)$$

Equations (2.2.5) and (2.2.6) involve two sensitivity coefficients which can be evaluated from Eq. (2.2.4),

$$\frac{\partial T_j}{\partial T_0} = 1, \quad \frac{\partial T_j}{\partial q} = -\frac{x_j}{k} \qquad (2.2.7)$$

Substituting Eqs. (2.2.4) and (2.2.7) into Eqs. (2.2.5) and (2.2.6), replacing q by its estimate $\hat{q}$, and simplifying, the following two normal equations are obtained:

$$T_0 \sum_{j=1}^{J} w_j^2 - \frac{\hat{q}}{k} \sum_{j=1}^{J} w_j^2 x_j = \sum_{j=1}^{J} w_j^2 Y_j \qquad (2.2.8)$$

$$\frac{-T_0}{k} \sum_{j=1}^{J} w_j^2 x_j + \frac{\hat{q}}{k^2} \sum_{j=1}^{J} w_j^2 x_j^2 = \frac{-1}{k} \sum_{j=1}^{J} w_j^2 Y_j x_j \qquad (2.2.9)$$

Solving this system of equations for the unknown heat flux $\hat{q}$ gives,

$$\hat{q} = -k \frac{\left(\sum_{j=1}^{J} w_j^2\right)\left(\sum_{j=1}^{J} w_j^2 x_j Y_j\right) - \left(\sum_{j=1}^{J} w_j^2 x_j\right)\left(\sum_{j=1}^{J} w_j^2 Y_j\right)}{\left(\sum_{j=1}^{J} w_j^2\right)\left(\sum_{j=1}^{J} w_j^2 x_j^2\right) - \left(\sum_{j=1}^{J} w_j^2 x_j\right)^2} \qquad (2.2.10)$$

It can be demonstrated that Eq. (2.2.10) reduces to Eq. (2.2.2) for $J=2$ and $w_1, w_2 \neq 1$. Again, note that the unknown heat flux is linear in the temperature measurements.

Equation (2.2.10) could also be developed by (1) determining the two constants a and b in Eq. (2.2.4) by fitting a weighted least squares curve to the experimental temperature data and (2) differentiating the curve to determine the heat flux (see Problem 2.1). However, the approach of using sensitivity coefficients $(\partial T/\partial q)$ is more consistent with the remainder of the text. Results similar to Eq. (2.2.10) can also be developed for cylindrical and spherical geometries (see Problems 2.2 and 2.3).

2.3 TRANSIENT ANALYSIS OF BODIES WITH SMALL INTERNAL THERMAL RESISTANCE

2.3.1 Exact Solution

For bodies in which the thermal conductivity is very large and/or the characteristic length scale $L(=\text{volume/surface area})$ is small, it is possible to ignore the internal thermal resistance. This analysis is referred to as the lumped (thermal) capacitance analysis. The energy balance on a body of arbitrary shape and of surface area A, volume V, and uniform temperature T is

$$q(t) = \rho c \frac{V}{A} \frac{dT}{dt} \qquad (2.3.1)$$

Note that $q(t)$ depends on the rate of change of temperature at time t and not on temperature data for all times past and future. If the heat flux is known, the temperature response of the body can be calculated by integrating Eq. (2.3.1)

$$T(t) = T_0 + \int_0^t q(\lambda) \frac{A}{\rho c V} d\lambda \qquad (2.3.2)$$

Note that the temperature at any time depends only on the past history and not on the future values of $q(t)$. Heat flux measurement devices that satisfy the assumption of negligible internal resistance are often referred to as slug calorimeters.

2.3.2 Approximate Solutions

Although Eq. (2.3.1) is an exact solution for the heat flux, practical application of this equation is generally with data available only at discrete times. Consequently, some difference approximations for dT/dt are needed. In the discussion that follows, the temperature data (Y_i) are taken at equally spaced time increments Δt. Three common two-point difference approximations to dT/dt are as follows:

$$\left.\frac{dT}{dt}\right|_{t_M} = \frac{Y_M - Y_{M-1}}{\Delta t} + O(\Delta t) \quad \text{Backward Difference (BD)} \qquad (2.3.3)$$

$$\left.\frac{dT}{dt}\right|_{t_M} = \frac{Y_{M+1} - Y_M}{\Delta t} + O(\Delta t) \quad \text{Forward Difference (FD)} \qquad (2.3.4)$$

$$\left.\frac{dT}{dt}\right|_{t_M} = \frac{Y_{M+1} - Y_{M-1}}{2\Delta t} + O(\Delta t^2) \quad \text{Central Difference (CD)} \qquad (2.3.5)$$

SEC. 2.3 TRANSIENT ANALYSIS OF BODIES

The symbol $O(\ )$ denotes order of magnitude of the truncation error, as determined by a Taylor series expansion. Note that the CD approximation is a higher-order approximation and thus for small Δt has a smaller truncation error than either BD or FD. The BD and FD results can be derived by passing a straight line through two points, and the CD results can be developed by passing a parabola through three points. Higher-order results can be obtained by increasing the number of points through which a polynomial is required to pass. For example, a fourth-order polynomial passing through five equally spaced points yields

$$\left.\frac{dT}{dt}\right|_{t_M} = \frac{Y_{M-2} - 8Y_{M-1} + 8Y_{M+1} - Y_{M+2}}{12\Delta t} + O(\Delta t^4) \qquad (2.3.6)$$

Since the truncation error becomes smaller with increasing polynomial order, one might be inclined to think that the five-point equation is superior to the three-point (CD) equation. This would be true if the data were errorless; however, it is demonstrated that errors in the temperature data play a crucial role in determining the accuracy of the computed heat flux. Instead of requiring a fourth-order polynomial to pass exactly through five data point, an alternative approach is to least-squares fit a straight line through five points and evaluate the slope at the center point. This approximation can be determined from Eq. (2.2.10) by choosing all the weighting factors equal to unity and considering the five equally spaced points of $x_i = -2\Delta t, -\Delta t, 0, \Delta t, 2\Delta t$; the result is

$$\left.\frac{dT}{dt}\right|_{t_M} = \frac{-2Y_{M-2} - Y_{M-1} + Y_{M+1} + 2Y_{M+2}}{10\Delta t} \qquad (2.3.7)$$

Note that when using any of the foregoing derivative approximations in conjunction with the exact solution Eq. (2.3.1), the heat flux is linear in the temperature measurements provided the volumetric heat capacity ρc is constant. All of these results have the appearance of a digital filter.

2.3.3 Temperature Errors and Approximate Solutions

Any measurement of temperature contains errors, and these errors have an impact on the computed heat flux. A convenient assumption to be used throughout this book is that the temperature errors are additive (see Section 1.4.2). This allows the measured temperature Y_i to be written as the sum of the true temperature T_i and an associated error δY_i

$$Y_i = T_i + \delta Y_i \qquad (2.3.8)$$

Let us determine the error in the derivative approximation caused by temperature errors. For the BD method,

$$\left.\frac{dT}{dt}\right|_{t_M} \approx \frac{T_M - T_{M-1}}{\Delta t} + \frac{\delta Y_M - \delta Y_{M-1}}{\Delta t} \qquad (2.3.9)$$

The effect of the temperature errors is contained in the term $(\delta Y_M - \delta Y_{M-1})/\Delta t$.

Note that the form of the error term is the same as the difference approximation itself. This will always occur in linear problems with additive errors. Similar results can be written down by inspection for the other difference approximations considered.

It is important to understand how a single temperature error affects the heat flux computed at the same time the error was made and how the error affects any subsequent calculations of heat flux. Suppose the backward difference approximation is used and all temperature measurement errors are identically zero except δY_M. From Eqs. (2.3.1) and (2.3.9),

$$q(t_M) = q_M = \rho c \frac{V}{A} \left(\frac{T_M - T_{M-1}}{\Delta t} + \frac{\delta Y_M}{\Delta t} \right) \qquad (2.3.10)$$

$$q_{M+1} = \rho c \frac{V}{A} \left(\frac{T_{M+1} - T_M}{\Delta t} - \frac{\delta Y_M}{\Delta t} \right) \qquad (2.3.11)$$

$$q_{M+2} = \rho c \frac{V}{A} \left(\frac{T_{M+2} - T_{M+1}}{\Delta t} \right) \qquad (2.3.12)$$

The single error δY_M affects only the calculations of q_M and q_{M+1} for the BD method, with all other q calculations being unaffected. For the FD method, only q_{M-1} and q_M are affected by the single error δY_M. At first glance, it might seem inappropriate to study only a single error because all temperature measurements are likely to have errors. However, the effects of all temperature errors can be superposed for linear problems (constant properties). If all errors δY_i are the same, there will not be any heat flux error because they cancel out at each time step. This can be understood by realizing that the slope of the temperature response curve does not change if all temperature measurements are shifted by the same amount. In general, all difference approximations can be written as

$$\left. \frac{dT}{dt} \right|_{t_M} \approx \sum_{i=1}^{n} a_i Y_{M-j+i} \qquad (2.3.13)$$

where n is the number of points used in the difference approximation; Eq. (2.3.13) is a compact way of writing Eqs. (2.3.3)–(2.3.7). For all of the difference approximations considered in this section, $\sum_{i=1}^{n} a_i = 0$; it is this characteristic that allows a uniform shift in all Y_i to have no effect on the slope calculation.

If certain statistical assumptions are made about the temperature errors, then it is possible to calculate the standard deviation of the error in the difference approximation. For example, if the temperature errors are independent, additive, and of constant variance (σ^2), the methods of Beck and Arnold[1] can be used to calculate the results in Table 2.1 (see also Section 1.4). These theoretical results favor the least squares approach because the standard deviation of the heat flux estimate is smallest.

SEC. 2.3 TRANSIENT ANALYSIS OF BODIES

TABLE 2.1 Standard Deviation of Error in Derivative Approximation for Temperature Errors that are Independent, Additive, and of Constant Variance (σ^2)

Difference Approximation	Equation Number	Standard Deviation Derivative Approximation
Backward	(2.3.3)	$\sqrt{2}\sigma/\Delta t = 1.414\sigma/\Delta t$
Forward	(2.3.4)	$\sqrt{2}\sigma/\Delta t = 1.414\sigma/\Delta t$
Central	(2.3.5)	$\sqrt{2}\sigma/(2\Delta t) = 0.707\sigma/\Delta t$
Five-point, fourth-order	(2.3.6)	$0.95\sigma/\Delta t$
Least-squares, straight-line, five-point	(2.3.7)	$\sigma/(\sqrt{10}\Delta t) = 0.316\sigma/\Delta t$

TABLE 2.2 Temperature Data for Example 2.1; from Beck and Arnold[1]

Observation Number	Time (s)	Y_i Temperature of Billet (°F)
9	768	191.65
10	864	184.44
11	960	177.64
12	1056	171.41
13	1152	165.04
14	1248	159.89
15	1344	155.19
16	1440	150.78
17	1536	146.68

EXAMPLE 2.1. A solid copper billet 0.0462 m (1.82 in.) long and 0.0254 m (1 in.) in diameter is heated in a furnace and then removed. Two thermocouples are attached to the billet. Some temperatures, Y_i, given by one of the thermocouples are listed in Table 2.2 as functions of time. See also the plot of Y_i versus time in Figure 2.2. For this test, $\rho c V/(A\Delta t) = 200$ W/m²-K (31 Btu/hr-ft²-F). Compare the five methods presented in this section for calculating the derivative dT/dt. These data are from an actual experiment presented on p. 243 of Beck and Arnold[1].

Solution. The results of Example 2.1 are summarized in Table 2.3 and Figure 2.3. Several conclusions can be drawn from this example.

1. Both FD and BD give the same numerical results but they are displaced in time by Δt.
2. All methods except BD use temperature measurements at times greater than the calculation time. This will be referred to as using "future (temperature) information."

CHAP. 2 SOLUTIONS OF THE INVERSE HEAT CONDUCTION PROBLEM

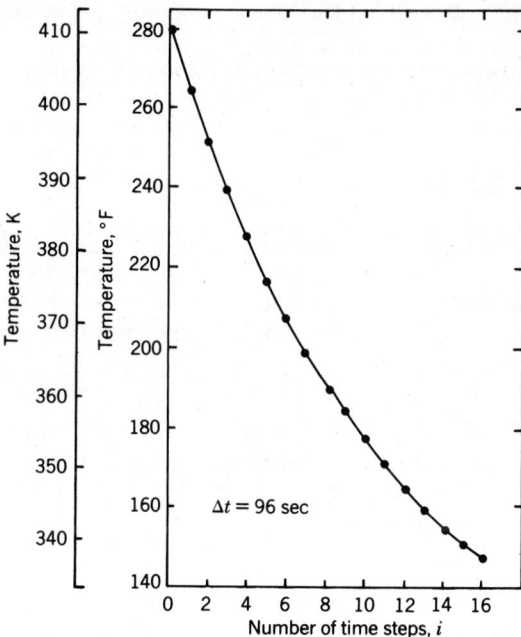

FIGURE 2.2 Temperatures for cooling billet example (Example 2.1).

TABLE 2.3 Average Temperature Difference ($\overline{\Delta T}$) for Example 2.1.

$$q_M = \frac{\rho c V}{A \Delta t} \overline{\Delta T}, \quad \overline{\Delta T} \text{ in } °F$$

M	$Y_M - Y_{M-1}$	$Y_{M+1} - Y_M$	$\frac{1}{2}(Y_{M+1} - Y_{M-1})$	Five-point, fourth order	Five-point linear least squares
9	—	−7.21	—	—	—
10	−7.21	−6.80	−7.01	—	—
11	−6.80	−6.23	−6.52	−6.47	−6.63
12	−6.23	−6.37	−6.30	−6.35	−6.17
13	−6.37	−5.05	−5.76	−5.81	−5.64
14	−5.05	−4.70	−4.93	−4.85	−5.11
15	−4.70	−4.41	−4.56	−4.54	−4.58
16	−4.41	−4.10	−4.26	—	—
17	−4.10	—	—	—	—

3. All methods except FD and BD show a continuous decay of slope with time. The physics of the problem dictates that the magnitude of the average temperature difference should decrease continuously with time.

4. By using more information (additional temperature measurement) to calculate heat flux at a given time, some values cannot be calculated at both early and late times.

SEC. 2.4 HEAT FLUX FROM MEASURED SURFACE TEMPERATURE HISTORY

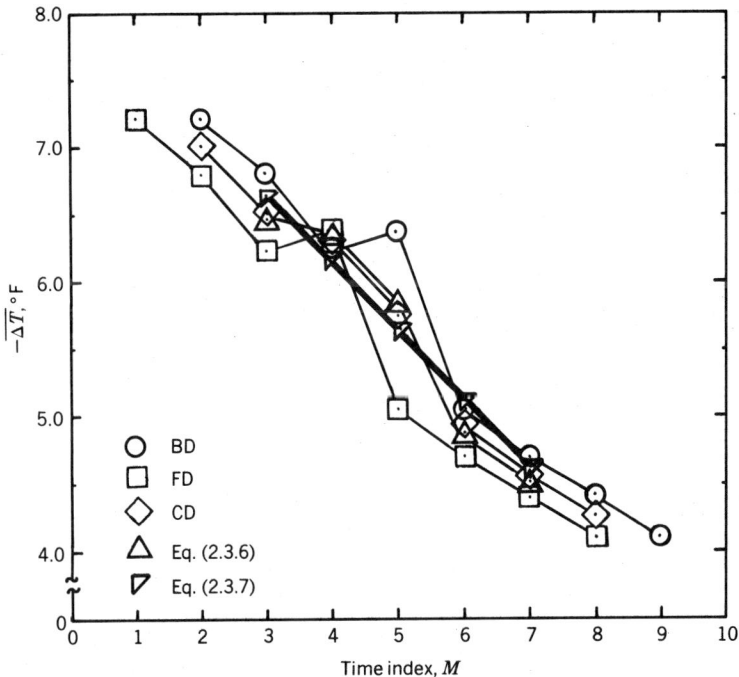

FIGURE 2.3 Average temperature difference $(\overline{\Delta T})$ for Example 2.1.

5. The linear least-squares result is appealing because the slope continuously decays with time in a smooth manner. □

2.4 HEAT FLUX FROM MEASURED SURFACE TEMPERATURE HISTORY

2.4.1 Exact Results for Continuous Surface Temperature History

If the surface temperature, $Y(t)$, of an object is known continuously as a function of time, some relatively simple exact solutions exist for the heat flux variation with time. This surface temperature specification yields a simpler inverse problem because the known surface temperature can be treated as a boundary condition in the traditional sense. One approach to this problem is to determine the temperature distribution within the body and then the temperature gradient at the surface is used to determine the heat flux. If the thermal properties are treated as constant, then Duhamel's theorem provides a convenient means of calculating the temperature field. The analysis starts with the temperature form of Duhamel's theorem[2,3,4]

$$T(x,t) = T_0 + \int_{t_0}^{t} u(x, t-\lambda) \frac{dY(\lambda)}{d\lambda} d\lambda + \sum_{i=0}^{N-1} u(x, t-\lambda_i) \Delta Y_i \qquad (2.4.1)$$

where $u(x,t)$ is the temperature response function for a body at zero initial temperature and subjected to a unit step in surface temperature, $Y(t)$ is the surface temperature variation with time, and T_0 is the uniform initial (for $t \leq t_0$) temperature. The integral in Eq. (2.4.1) allows a continuous surface temperature in time, and the summation term allows N discrete steps in surface temperature occurring at $\lambda_i = i\Delta t$. Some understanding of Duhamel's theorem can be gained by considering the discrete version of Eq. (2.4.1). If a series of steps in the surface temperature occurs as shown in Figure 2.4, the temperature at position x and at time t for $2\Delta t < t < 3\Delta t$ is given by

$$T(x, t) = T_0 + u(x, t)\Delta Y_0 + u(x, t - \Delta t)\Delta Y_1 + u(x, t - 2\Delta t)\Delta Y_2 \quad (2.4.2)$$

The magnitude of the actual step in surface temperature is multiplied by the response due to a unit step in surface temperature, $u(x, t)$. The unit step response, u, must be shifted in time to correspond to the time when the temperature step actually occurs. Additional information on Duhamel's theorem can be found in Carslaw and Jaeger[2], Schneider[3], Myers,[4] and Chapter 3.

If only the heat flux is of interest, it is not necessary to calculate the entire temperature field. The heat flux at the active surface can be determined from Fourier's law by differentiating Eq. (2.4.1) to get [for $Y(t)$ continuous in time t],

$$q(t) = -k \left.\frac{\partial T}{\partial x}\right|_{x=0} = -k \int_{t_0}^{t} \left.\frac{\partial u(x, t - \lambda)}{\partial x}\right|_{x=0} Y'(\lambda) d\lambda \quad (2.4.3)$$

where $Y'(\lambda) = dY/d\lambda$. Equation (2.4.3) is exact within the restrictions of Duhamel's theorem. In order to apply Eq. (2.4.3), the derivative of the unit step response must be known. The simplest case of a unit step function is for a semi-infinite planar solid.

$$u(x, t) = 1 - \mathrm{erf}\left(\frac{x}{2\sqrt{\alpha t}}\right), \quad \left.\frac{\partial u}{\partial x}\right|_{x=0} = \frac{-1}{\sqrt{\pi \alpha t}} \quad (2.4.4)$$

Another simple case is the infinite plate of thickness L where the step change in surface temperature occurs at $x = 0$ and the "inactive surface" is perfectly insulated.

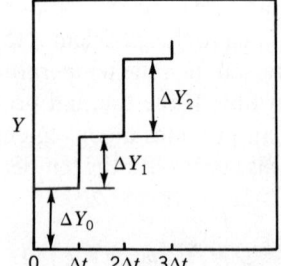 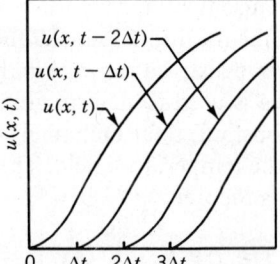

FIGURE 2.4 Illustration of temperature form of Duhamel's theorem.

SEC. 2.4 HEAT FLUX FROM MEASURED SURFACE TEMPERATURE HISTORY

$$u(x,t) = 1 - 2\sum_{n=0}^{\infty} \frac{1}{\gamma_n} e^{-\alpha\gamma_n^2 t/L^2} \sin\frac{\gamma_n x}{L}$$

$$\left.\frac{\partial u}{\partial x}\right|_{x=0} = -\frac{2}{L}\sum_{n=0}^{\infty} e^{-\alpha\gamma_n t/L^2} \qquad (2.4.5)$$

$$\gamma_n = (2n+1)\frac{\pi}{2}, \quad n=0,1,2,\ldots$$

Other step function solutions for various geometries and/or "inactive surface" boundary conditions are available in the literature. The analysis that follows is restricted to the semi-infinite planar solid; for this case, Eq. (2.4.3) becomes

$$q(t) = \sqrt{\frac{k\rho c}{\pi}} \int_{t_0}^{t} \frac{Y'(\lambda)}{\sqrt{t-\lambda}} d\lambda \qquad (2.4.6)$$

$$= \sqrt{\frac{k\rho c}{\pi}} \left\{ \frac{1}{\sqrt{t}} [Y(t) - Y(t_0)] + \frac{1}{2} \int_{t_0}^{t} \frac{Y(t) - Y(\lambda)}{(t-\lambda)^{3/2}} d\lambda \right\} \qquad (2.4.7)$$

Equation (2.4.7) comes from integration by parts of Eq. (2.4.6). Note that both of these forms have singularities at $\lambda = t$. If a purely numerical procedure is used, these singularities can cause some difficulty. Therefore, an analytical integration procedure is considered. The reader is reminded that Eqs. (2.4.6) and (2.4.7) presume that the body is at a uniform temperature $Y(t_0)$ for $t \leq t_0$.

It may be possible to use a gaussian quadrature[5] to treat this singularity; however, such an approach would not allow the experimenter freedom in choosing times at which the surface temperature is sampled. In the next section, a combined analytical numerical procedure will be explored.

2.4.2 Approximate Results for Semi-Infinite Body with Surface Temperature Measured at Discrete Times

Assume that the surface temperature Y_j is measured only at discrete times t_j. Between successive times, the surface temperature is assumed to vary linearly with time. For these assumptions, Eq. (2.4.6) can be integrated analytically to give

$$\hat{q}(t_M) = \hat{q}_M = 2\sqrt{\frac{k\rho c}{\pi}} \sum_{i=1}^{M} \left(\frac{Y_i - Y_{i-1}}{t_i - t_{i-1}}\right)\left(\sqrt{t_M - t_{i-1}} - \sqrt{t_M - t_i}\right) \qquad (2.4.8a)$$

$$= 2\sqrt{\frac{k\rho c}{\pi}} \sum_{i=1}^{M} \frac{Y_i - Y_{i-1}}{\sqrt{t_M - t_i} + \sqrt{t_M - t_{i-1}}} \qquad (2.4.8b)$$

where the $\hat{}$ symbol denotes an estimated value.

The important material property group is $(k\rho c)^{1/2}$; values of this parameter are presented in Table 2.4. For a given heat flux into a semi-infinite body, a

TABLE 2.4 Thermal Property $\sqrt{k\rho c}$

Material	(J/m²-K-$\sqrt{s}$)	(Btu/ft²-F-$\sqrt{hr}$)
Copper, pure	1150	106.8
Silver, pure	1000	92.9
Aluminum, pure	703	65.3
Steel, low carbon	442	41.0
Steel, 20% Cr	282	26.2
Steel, 40% Ni	198	18.4
Glass, plate	42	3.9

body with a small value of $k\rho c$ has a larger surface temperature rise than bodies with large $k\rho c$. Physically, this occurs because a large-k body is able to remove the heat from the surface more rapidly. Note that the heat flux is linear in the temperature measurements, provided $(k\rho c)^{1/2}$ is independent of temperature.

EXAMPLE 2.2. In order to demonstrate the method given by Eq. (2.4.8), an example problem is selected for which an exact solution is available. Consider a heat flux with a parabolic variation with time

$$\frac{q}{q_{max}} = 4 \frac{t}{t_{max}} \left(1 - \frac{t}{t_{max}}\right) \qquad (2.4.9)$$

Using the analytical solution of Carslaw and Jaeger[2] for $q \sim t^n$ and superposition, the surface temperature variation is given by (see also Problem 1.4).

$$T_s(t) - T_0 = \frac{4q_{max}}{\sqrt{\frac{k\rho c}{t_{max}}}} \left[\frac{\Gamma(2)}{\Gamma(5/2)} \left(\frac{t}{t_{max}}\right)^{3/2} - \frac{\Gamma(3)}{\Gamma(\frac{7}{2})} \left(\frac{t}{t_{max}}\right)^{5/2}\right] \qquad (2.4.10)$$

where $\Gamma(x)$ is the Gamma function. [$\Gamma(2)=1$, $\Gamma(3)=2$, $\Gamma(5/2)=3\pi^{1/2}/4$ and $\Gamma(7/2)=15\pi^{1/2}/8$.] The results of the foregoing analytical solution are given in Figure 2.5 for the following parameters:

$$q_{max} = 12{,}000 \text{ Btu/ft}^2\text{-sec}, \quad T_0 = 540°\text{R}$$

$$k\rho c = 3.2855 \left(\frac{\text{Btu}}{\text{ft}^2\text{s-R}}\right)^2\text{-s} \quad \text{(Copper at elevated temperature)}$$

These conditions are intended to be representative of those encountered in traversing a copper calorimeter across a high heat flux arc jet. The exact temperature data calculated from Eq. (2.4.10) for a time step of $\Delta t = 0.001$ s applied to the inverse solution given by Eq. (2.4.8), yield the results shown in Figure 2.6. At early times, the percent error in the computed heat flux is in excess of -15%; after a time of 0.01 s the heat flux error remains nearly constant at approximately -2% until approximately 0.065 s. As the final time of 0.07 s is approached, the errors become large and positive. This is because the heat flux is approaching zero at 0.07 s. □

SEC. 2.4 HEAT FLUX FROM MEASURED SURFACE TEMPERATURE HISTORY 63

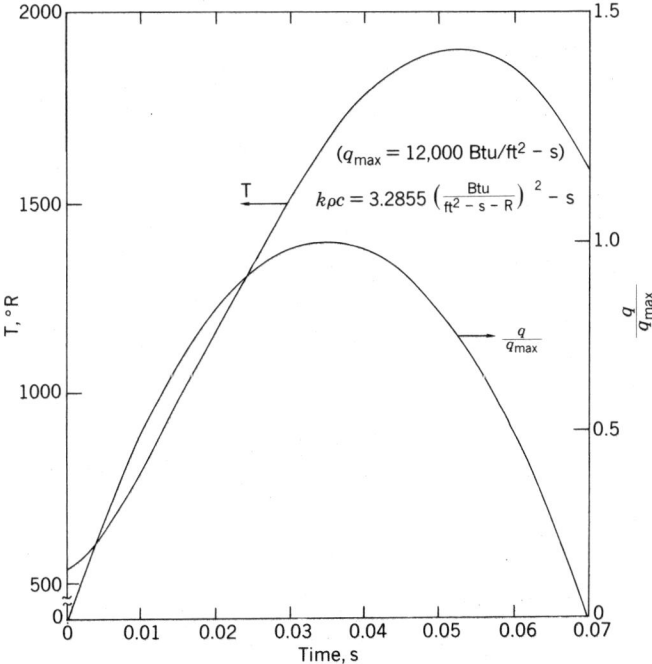

FIGURE 2.5 Parabolic heat flux history and corresponding surface temperature variation.

Although the results for the preceding example are good, the use of exact temperature data is not a very severe test of any inverse method. In order to simulate the effect of temperature errors on the computed heat flux history, a random error term was added to the results computed from Eq. (2.4.10)

$$Y_m = T_s(t_m) + \theta \cdot \Delta T$$

where θ is a random variable of uniform distribution with values in the range $[-1, 1]$ and ΔT is the maximum magnitude of the temperature error. For the calculations that follow, it is assumed that $\Delta T = 5R$; the calculated errors are shown in Figure 2.6. In general, the results using simulated temperature errors scatter about the results using errorless temperature data. However, the realistic temperature data produce a much wider variation in the heat flux error.

2.4.3 Temperature Error Propagation in Eq. (2.4.8)

It was demonstrated in Section 2.3.3 that if an error δY_i is made in a single temperature measurement Y_i, the corresponding heat flux error becomes identically zero after n time steps where n is the number of measurement points used in approximating the derivative dT/dt. For a semi-infinite body that has a

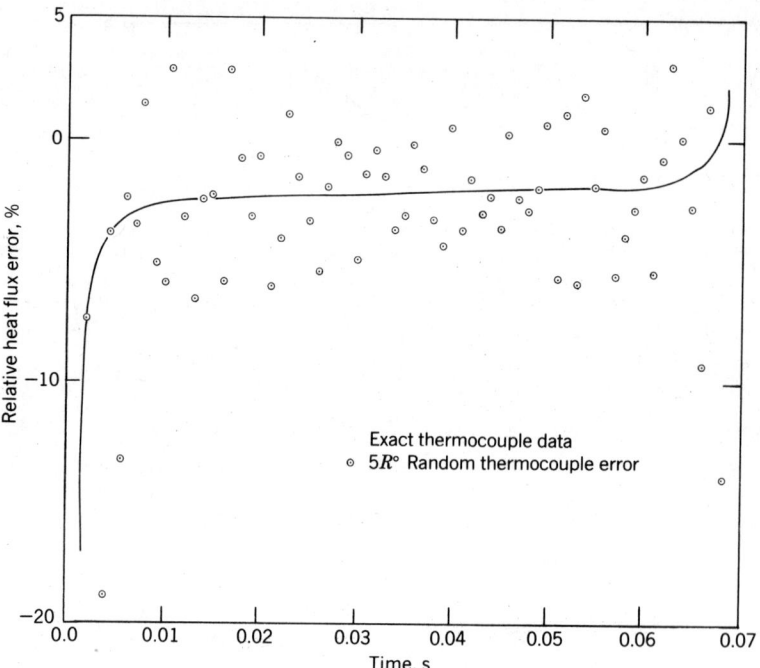

FIGURE 2.6 Relative heat flux error (%) using exact and inexact thermocouple data.

finite internal thermal resistance, the heat flux error corresponding to a single temperature error takes an infinite number of time steps to decay to zero. This can be demonstrated by considering the first few terms of Eq. (2.4.8),

$$\hat{q}_M = 2\sqrt{\frac{k\rho c}{\pi \Delta t}} [(Y_1 - Y_0)(\sqrt{M} - \sqrt{M-1}) + (Y_2 - Y_1)(\sqrt{M-1} - \sqrt{M-2}) + \cdots],$$
$$M = 1, 2, \ldots \qquad (2.4.11)$$

where the data were taken with equal increments of Δt. Note that the initial temperature measurement Y_0 appears in only one place in Eq. (2.4.11). Suppose there is an error δY_0 in the temperature $Y_0 (= T_0 + \delta Y_0)$ but all other temperature measurements are exact. The corresponding error in heat flux can be determined from Eq. (2.4.11),

$$\hat{q}_M = 2\sqrt{\frac{k\rho c}{\pi \Delta t}} [(Y_1 - T_0)(\sqrt{M} - \sqrt{M-1}) + (Y_2 - Y_1)(\sqrt{M-1} - \sqrt{M-2}) + \cdots]$$
$$-2\sqrt{\frac{k\rho c}{\pi \Delta t}} \delta Y_0 (\sqrt{M} - \sqrt{M-1}) \qquad (2.4.12)$$

The term inside the brackets in Eq. (2.4.12) is the heat flux for errorless data;

SEC. 2.4 HEAT FLUX FROM MEASURED SURFACE TEMPERATURE HISTORY

the term in Eq. (2.4.12) that contains the temperature error δY_0 represents the corresponding heat flux error. Let $\delta q_M^{(0)}$ be the error in $\hat{q}_M$ corresponding to the temperature error δY_0; the superscript on δq_M is important because it indicates the time at which the temperature error occurred. Then

$$\frac{-1}{2\sqrt{\frac{k\rho c}{\pi \Delta t}}} \frac{\delta q_M^{(0)}}{\delta Y_0} = \sqrt{M} - \sqrt{M-1}, \quad M = 1, 2, \ldots \quad (2.4.13)$$

represents the dimensionless heat flux error and shows how it decays for subsequent times. The normalization is conveniently chosen so that the first value is unity. Table 2.5 presents results showing how the dimensionless heat flux error due to an initial temperature error, δY_0, decays with time. The results of Table 2.5 indicate that a positive temperature error δY_0 causes a negative error in heat flux and this error damps very slowly with time. Even after 100 time steps, the dimensionless heat flux error is still 5% of its initial value. If the initial heat flux error was large, then this heat flux may be very inaccurate. For a given value of temperature error δY_0, the heat flux error is proportional to $(k\rho c)^{1/2}$ and inversely proportional to $1/(\Delta t)^{1/2}$; small time steps and large values of $k\rho c$ both produce large heat flux errors.

Equation (2.4.11) reveals that an error in Y_1 may be quite different from an error in Y_0 because Y_1 appears in two places in the heat flux equation. If a single temperature error δY_M occurs, the resulting heat flux error and its decay with

TABLE 2.5 Heat Flux Error for δY_0 at Initial Time

M (time index)	$\dfrac{-1}{2\sqrt{\frac{k\rho c}{\pi \Delta t}}} \dfrac{\delta q_M^{(0)}}{\delta Y_0}$
1	1.000
2	0.414
3	0.318
4	0.268
5	0.236
6	0.213
7	0.196
8	0.183
9	0.172
10	0.162
100	0.050

subsequent time steps are as follows:

$$\frac{1}{2\sqrt{\frac{k\rho c}{\pi \Delta t}}} \frac{\delta q_M^{(M)}}{\delta Y_M} = 1.0 \qquad (2.4.14a)$$

$$\frac{1}{2\sqrt{\frac{k\rho c}{\pi \Delta t}}} \frac{\delta q_{M+i}^{(M)}}{\delta Y_M} = \sqrt{i+1} - 2\sqrt{i} + \sqrt{i-1} \quad \begin{array}{l} i = 1, 2, \ldots \\ M = 1, 2, \ldots \end{array} \qquad (2.4.14b)$$

The subscript on q represents the computation time $t_M = M\Delta t$, and the superscript indicates the time at which the temperature error occurred. Note that the dimensionless heat flux error in Eq. (2.4.14) is independent of the time (M) at which it occurred, provided $M \neq 0$. The results of Eq. (2.4.14) are given in Table 2.6.

The heat flux error corresponding to $\delta Y_M (M \neq 0)$ decays considerably faster than the error corresponding to δY_0. This implies that greater care should be taken in measuring Y_0 than other values of Y.

It should be reiterated that all temperature measurements will contain errors. A complete error analysis can be accomplished by superposing the heat flux error calculations for a single temperature error because the problem under consideration is linear.

TABLE 2.6 Decay of Heat Flux Error Resulting From Temperature Error δY_M at Time $t_M = M\Delta t$

i (time index)	$\dfrac{1}{2\sqrt{\dfrac{k\rho c}{\pi \Delta t}}} \dfrac{\delta q_{M+i}^{(M)}}{\delta Y_M}$
0	1.00000
1	−0.58579
2	−0.09638
3	−0.04989
4	−0.03188
5	−0.02265
6	−0.01716
7	−0.01359
8	−0.01110
9	−0.00930
10	−0.00793
100	−0.00025

2.5 EXACT SOLUTIONS OF INVERSE HEAT CONDUCTION PROBLEMS

2.5.1 Literature Review

Few exact solutions to the inverse problem of heat conduction for which the temperature sensor is at an arbitrary location are available in the literature. This is in contrast to the direct problem of heat conduction for which a wide range of solutions is available. Burggraf[6] presented one of the earliest exact solutions. He approached the problem by assuming that both the temperature $Y(t)$ and heat flux $q_E(t)$ were known at a sensor location. The temperature field was developed in terms of an infinite series of all-order derivatives of both $Y(t)$ and $q_E(t)$. If the temperature sensor was located at the center of a solid cylinder or sphere, then $q_E(t)$ was identically zero (for one-dimensional radial heat conduction). Langford[7] independently developed results similar to the Burggraf solution. Kover'yanov[9] developed results for hollow cylinders and spheres. The heat flux at the exposed surface was determined by differentiating the temperature field. Imber and Khan[8] obtained an exact solution for the temperature field using Laplace transforms when the temperature was known at two distinct interior points. Their temperature solution can be extrapolated in both directions toward the boundaries. The extrapolation distance is limited to the distance between the two temperature sensors. No computational results were presented for the more difficult problem of calculating heat flux at the exposed surface.

2.5.2 Derivation of Exact Solution for Planar Geometry

The analysis that follows closely parallels that of Burggraf.[6] The body is divided into (1) an inverse region and (2) a direct region as indicated in Figure 1.4. The direct region has conventional boundary conditions: specified temperature $Y(t)$ at the left face and arbitrary boundary conditions at the "inactive surface," L. By some means, it is necessary to solve for the temperature field in region 2. Next, the solution is differentiated at the location of the temperature sensor in order to calculate the heat flux q_E at $x_1 = E$. Once the heat flux at the sensor location is calculated, the inverse problem has two boundary conditions specified at the same boundary. The Burggraf solution requires that $q_E(t)$ and all of its derivatives are known.

Starting with the constant-property form of the energy equation,

$$\frac{\partial T}{\partial t} = \alpha \nabla^2 T \qquad (2.5.1)$$

and differentiating it with respect to time yields,

CHAP. 2 SOLUTIONS OF THE INVERSE HEAT CONDUCTION PROBLEM

$$\frac{\partial^2 T}{\partial t^2} = \frac{\partial}{\partial t}(\alpha\nabla^2 T) = \alpha\nabla^2 \frac{\partial T}{\partial t} = \alpha\nabla^2(\alpha\nabla^2 T)$$

$$\frac{\partial^2 T}{\partial t^2} = \alpha^2 \nabla^4 T \qquad (2.5.2)$$

Generalizing for an arbitrary time derivative gives

$$\frac{\partial^n T}{\partial t^n} = \alpha^n \nabla^{2n} T \qquad (2.5.3)$$

The same type of procedure is applied to Fourier's law:

$$q = -k\nabla T \qquad (2.5.4)$$

$$\frac{\partial q}{\partial t} = -k\nabla \frac{\partial T}{\partial t} = -k\nabla(\alpha\nabla^2 T) = \alpha\nabla^2(-k\nabla T) = \alpha\nabla^2 q \qquad (2.5.5)$$

For time derivatives of the heat flux of arbitrary order, the derivative is

$$\frac{\partial^n q}{\partial t^n} = \alpha^n \nabla^{2n} q = -k\alpha^n \nabla^{2n+1} T \qquad (2.5.6)$$

The temperature field is assumed to be an infinite series involving the temperature gradients at the temperature sensor location $r = E$,

$$T(r, t) = \sum_{n=0}^{\infty} H_n(r) \nabla^n T \bigg|_{r=E} \qquad (2.5.7)$$

In the analysis that follows, the geometry is restricted to one dimension such that

$$\nabla^2 T = \frac{1}{r^m} \frac{d}{dr}\left(r^m \frac{dT}{dr}\right) \quad \begin{array}{l} m=0 \text{ planar} \\ m=1 \text{ cylindrical} \\ m=2 \text{ spherical} \end{array} \qquad (2.5.8)$$

It is convenient to divide the series in Eq. (2.5.6) into even and odd terms,

$$T(r, t) = \sum_{i=0}^{\infty} H_{2i}(r)\nabla^{2i} T \bigg|_{r=E} + \sum_{i=0}^{\infty} H_{2i+1}(r)\nabla^{2i+1} T \bigg|_{r=E} \qquad (2.5.9)$$

Substituting Eqs. (2.5.3) and (2.5.5) into Eq. (2.5.8) yields

$$T(r, t) = \sum_{i=0}^{\infty} H_{2i}(r) \frac{1}{\alpha^i} \frac{\partial^i T}{\partial t^i}\bigg|_{r=E} + \sum_{i=0}^{\infty} H_{2i+1}(r) \frac{1}{(-k\alpha^i)} \frac{\partial^i q}{\partial t^i}\bigg|_{r=E}$$

$$= \sum_{i=0}^{\infty} f_i(r) \frac{d^i Y}{dt^i} - \frac{1}{k}\sum_{i=0}^{\infty} g_i(r) \frac{d^i q_E}{dt^i} \qquad (2.5.10)$$

where

$$f_i(r) = \frac{H_{2i}(r)}{\alpha^i}, \quad g_i(r) = \frac{H_{2i+1}(r)}{\alpha^i}, \quad \frac{\partial^i T}{\partial t^i}\bigg|_{r=E} = \frac{\partial^i Y}{\partial t^i}, \quad \frac{\partial^i q}{\partial t^i}\bigg|_{r=E} = \frac{\partial^i q_E}{\partial t^i} \qquad (2.5.11)$$

SEC. 2.5 EXACT SOLUTIONS OF INVERSE HEAT CONDUCTION PROBLEMS

Equation (2.5.10) is the general solution for the temperature field in the inverse region.

The remaining problem is to determine the functions $f(r)$ and $g(r)$. These functions are found by substituting Eq. (2.5.10) into the differential equation, Eq. (2.5.1),

$$\frac{\partial}{\partial t}\left[\sum_{n=0}^{\infty} f_n(r)\frac{d^n Y}{dt^n} - \frac{1}{k}\sum_{n=0}^{\infty} g_n(r)\frac{d^n q_E}{dt^n}\right]$$

$$= \alpha \nabla^2 \left[\sum_{n=0}^{\infty} f_n(r)\frac{d^n Y}{dt^n} - \frac{1}{k}\sum_{n=0}^{\infty} g_n(r)\frac{d^n q_E}{dt^n}\right]$$

Collecting like powers of $(d^n Y)/dt^n$ and $(d^n q_E)/dt^n$ gives,

$$\sum_{n=0}^{\infty} (f_{n-1} - \alpha\nabla^2 f_n)\frac{d^n Y}{dt^n} = \frac{1}{k}\sum_{n=0}^{\infty} (g_{n-1} - \alpha\nabla^2 g_n)\frac{d^n q_E}{dt^n} \quad (2.5.12)$$

A solution is obtained by requiring that each term inside the brackets of Eq. (2.5.12) is identically zero,

$$\begin{aligned}\nabla^2 f_0 &= 0 \qquad \nabla^2 g_0 = 0 \\ \nabla^2 f_n &= \frac{1}{\alpha} f_{n-1} \\ \nabla^2 g_n &= \frac{1}{\alpha} g_{n-1}\end{aligned}\quad n = 1, 2, \ldots \qquad (2.5.13)$$

The boundary conditions on the f and g functions are determined from the requirement that the solution exactly matches the temperature data $Y(t)$,

$$T(E, t) = Y(t) = \sum_{n=0}^{\infty} f_n(E)\frac{d^n Y}{dt^n} - \frac{1}{k}\sum_{n=0}^{\infty} g_n(E)\frac{d^n q_E}{dt^n}$$

or

$$f_0(E) = 1, \quad g_0(E) = 0, \quad f_n(E) = g_n(E) = 0, \quad n = 1, 2, \ldots \qquad (2.5.14)$$

Also the heat flux is given by

$$q_E = -k\frac{\partial T}{\partial r}\bigg|_{r=E} = -k\sum_{n=0}^{\infty} f'_n(E)\frac{d^n Y}{dt^n} + \sum_{n=0}^{\infty} g'_n(E)\frac{d^n q_E}{dt^n}$$

$$g'_0(E) = 1, \quad f'_0(E) = 0, \quad f'_n(E) = g'_n(E) = 0, \quad n = 1, 2, \ldots \qquad (2.5.15)$$

The solution to Eq. (2.5.13) subject to the boundary conditions given by Eqs. (2.5.14) and (2.5.15) completely determines the f and g functions. Note that these functions must be determined in a sequential manner starting with f_0 and g_0.

CHAP. 2 SOLUTIONS OF THE INVERSE HEAT CONDUCTION PROBLEM

Observations

1. For an insulated surface at $r=E$, the f-series alone determines the temperature profile.
2. For an isothermal surface at $r=E$, the g-series alone determines the temperature profile.
3. The functions f_0 and g_0 represent steady-state solutions.
4. All-order derivatives of $Y(t)$ and $q_E(t)$ must exist.

Solution for Planar Geometry $(r=x)$. By direct substitution, it is shown that

$$f_n = \frac{1}{(2n)!} \frac{(E-x)^{2n}}{\alpha^n}, \quad g_n = \frac{-1}{(2n+1)!} \frac{(E-x)^{2n+1}}{\alpha^n} \tag{2.5.16}$$

is a solution to Eqs. (2.4.12)–(2.4.14). The temperature distribution is written as

$$T(x,t) = Y(t) + \sum_{n=1}^{\infty} \frac{1}{(2n)!} \frac{(E-x)^{2n}}{\alpha^n} \frac{d^n Y}{dt^n}$$

$$+ \frac{(E-x)}{k}\left[q_E(t) + \sum_{n=1}^{\infty} \frac{1}{(2n+1)!} \frac{(E-x)^{2n}}{\alpha^n} \frac{d^n q_E}{dt^n}\right] \tag{2.5.17}$$

The planar geometry solution is similar in appearance to a Taylor series expansion about the temperature sensor depth.

The heat flux at the active surface is of primary interest; it is determined by using Fourier's law and Eq. (2.5.17) and evaluating at $x=0$:

$$q(t) = q_E + k \sum_{n=1}^{\infty} \frac{E^{2n-1}}{(2n-1)!} \frac{1}{\alpha^n} \frac{d^n Y}{dt^n} + \sum_{n=1}^{\infty} \frac{E^{2n}}{(2n)!} \frac{d^n q_E}{dt^n} \tag{2.5.18}$$

It is convenient to normalize the heat fluxes and introduce a dimensionless time as follows:

$$Q = \frac{qE}{k}, \quad Q_E = \frac{q_E E}{k}, \quad \tau_E = \frac{\alpha t}{E^2} \tag{2.5.19}$$

Note that Q has the units of temperature. The normalized heat flux is written as

$$Q = Q_E + \sum_{n=1}^{\infty} \frac{1}{(2n-1)!} \frac{d^n Y}{d\tau_E^n} + \sum_{n=1}^{\infty} \frac{1}{(2n)!} \frac{d^n Q_E}{d\tau_E^n} \tag{2.5.20}$$

This is a very important result because it is exact for continuous temperature measurements and has a simple form. The solution clearly shows the dependence of the surface heat flux on all orders of time derivatives of the measured temperature and the related heat flux, both at $x=E$. The temperature level itself is not significant since only derivatives are needed. It is surprising that the solution given by Eq. (2.5.20) shows no explicit dependence on the initial temperature distribution in the body. Burggraf[6] has pointed out, however, that a polynomial fit of finite order to the experimental temperature data, $Y(t)$, implies that the initial temperature profile must be nonuniform.

SEC. 2.5 EXACT SOLUTIONS OF INVERSE HEAT CONDUCTION PROBLEMS

Since the inverse solution given by Eq. (2.5.20) is exact, one might ask if there is any need for further investigation. The answer to this is a definite yes because the exact solution has several practical limitations:

1. High-order derivatives of discrete temperature data $Y(t)$ and heat flux Q_E must be evaluated numerically.
2. It is awkward for composite bodies and not appropriate for temperature-dependent properties.
3. It may require a numerical procedure to solve the direct problem and determine $Q_E(t)$.
4. It is not applicable to the overspecified problem of more than one interior temperature sensor.
5. The method does not lend itself readily to the case of multiple heat flux determination such as the two-dimensional case.

Although the above exact solution has limited applicability in a practical sense, it is extremely important because of the insights provided.

2.5.3 Expressions for Cylinders and Spheres

Solid cylinders and spheres in which the heat flux is one dimensional radially and the temperature sensor is located at the center have simple exact solutions for the IHCP. Consequently, the heat flux at the exterior surface depends only on the temperature response and its derivatives. From Burggraf[6] and Langford,[7] the temperature fields for solid cylinders and spheres are given by

$$T(x, t) = Y(t) + \sum_{n=1}^{\infty} \frac{(R-x)^{2n}}{2^{2n}(n!)^2 \alpha^n} \frac{d^n Y}{dt^n} \quad \text{(cylinder)} \qquad (2.5.21)$$

$$T(x, t) = Y(t) + \sum_{n=1}^{\infty} \frac{(R-x)^{2n}}{(2n+1)! \alpha^n} \frac{d^n Y}{dt^n} \quad \text{(sphere)} \qquad (2.5.22)$$

Note that the coordinate x is attached to the exterior surface and that R is the external radius. The heat flux is evaluated from Fourier's law and Eqs. (2.5.21) and (2.5.22); the results are

$$q = -k \left. \frac{\partial T}{\partial x} \right|_{x=0} = k \sum_{n=1}^{\infty} \frac{n R^{2n-1}}{2^{2n-1}(n!)^2 \alpha^n} \frac{d^n Y}{dt^n} \quad \text{(cylinder)} \qquad (2.5.23)$$

$$q = -k \left. \frac{\partial T}{\partial x} \right|_{x=0} = k \sum_{n=1}^{\infty} \frac{2n R^{2n-1}}{(2n+1)! \alpha^n} \frac{d^n Y}{dt^n} \quad \text{(sphere)} \qquad (2.5.24)$$

Comments on hollow cylinders and spheres can be found in Burggraf,[6] Langford,[7] and Kover'yanov.[9]

2.5.4 Example Results for Planar Geometry

In order to apply the exact solution, the time derivatives must be approximated numerically. For simplicity, suppose the temperature sensor is attached at a perfectly insulated boundary and central differences are used. Only the first five derivatives are considered; in order for the method to be of practical utility, the series must be truncated at a reasonable limit. Applying Eq. (2.5.20),

$$Q = \sum_{n=1}^{\infty} \frac{1}{(2n-1)!} \frac{d^n Y}{d\tau_E^n} \qquad (2.5.25)$$

The first five derivatives are approximated as follows:

$$\frac{dY_j}{d\tau_E} \approx \frac{Y_{j+1} - Y_{j-1}}{2\Delta\tau_E} \qquad (2.5.26)$$

$$\frac{d^2 Y_j}{d\tau_E^2} \approx \frac{Y_{j-1} - 2Y_j + Y_{j+1}}{\Delta\tau_E^2} \qquad (2.5.27)$$

$$\frac{d^3 Y_j}{d\tau_E^3} \approx \frac{-Y_{j-2} + 2Y_{j-1} - 2Y_{j+1} + Y_{j+2}}{2\Delta\tau_E^3} \qquad (2.5.28)$$

$$\frac{d^4 Y_j}{d\tau_E^4} \approx \frac{Y_{j-2} - 4Y_{j-1} + 6Y_j - 4Y_{j+1} + Y_{j+2}}{\Delta\tau_E^4} \qquad (2.5.29)$$

$$\frac{d^5 Y_j}{d\tau_E^5} \approx \frac{-Y_{j-3} + 4Y_{j-2} - 5Y_{j-1} + 5Y_{j+1} - 4Y_{j+2} + Y_{j+3}}{2\Delta\tau_E^5} \qquad (2.5.30)$$

Some insight is gained by looking at the coefficients of the various order differences; the appropriate term to consider is

$$\frac{1}{(2n-1)!\Delta\tau_E^n} \qquad (2.5.31)$$

Equation (2.5.31) is tabulated in Table 2.7 for various values of the dimensionless time step. For each value of $\Delta\tau_E$, the maximum coefficient is underlined. Note that as $\Delta\tau_E$ becomes smaller, higher-order derivatives appear to become more important. If $\Delta\tau_E$ is large ($>10^2$) then the first derivative is the dominant term; if $\Delta\tau_E$ is much smaller than unity, then several high-order terms appear to be significant.

Equation (2.5.25) can be written in an alternative form by using the difference approximations given by Eqs. (2.5.26)–(2.5.30) and simplifying to obtain

$$Q_j = D_{-3} Y_{j-3} + D_{-2} Y_{j-2} + D_{-1} Y_{j-1} + D_0 Y_j + D_1 Y_{j+1} + D_2 Y_{j+2} + D_3 Y_{j+3}$$

$$(2.5.32)$$

TABLE 2.7 Difference Coefficients in Burggraf's Exact Solution, $\frac{1}{(2n-1)!}(\Delta\tau_E)^n$

n	$\Delta\tau_E=10^2$	$\Delta\tau_E=1$	$\Delta\tau_E=0.5$	$\Delta\tau_E=0.25$	$\Delta\tau_E=0.125$	$\Delta\tau_E=0.0625$	$\Delta\tau_E=0.03125$	$\Delta\tau_E=0.01$
1	0.01	1.0	2.0	4.0	8.0	16.0	32.0	100
2	1.6667×10^{-5}	0.166667	0.666667	2.666667	10.666667	42.666667	170.666667	1.66667×10^3
3	8.3333×10^{-9}	0.008333	0.066667	0.533333	4.266667	34.133333	273.066667	8.3333×10^3
4	1.9841×10^{-12}	0.000198	0.003175	0.050794	0.812698	13.003175	208.050794	1.9841×10^4
5	2.7557×10^{-16}	0.000003	0.000088	0.002822	0.090300	2.889594	92.467019	2.7557×10^4
6			0.000002	0.000103	0.006567	0.420305	26.899497	2.5052×10^4
7				0.000003	0.000337	0.043108	5.517845	1.6059×10^4

TABLE 2.8 Temperature Weighting Coefficients for Eq. (2.5.32); Only First Five Derivatives of Eq. (2.5.25) Are Included

$\Delta\tau_E$	D_{-3}	D_{-2}	D_{-1}	D_0	D_1	D_2	D_3
10^2	1.3779×10^{-16}	-4.1647×10^{-9}	-4.9833×10^{-3}	-3.3333×10^{-5}	5.0167×10^{-3}	4.1687×10^{-9}	1.3779×10^{-16}
8	4.2049×10^{-11}	-8.0894×10^{-6}	-5.9880×10^{-2}	-5.2080×10^{-3}	6.5088×10^{-2}	8.1863×10^{-6}	4.2049×10^{-11}
4	1.3456×10^{-9}	-6.4324×10^{-5}	-1.1446×10^{-1}	-2.0829×10^{-2}	1.3528×10^{-1}	6.5874×10^{-5}	1.3456×10^{-9}
2.0	4.3058×10^{-8}	-5.0826×10^{-4}	-2.0734×10^{-1}	-8.3259×10^{-2}	2.9058×10^{-1}	5.3306×10^{-4}	4.3058×10^{-8}
1.0	1.3779×10^{-6}	-3.9627×10^{-3}	-3.2580×10^{-1}	-3.3214×10^{-1}	6.5755×10^{-1}	4.3596×10^{-3}	1.3779×10^{-6}
0.5	4.4092×10^{-5}	-2.9982×10^{-2}	-2.7959×10^{-1}	-1.3143×10^{0}	1.5875×10^{0}	3.6332×10^{-2}	4.4092×10^{-5}
0.25	1.4109×10^{-3}	-2.1023×10^{-1}	9.8977×10^{-1}	-5.0286×10^{0}	3.9372×10^{0}	3.1182×10^{-1}	1.4109×10^{-3}
0.125	4.5150×10^{-2}	-1.1400×10^{0}	7.4568×10^{0}	-1.6457×10^{1}	7.3750×10^{0}	2.7654×10^{0}	4.5150×10^{-2}
0.0625	1.4448×10^{0}	1.7157×10^{0}	9.5633×10^{0}	-7.3143×10^{0}	-2.8255×10^{1}	2.4291×10^{1}	1.4448×10^{0}
0.03125	4.6234×10^{1}	2.5645×10^{2}	-6.3564×10^{2}	9.0697×10^{2}	-6.8744×10^{2}	1.5965×10^{2}	4.6234×10^{1}
10^{-3}	1.3779×10^{9}	5.7057×10^{9}	-7.6745×10^{9}	1.1901×10^{9}	6.0875×10^{9}	-5.3089×10^{9}	1.3779×10^{9}

where the D's are independent of the time index j and can be written as

$$D_{-3} = \frac{1}{2 \cdot 9! \Delta \tau_E^5}$$

$$D_{-2} = \frac{4}{2 \cdot 9! \Delta \tau_E^5} + \frac{1}{7! \Delta \tau_E^4} - \frac{1}{2 \cdot 5! \Delta \tau_E^3}$$

$$D_{-1} = \frac{-5}{2 \cdot 9! \Delta \tau_E^5} - \frac{4}{7! \Delta \tau_E^4} + \frac{1}{5! \Delta \tau_E^3} + \frac{1}{3! \Delta \tau_E^2} - \frac{1}{2 \Delta \tau_E}$$

$$D_0 = \frac{6}{7! \Delta \tau_E^4} - \frac{2}{3! \Delta \tau_E^2} \qquad (2.5.33)$$

$$D_1 = \frac{5}{2 \cdot 9! \Delta \tau_E^5} - \frac{4}{7! \Delta \tau_E^4} - \frac{1}{5! \Delta \tau_E^3} + \frac{1}{3! \Delta \tau_E^2} + \frac{1}{2 \Delta \tau_E}$$

$$D_2 = \frac{-4}{2 \cdot 9! \Delta \tau_E^5} + \frac{1}{7! \Delta \tau_E^4} + \frac{1}{2 \cdot 5! \Delta \tau_E^3}$$

$$D_3 = \frac{1}{2 \cdot 9! \Delta \tau_E^5}$$

Equation (2.5.32) has the appearance of a seven-point moving average filter, with the sum of the coefficients equal to zero. Again, note that q is linear in the temperature measurements. The coefficients are not symmetrical; that is, $D_{-i} \neq D_i$ except for certain values of i. Table 2.8 presents the temperature coefficients defined in Eq. (2.5.33) as functions of the dimensionless time step $\Delta \tau_E$. For large dimensionless time steps ($> 10^2$), D_1 and D_{-1} are the dominant terms; only D_{-1} and D_1 contain $1/\Delta \tau_E$ terms. This suggests, in turn, that for large time steps, a two-point central difference might be adequate. The opposite is true for small time steps; all of the D's appear to be approximately the same order of magnitude. Thus, the seven-point formula considered in this example might not be adequate for small time steps.

In summary, an exact solution has been presented for the inverse problem of heat conduction, provided the properties are constant. The solution requires that infinite-order derivatives of the experimental data $Y(t)$ must exist. If small dimensionless time steps are used, high-order derivatives can dominate. From a practical point of view, the utility of the Burggraf solution is to provide beneficial insight into inverse heat conduction problems.

REFERENCES

1. Beck, J. V. and Arnold, K. J., *Parameter Estimation in Engineering and Science*, Wiley, New York, 1977.
2. Carslaw, H. S. and Jaeger, J. C., *Conduction of Heat in Solids*, 2nd ed., Oxford University Press, London, 1959.
3. Schneider, P. J., *Conduction Heat Transfer*, Addison-Wesley, Reading, MA, 1955.

4. Meyers, G. E., *Analytical Methods in Conduction Heat Transfer*, McGraw-Hill, New York, 1971.
5. Carnahan, B., Luther, H. A., and Wilkes, J. O., *Applied Numerical Methods*, Wiley, New York, 1969.
6. Burggraf, O. R., An Exact Solution of the Inverse Problem in Heat Conduction Theory and Applications, *ASME J. Heat Transfer*, **86**C, 373–382, August 1964.
7. Langford, D., New Analytical Solutions of the One-Dimensional Heat Equation for Temperature and Heat Flow Rate Both Prescribed at the Same Fixed Boundary (with applications to the phase change problem), *Q. Appl. Math.* **24** (4), 315–322 (1976).
8. Imber, M. and Khan, J., Prediction of Transient Temperature Distributions with Embedded Thermocouples, *AIAA J.* **10** (6), 784–789 (1972).
9. Kover'yanov, V. A., Inverse Problem of Nonsteady-State Thermal Conductivity, *Teplofizika Vysokikh Temperatur*, **5** (1), 141–143 (1967).

PROBLEMS

2.1. Develop Eq. (2.2.10) by (a) using a weighted least-squares procedure that fits a linear equation to the experimental temperature data and (b) differentiating the curve fit to determine the average temperature gradient, and (c) use of Fourier's law to obtain the heat flux.

2.2. For cylindrical geometries, the temperature profile is linear in the variable $\ln r$; $T - T_i = [-Q \ln(r/r_i)]/(2\pi kL)$, $Q = q(2\pi r L)$ where Q is constant. Develop a heat flux estimating equation analogous to Eq. (2.2.10) that minimizes the weighted least-squares error between the computed and experimental temperatures. The heat flux is to be found at $r = r_i$. Is this result the same as the one obtained for a temperature profile linear in r?

2.3. For spherical geometries, the temperature profile is linear in the variable $1/r$; $T - T_i = (Q/4\pi k)(1/r - 1/r_i)$, $Q = q(4\pi r^2) = $ constant. Develop a heat flux equation analogous to Eq. (2.2.10) that minimizes the weighted least-squares error between the computed and experimental temperatures. Is this result the same as that obtained for a temperature profile linear in r?

2.4. Develop Eq. (2.3.7). It is acceptable to start with Eq. (2.2.10) as suggested in the text immediately above Eq. (2.3.7).

2.5. Demonstrate that if five temperature-time points are used to determine a least-squares straight line [e.g., Eq. (2.3.7)], then a single temperature error ε_M will cause errors in the calculation of q_{M-2}, q_{M-1}, q_M, q_{M+1}, and q_{M+2}. What are the heat flux errors?

2.6. Verify the numerical results in Table 2.3.

2.7. Derive Eq. (2.4.7) from Eq. (2.4.6) by integration by parts.
   Hint: rewrite the integrand as

PROBLEMS

$$\frac{Y'(\lambda) + Y'(t) - Y'(t)}{\sqrt{t-\lambda}}$$

and express in terms of two integrals.

2.8. Verify that Eq. (2.5.16) is a solution to Eqs. (2.5.13, 14, 15).

2.9. Prove that the sum of terms in Eq. (2.4.14a, b) adds to zero. What is the physical significance of this result?

2.10. For the spherical heat flux equation given by Eq. (2.5.24), generate a table analogous to Table 2.7. What conclusions can be made?

2.11. Starting with Eq. (2.4.7) and the assumption that the surface temperature response varies piecewise linearly with time, show that the heat flux can be expressed as

$$\hat{q}(t_M) = \hat{q}_M = \left(\frac{k\rho c}{\pi}\right)^{1/2} \left\{ \frac{Y_M - Y_0}{(t_M - t_0)^{1/2}} + \frac{Y_M - Y_{M-1}}{(t_M - t_{M-1})^{1/2}} \right.$$
$$\left. + \sum_{i=1}^{M+1} \left[\frac{Y_M - Y_i}{(t_M - t_i)^{1/2}} - \frac{Y_M - Y_{i-1}}{(t_M - t_{i-1})^{1/2}} + \frac{2(Y_i - Y_{i-1})}{(t_M - t_i)^{1/2} + (t_M - t_{i-1})^{1/2}} \right] \right\}$$

2.12. The result given in Problem 2.11 has been given in numerous places in the literature and is algebraically more complicated than Eq. (2.4.8). Show that the above expression can be simplified to Eq. (2.4.8b). Hint: Expand the first two summations and show that most of the terms cancel.

CHAPTER 3

APPROXIMATE METHODS FOR DIRECT HEAT CONDUCTION PROBLEMS

3.1 INTRODUCTION

In this chapter some approximate methods are given for solving direct transient heat conduction problems. These direct techniques are the first stage of solution procedures for the inverse heat conduction problems. The second stage, which involves specific algorithms for the IHCP, is developed in Chapter 4. The approximate heat conduction solution procedures of this chapter are combined in Chapters 5 and 6 with the IHCP algorithms of Chapter 4.

3.1.1 Various Numerical Approaches

Partial differential equations including the transient heat conduction equation can be solved using a variety of methods including exact and numerical procedures. The exact methods include the classical methods of separation of variables and Laplace transforms.

Two types of numerical procedures are discussed in this chapter. One is based on an integral formulation of the mathematical model and the other on a differential form of the model.

The transient heat conduction equation can be either linear or nonlinear. For the linear case the partial differential equation formulation can be equivalently presented by an integral equation. The advantage of the integral form is that the space dependence can be represented exactly leaving only the time dependence to be approximated.

Two different approaches that employ integral equations for solving transient

SEC. 3.1 INTRODUCTION

heat conduction problems are those using Duhamel's theorem and Green's functions both of which yield convolution-type integrals. Since these methods are very similar, only Duhamel's theorem is given. A numerical approximation of the integral is also given. The heat-conducting body can be of arbitrary shape and of nonhomogeneous material. Although the heat transfer may be two- or three-dimensional, in this chapter there is a single forcing input which is only time dependent.

A new technique which utilizes Duhamel's theorem for solving connected basic geometries is called the unsteady surface element method.[1-5] It also involves solving an inverse problem but there are no measurement errors and the matching conditions are at *surfaces* rather than at interior points of the bodies.

The solution of a nonlinear transient heat conduction problem requires an approach that discretizes the partial differential equation. Two methods for solving the nonlinear transient heat conduction equation are the finite difference and finite element (FE) methods. An alternative method is used in this chapter in which conservation of energy is applied directly to finite control volumes; it is called the finite control volume procedure.

A nonlinear problem is one for which the heat conduction equation or a boundary condition is a nonlinear function of temperature. One example of a nonlinear partial differential equation is:

$$\frac{\partial}{\partial x}\left(k\frac{\partial T}{\partial x}\right) = \rho c \frac{\partial T}{\partial t} \qquad (3.1.1)$$

where k and/or ρc are functions of temperature T. If k and ρc are functions of position, x, only, then Eq. (3.1.1) is linear. An example of a nonlinear boundary condition is the radiation condition,

$$-k\frac{\partial T}{\partial x}\bigg|_{x=0} = \sigma\varepsilon[T_\infty^4 - T^4(0, t)] \qquad (3.1.2)$$

3.1.2 Scope of Chapter

This chapter gives a derivation of Duhamel's integral (Section 3.2), and a numerical approximation thereof. Section 3.3 discusses finite control volume solutions of the transient heat conduction equation. Both linear and nonlinear cases are included.

Section 3.3.5 demonstrates that the linear finite control volume approach can yield expressions that are analogous to those obtained from the Duhamel's theorem method. This is important because the same linear IHCP algorithms can be employed using many different methods of approximating the heat conduction equation and boundary conditions; such methods include a numerical form of Duhamel's theorem, finite differences, and finite elements.

3.2 DUHAMEL'S THEOREM

3.2.1 Derivation of Duhamel's Theorem

Duhamel's theorem can be considered to be a result of the principle of superposition and thus it is only valid for linear cases. There are several ways of deriving it, one of which uses Laplace transforms. See Ozisik (Reference 6, p. 197) and Luikov (Reference 7, p. 344). Another derivation uses the concept of superposition; that is the method used in Arpaci (Reference 8, p. 307) and Myers (Reference 9, p. 155). The following derivation employs the principle of superposition and it also produces a convenient numerical approximation.

Duhamel's theorem employs a "building block" solution which is used with the superposition principle to obtain the temperature at any point and time. One such solution is $\phi(\mathbf{r}, t)$ which is for the temperature rise at a point $\mathbf{r}$ in a heat-conducting body due to the heat flux,

$$q(t) = \begin{cases} 0, & t < 0 \\ 1, & t > 0 \end{cases} \quad (3.2.1)$$

This is sometimes called a unit step heat flux or even a constant unit heat flux. The thermal properties of the body are independent of temperature but can be functions of position, and the temperature distribution need not be one-dimensional. For the derivation herein the only requirement is that the prescribed surface heat flux is given by the product,

$$q(s, t) = f(t)g(s)$$

where s is a coordinate measured along the surface. For convenience, the heat flux is written as a function of time only.

The surface heat flux is approximated in the manner shown in Figure 3.1. The heat fluxes at times

$$\lambda_{1/2}, \ \lambda_{3/2}, \ldots, \lambda_{M-1/2}$$

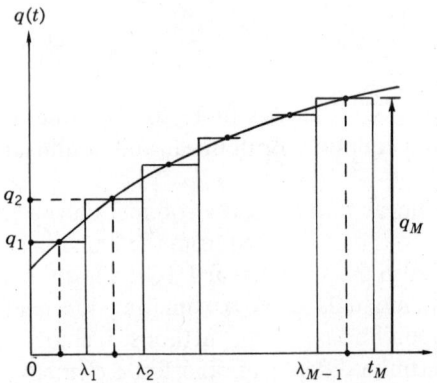

FIGURE 3.1 Approximate representation of $q(t)$ by M steps in heat flux.

SEC. 3.2 DUHAMEL'S THEOREM

are used to represent the heat fluxes between times of, respectively,

$$0 \text{ to } \lambda_1, \quad \lambda_1 \text{ to } \lambda_2, \ldots, \quad \lambda_{M-1} \text{ to } t_M$$

The corresponding heat fluxes are denoted $q_1, q_2, \ldots, q_M$ or in general

$$q_M = q(\lambda_{M-1/2}) = q[(M-\tfrac{1}{2})\Delta\lambda] \tag{3.2.2}$$

The initial temperature distribution in the body is the constant value of T_0. Then using the principle of superposition, the temperature at location $\mathbf{r}$, at time t_M, is composed of contributions due to the heat flux components q_1 through q_M inclusive:

$$\begin{aligned}
T(\mathbf{r}, t_M) = T_0 &+ q_1[\phi(\mathbf{r}, t_M - \lambda_0) - \phi(\mathbf{r}, t_M - \lambda_1)] \\
&+ q_2[\phi(\mathbf{r}, t_M - \lambda_1) - \phi(\mathbf{r}, t_M - \lambda_2)] \\
&\vdots \\
&+ q_M[\phi(\mathbf{r}, t_M - \lambda_{M-1}) - \phi(\mathbf{r}, t_M - \lambda_M)]
\end{aligned} \tag{3.2.3}$$

where $\phi(\mathbf{r}, t_M - \lambda_M) = \phi(\mathbf{r}, 0) = 0$.

Using equal time steps and times of

$$\lambda_i = i\Delta\lambda, \quad i = 0, 1, \ldots, M \tag{3.2.4}$$

yields:

$$\lambda_j - \lambda_i = j\Delta\lambda - i\Delta\lambda = (j-i)\Delta\lambda = \lambda_{j-i} \tag{3.2.5}$$

Then Eq. (3.2.3) can be written as

$$T(\mathbf{r}, t_M) = T_0 + \sum_{n=1}^{M} q_n \frac{\phi(\mathbf{r}, t_M - \lambda_{n-1}) - \phi(\mathbf{r}, t_M - \lambda_n)}{\Delta\lambda} \Delta\lambda \tag{3.2.6}$$

For the limiting case of $\Delta\lambda \to 0$ and for a fixed time $t_M = t$, Eq. (3.2.6) becomes the integral

$$T(\mathbf{r}, t) = T_0 + \int_0^t q(\lambda) \left[-\frac{\partial \phi(\mathbf{r}, t-\lambda)}{\partial \lambda} \right] d\lambda \tag{3.2.7}$$

From the relation

$$-\frac{\partial \phi(\mathbf{r}, t-\lambda)}{\partial \lambda} = \frac{\partial \phi(\mathbf{r}, t-\lambda)}{\partial t} \tag{3.2.8}$$

Duhamel's theorem (sometimes called Duhamel's integral) becomes

$$T(\mathbf{r}, t) = T_0 + \int_0^t q(\lambda) \frac{\partial \phi(\mathbf{r}, t-\lambda)}{\partial t} d\lambda \tag{3.2.9}$$

Equation (3.2.9) is a heat flux form of Duhamel's theorem; it is a convolution because there is a product of two functions, one of λ and the other of $t-\lambda$; that is, there is folding or convoluting of one function with respect to the other. For a temperature-based form of Duhamel's theorem, see Eq. (2.4.1) and References 6–10.

3.2.2 Numerical Approximation of Duhamel's Theorem

A numerical approximation for Eq. (3.2.9) is obtained from Eq. (3.2.6). Note that

$$\phi(\mathbf{r}, t_M - \lambda_{n-1}) - \phi(\mathbf{r}, t_M - \lambda_n) = \phi(\mathbf{r}, t_{M-n+1}) - \phi(\mathbf{r}, t_{M-n}) = \Delta\phi(\mathbf{r}, t_{M-n}) \quad (3.2.10)$$

and thus Eq. (3.2.6) can be written as

$$T(\mathbf{r}, t_M) = T_0 + \sum_{n=1}^{M} q_n \Delta\phi(\mathbf{r}, t_{M-n}) \quad (3.2.11)$$

The notation can be simplified by omitting the $\mathbf{r}$ dependence and writing

$$\boxed{T_M = T_0 + \sum_{n=1}^{M} q_n \Delta\phi_{M-n}, \quad \Delta\phi_i = \phi_{i+1} - \phi_i} \quad (3.2.12)$$

where q_n is best evaluated at time $(n-\tfrac{1}{2})\Delta t$ as indicated by Eq. (3.2.2). If the actual heat flux is constant over each time step, Eq. (3.2.12) gives an exact expression for the temperature, T_M; on the other hand, Eq. (3.2.12) yields approximate results if the true heat flux varies over the time steps. This equation is quite important in investigation of the IHCP because it gives a convenient expression for the temperature in terms of the heat flux components. The observation that the sum of the subscripts on the right side of Eq. (3.2.12) is M can aid in memorizing this equation.

EXAMPLE 3.1. Calculate the temperature at 1 cm inside a thick steel plate ($k = 40$ W/m-C, $\alpha = 10^{-5}$ m²/s) that is initially at 30°C with an applied heat flux of

$$q = \dot{q}_0 t, \quad \dot{q}_0 = 10^6 \text{ W/m}^2\text{-s}, \quad t \text{ in s}$$

starting at time zero. Find values of the temperature at times 1, 2, and 3 s. Use Eq. (3.2.12) to approximate each temperature and compare it with the exact value.

Solution. The temperatures are denoted T_1, T_2, and T_3 and correspond to $t = 1, 2,$ and 3 s; from Eq. (3.2.12) they are given by

$$T_1 = T_0 + q_1 \Delta\phi_0 \quad \text{(a)}$$

$$T_2 = T_0 + q_1 \Delta\phi_1 + q_2 \Delta\phi_0 \quad \text{(b)}$$

$$T_3 = T_0 + q_1 \Delta\phi_2 + q_2 \Delta\phi_1 + q_3 \Delta\phi_0 \quad \text{(c)}$$

The ϕ_i values are for a semi-infinite body with a unit step increase in surface heat flux; ϕ_i values can be found using entries in Table 1.2 for T^+ defined by Eq. (1.6.18c) and letting $q_c = 1$, $T(x, t) = \phi$, and $T_0 = 0$; hence

$$\phi = \left(\frac{x}{k}\right) T^+ = \left(\frac{0.01}{40}\right) T^+$$

Also the time t is related to t_x^+ by Eq. (1.6.18d) which gives

$$t_x^+ = \frac{\alpha t}{x^2} = \frac{10^{-5}}{10^{-4}} = 0.1 t$$

SEC. 3.2 DUHAMEL'S THEOREM

Then the T^+ values at $t_x^+ = 0.1, 0.2,$ and 0.3 are used to obtain the $\Delta\phi_i$ values,

$$\Delta\phi_0 = \phi_1 = \frac{x}{k}\phi_1^+ = \frac{0.01}{40}(0.003943) = 9.8575E\text{-}7 \text{ C-m}^2/\text{W}$$

$$\Delta\phi_1 = \phi_2 - \phi_1 = \frac{x}{k}(\phi_2^+ - \phi_1^+) = 6.69725E\text{-}6 \text{ C-m}^2/\text{W}$$

$$\Delta\phi_2 = \phi_3 - \phi_2 = \frac{x}{k}(\phi_3^+ - \phi_2^+) = 1.029025E\text{-}5 \text{ C-m}^2/\text{W}$$

The q values are evaluated at $t = 0.5, 1.5,$ and 2.5 s so that

$$q_1 = 0.5E6, \quad q_2 = 1.5E6, \quad q_3 = 2.5E6 \text{ W/m}^2$$

and thus the temperatures found from Eqs. (a), (b), and (c) are

$$T_1 = 30 + 0.5E6(9.8575E\text{-}7) = 30.493°C$$

$$T_2 = 30 + 0.5E6(6.69725E\text{-}6) + 1.5E6(9.8575E\text{-}7)$$
$$= 34.827°C$$

$$T_3 = T_0 + 0.5E6(1.029025E\text{-}5) + 1.5E6(6.69725E\text{-}6)$$
$$+ 2.5E6(9.8575E\text{-}7) = 47.655°C$$

The exact temperatures are obtained from Carslaw and Jaeger (Reference 10, p. 77),

$$T = T_0 + \left(\frac{\dot{q}_0}{\alpha k}\right)(4\alpha t)^{3/2} i^3\text{erfc}[(4t_x^+)^{-1/2}] \quad (3.2.13)$$

where

$$i^3\text{erfc}(z) = [(1+z^2)\pi^{-1/2}e^{-z^2} - z(1.5+z^2)\text{erfc}(z)]/6 \quad (3.2.14)$$

The temperatures T_1, T_2, and T_3 are obtained from Eq. (3.2.13) at times $t = 1, t = 2,$ and $t = 3$ s or equivalently for $t_x^+ = 0.1, 0.2,$ and 0.3. The resulting exact values are

$$T_1 = 30.1876°C, \quad T_2 = 34.0743°C, \quad T_3 = 46.7255°C$$

Hence the errors in the approximate temperatures are +0.305, 0.753, and 0.930, respectively. These errors are increasing, but relative to the exact temperature *rises* they are decreasing; namely, 163%, 18.5%, and 5.6%, respectively. □

3.2.3 Matrix Form of Duhamel's Theorem

It is frequently advantageous to perform algebraic manipulations utilizing a matrix form of the model. This subsection displays a matrix form for the numerical statement of Duhamel's theorem.

An expansion of the constant heat flux numerical approximation of Duhamel's integral, Eq. (3.2.12), with M replaced by 1 to $M + r - 1$, is

$$T_1 = T_0 + q_1\Delta\phi_0$$
$$T_2 = T_0 + q_1\Delta\phi_1 + q_2\Delta\phi_0$$

$$T_3 = T_0 + q_1 \Delta\phi_2 + q_2 \Delta\phi_1 + q_3 \Delta\phi_0$$
$$\vdots$$
$$T_M = T_0 + q_1 \Delta\phi_{M-1} + q_2 \Delta\phi_{M-2} + \cdots + q_{M-1}\Delta\phi_1 + q_M \Delta\phi_0$$
$$T_{M+1} = T_0 + q_1 \Delta\phi_M + q_2 \Delta\phi_{M-1} + \cdots + q_M \Delta\phi_1 + q_{M+1}\Delta\phi_0$$
$$\vdots$$
$$T_{M+r-1} = T_0 + q_1 \Delta\phi_{M+r-2} + \cdots + q_{M+r-2}\Delta\phi_1 + q_{M+r-1}\Delta\phi_0 \quad (3.2.15)$$

which can be written in expanded matrix form as

$$\begin{bmatrix} T_1 \\ T_2 \\ T_3 \\ \vdots \\ T_M \\ T_{M+1} \\ \vdots \\ T_{M+r-1} \end{bmatrix} = \begin{bmatrix} \Delta\phi_0 & & & & & & & \\ \Delta\phi_1 & \Delta\phi_0 & & & & & & \\ \Delta\phi_2 & \Delta\phi_1 & \Delta\phi_0 & & & & & \\ \vdots & \vdots & \vdots & & & & & \\ \Delta\phi_{M-1} & \Delta\phi_{M-2} & \Delta\phi_{M-3} & \cdots & \Delta\phi_0 & & & \\ \Delta\phi_M & \Delta\phi_{M-1} & \Delta\phi_{M-2} & \cdots & \Delta\phi_1 & \Delta\phi_0 & & \\ \vdots & \vdots & \vdots & & \vdots & \vdots & & \\ \Delta\phi_{M+r-2} & \Delta\phi_{M+r-3} & \Delta\phi_{M+r-4} & \cdots & \Delta\phi_{r-1} & \Delta\phi_{r-2} & \cdots & \Delta\phi_0 \end{bmatrix}$$

$$\times \begin{bmatrix} q_1 \\ q_2 \\ q_3 \\ \vdots \\ q_M \\ q_{M+1} \\ \vdots \\ q_{M+r-1} \end{bmatrix} + T_0 \mathbf{1} \quad (3.2.16)$$

where **1** is a vector of ones. In a more compact form Eq. (3.2.16) is

$$\mathbf{T} = \mathbf{X}\mathbf{q} + T_0 \mathbf{1} \quad (3.2.17)$$

where **T** and **q** are the appropriate vectors displayed in Eq. (3.2.16) and **X** is the lower triangular matrix,

$$\mathbf{X} = \begin{bmatrix} \Delta\phi_0 & & & & & & \\ \Delta\phi_1 & \Delta\phi_0 & & & & & \\ \Delta\phi_2 & \Delta\phi_1 & \Delta\phi_0 & & & & \\ \vdots & \vdots & \vdots & & & & \\ \Delta\phi_{M-1} & \Delta\phi_{M-2} & \Delta\phi_{M-3} & \cdots & \Delta\phi_0 & & \\ \Delta\phi_M & \Delta\phi_{M-1} & \Delta\phi_{M-2} & \cdots & \Delta\phi_1 & \Delta\phi_0 & \\ \vdots & \vdots & \vdots & & \vdots & \vdots & \\ \Delta\phi_{M+r-2} & \Delta\phi_{M+r-3} & \Delta\phi_{M+r-4} & \cdots & \Delta\phi_{r-1} & \Delta\phi_{r-2} & \cdots & \Delta\phi_0 \end{bmatrix}$$

(3.2.18a)

The **X** matrix is called the pulse sensitivity coefficient matrix for **q**. Its structure is lower triangular with $\Delta\phi_0$'s along the main diagonal, $\Delta\phi_1$'s along the diagonal

SEC. 3.2 DUHAMEL'S THEOREM

just below the main one, and so on. The components of the **X** matrix are called sensitivity coefficients because they are equal to the first derivatives of the temperature with respect to heat flux components. From Eq. (3.2.15), one can verify that the ijth component of **X** is

$$X_{ij} = \frac{\partial T_i}{\partial q_j} = \begin{cases} \Delta\phi_{i-j}, & i \geq j \\ 0, & i < j \end{cases} \qquad (3.2.18b)$$

which shows that X_{ij} depends only on the difference between i and j for linear problems.

In the sequential IHCP methods investigated in Chapter 4, the heat flux components, $q_1, q_2, \ldots, q_{M-1}$, are considered previously estimated and are denoted $\hat{q}_1, \ldots, \hat{q}_{M-1}$. Estimates of $q_M, q_{M+1}, \ldots, q_{M+r-1}$ are needed. Moreover, there may be a known heat flux at $x = L$, a known convection condition at $x = L$, or known internal volumetric heating. For these cases it is convenient to use the modification of Eq. (3.2.17),

$$\mathbf{T} = \mathbf{Xq} + \hat{\mathbf{T}}|_{q=0} \qquad (3.2.19)$$

This is called the standard form for the linear IHCP. This equation is not restricted to representing a numerical form of Duhamel's Theorem. It can be derived using Taylor series for the linear inverse heat conduction problem. The various components of Eq. (3.2.19) are

$$\mathbf{T} = \begin{bmatrix} T_M \\ T_{M+1} \\ \vdots \\ T_{M+r-1} \end{bmatrix}, \quad \mathbf{q} = \begin{bmatrix} q_M \\ q_{M+1} \\ \vdots \\ q_{M+r-1} \end{bmatrix} \qquad (3.2.20a,b)$$

$$\mathbf{X} = \begin{bmatrix} \Delta\phi_0 & & & \\ \Delta\phi_1 & \Delta\phi_0 & & \\ \vdots & \vdots & & \\ \Delta\phi_{r-1} & \Delta\phi_{r-2} & \cdots & \Delta\phi_0 \end{bmatrix} \qquad (3.2.21)$$

$$\hat{\mathbf{T}}|_{q=0} = \begin{bmatrix} \hat{T}_M|_{q_M=0} \\ \hat{T}_{M+1}|_{q_M=q_{M+1}=0} \\ \vdots \\ \hat{T}_{M+r-1}|_{q_M=q_{M+1}=\cdots=q_{M+r-1}=0} \end{bmatrix} \qquad (3.2.22)$$

Equations (3.2.19)–(3.2.22) apply for the numerical convolution as well as for the finite difference methods. For the case of known heating or energy sources other than where q is to be estimated, **X** and $\hat{\mathbf{T}}|_{q=0}$ must be properly interpreted.

The quantity ϕ_i needed in **X** is the temperature rise at the sensor location for a unit step change in the surface heat flux at $t = 0$ for the same partial differential equation and boundary conditions as the original problem except the differential equation and boundary conditions (other than at $x = 0$) are homogeneous. An example is the problem for a hollow cylinder,

$$\frac{k}{r}\frac{\partial}{\partial r}\left(r\frac{\partial T}{\partial r}\right) = \rho c \frac{\partial T}{\partial t} - \frac{g(r,t)}{k}, \quad a < r < b \qquad (3.2.23)$$

$$-k\frac{\partial T}{\partial r}\bigg|_{r=a} = q(t), \quad \text{unknown heat flux} \qquad (3.2.24a)$$

$$-k\frac{\partial T}{\partial r}\bigg|_{r=b} = h[T(b,t) - T_\infty(t)] \qquad (3.2.24b)$$

$$T(r,0) = F(r) \qquad (3.2.25)$$

The $\phi(r,t)$ problem is the solution of the problem of $\partial T/\partial q_c$ where $q(t) = q_c$. See Section 1.6. The $\phi(r,t)$ problem is

$$\frac{k}{r}\frac{\partial}{\partial r}\left(r\frac{\partial \phi}{\partial r}\right) = \rho c \frac{\partial \phi}{\partial t} \qquad (3.2.26)$$

$$-k\frac{\partial \phi}{\partial r}\bigg|_{r=a} = 1 \qquad (3.2.27a)$$

$$-k\frac{\partial \phi}{\partial r}\bigg|_{r=b} = h\phi(b,t) \qquad (3.2.27b)$$

$$\phi(r,0) = 0 \qquad (3.2.28)$$

Note that the only nonhomogeneous term is at the heated surface, $r = a$. The ϕ_i value is the solution of Eqs. (3.2.26)–(3.2.28) for the sensor location and time t_i.

The symbol $\hat{T}_M|_{q_M = 0}$ represents the calculated temperature for the model at time t_M for the estimated heat flux components $\hat{q}_1, \hat{q}_2, \ldots, \hat{q}_{M-1}$, but q_M is set equal to zero. For example, for the model given by Eqs. (3.2.23)–(3.2.25), known values of $g(r,t)$ are used in Eq. (3.2.23), known values of $T_\infty(t)$ are used in Eq. (3.2.24b), and the known initial temperature distribution, $F(r)$, is used; the temperature at the sensor location is calculated at time t_M with these conditions and with the heat flux, $q(t)$, equal to $\hat{q}_1$ for $0 < t < \Delta t$, $\hat{q}_2$ for Δt to $2\Delta t, \ldots, \hat{q}_{M-1}$ for $t_{M-2} < t < t_{M-1}$, and $q = 0$ for $t_{M-1} < t < t_M$. The temperature $\hat{T}_{M+1}|_{q_M = q_{M+1} = 0}$ is similarly found at the sensor location and at t_{M+1} with the known $\hat{q}_i$ values for $i = 1, 2, \ldots, M-1$ and also $q_M = q_{M+1} = 0$. For the case $g(r,t) = 0$ in Eq. (3.2.23), $T_\infty(t) = T_0$ in Eq. (3.2.24b) and $T(r,0) = T_0$ in Eq. (3.2.25), the $\hat{T}|_{q=0}$ vector is simply given by

$$\hat{T}|_{q=0} = \begin{bmatrix} \sum_{i=1}^{M-1} \hat{q}_i \Delta\phi_{M-i} + T_0 \\ \sum_{i=1}^{M-1} \hat{q}_i \Delta\phi_{M-i+1} + T_0 \\ \vdots \\ \sum_{i=1}^{M-1} \hat{q}_i \Delta\phi_{M-i+r-1} + T_0 \end{bmatrix} \qquad (3.2.29)$$

SEC. 3.3 DIFFERENCE METHODS

It is appropriate to compare Eqs. (3.2.17) and (3.2.19). Although the symbolism used in these two equations is very similar, the two equations are really quite different. Equation (3.2.17) is displayed more clearly by Eq. (3.1.16) for which temperatures are given starting at time t_1 and continuing up to time t_{M+r-1}. Moreover, there is only one heat input to the body; namely, the unknown surface heat flux, and also the initial temperature is the uniform value of T_0. Equation (3.2.19) is much more general and can include the conditions for Eq. (3.2.16) or Eq. (3.2.17) by simply setting $M=1$ and $\hat{T}_1|_{q_1=0}=T_0$. Equation (3.2.19) can also cover the case of an arbitrary temperature distribution at time t_{M-1} and a known heat flux at $x=L$ [or other nonhomogeneous conditions as shown in Eqs. (3.2.23) and (3.2.24)].

3.3 DIFFERENCE METHODS

Duhamel's theorem is a powerful technique for solving a wide variety of linear heat conduction problems. Unfortunately, many situations exist in which either (1) this technique becomes cumbersome or (2) it is not applicable. The most severe limitation is for nonlinear problems such as for bodies with temperature-dependent thermal properties. In many heat transfer problems, the temperature change of a body exposed to a heat flux is sufficiently large that the changes in thermal properties are appreciable. Fortunately, numerical methods can be used to convert the nonlinear partial differential equation of heat conduction into a system of linear algebraic equations involving the temperature at discrete locations.

3.3.1 Finite Control Volume Procedure for Constant Property Planar Geometries

Many methods exist for discretizing partial differential equations. Two popular ones are (1) finite differences and (2) finite elements. An alternative approach in which conservation of energy is applied directly to control volumes is adopted here. The finite control volume (FCV) procedure uses a control volume of arbitrary and finite size as in Figure 3.2 and the integral form of the energy equation is applied to this control volume. Conservation of energy for a stationary solid requires that the sum of the rate of heat flow across the control volume boundary and the time rate of change of energy within the control volume is equal to the rate at which energy is produced within the control volume. In equation form, the conservation of energy can be written as

$$\iint_A \mathbf{q} \cdot d\mathbf{A} + \frac{\partial}{\partial t} \iiint_\mathcal{V} \rho e \, d\mathcal{V} = \iiint_\mathcal{V} e''' \, d\mathcal{V} \qquad (3.3.1)$$

where A and $\mathcal{V}$ are the closed surface area and volume respectively of the control volume. $\mathbf{q}$ is the heat flux vector leaving the control volume surface, e is the

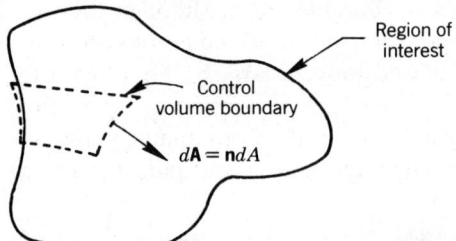

FIGURE 3.2 Typical finite control volume.

specific energy (energy per unit mass), and e''' is the energy production rate per unit volume. In the control volume shown in Figure 3.2, a part of the control volume boundary coincides with the boundary of the body, and the remaining control volume boundary is completely within the body. Consequently, the surface integral of the heat flux can involve both heat flux from external sources (boundary conditions) and heat conducted to the given control volume from all neighboring control volumes.

In its present form, Eq. (3.3.1) is too general to demonstrate the details of the FCV procedure. Let us restrict our attention to one-dimensional planar problems and divide the body of interest into increments of equal size (called elements) as shown in Figure 3.3. Each element has its own thermal properties and an element boundary can also correspond to a material interface. The control volume boundaries are chosen to lie at the midplane of each element; this selection of control volume boundary is somewhat arbitrary. The choice for control volume boundaries permits two different materials to be present within a single control volume. Each element contains two nodes, one at each boundary of the element which are identified by a (•) in Figure 3.3.

First, the general conservation of energy Eq. (3.3.1) is applied to the control volume surrounding node 1; Figure 3.4a provides an expanded view of this control volume. The result is

$$-k\frac{\partial T}{\partial x}\bigg|_{\Delta x/2} - q(t) + \frac{d}{dt}\int_0^{\Delta x/2} \rho c T\, dx = 0 \qquad (3.3.2)$$

where the energy content is given by $e = cT$ and $e''' \equiv 0$. For the arbitrary interior node j and the backface node N (see Figure 3.4b and c, respectively), application of the conservation of energy equation yields

$$-k\frac{\partial T}{\partial x}\bigg|_{x_j+\Delta x/2} + k\frac{\partial T}{\partial x}\bigg|_{x_j-\Delta x/2} + \frac{d}{dt}\int_{x_j-\Delta x/2}^{x_j+\Delta x/2} \rho c T\, dx = 0 \qquad (3.3.3)$$

$$q_N(t) + k\frac{\partial T}{\partial x}\bigg|_{x_N-\Delta x/2} + \frac{d}{dt}\int_{x_N-\Delta x/2}^{x_N} \rho c T\, dx = 0 \qquad (3.3.4)$$

This control volume approach insures that the heat conducted out of one control volume is conducted into the adjacent control volume. In order to evaluate the integrals and derivatives in Eqs. (3.3.2)–(3.3.4) and convert them

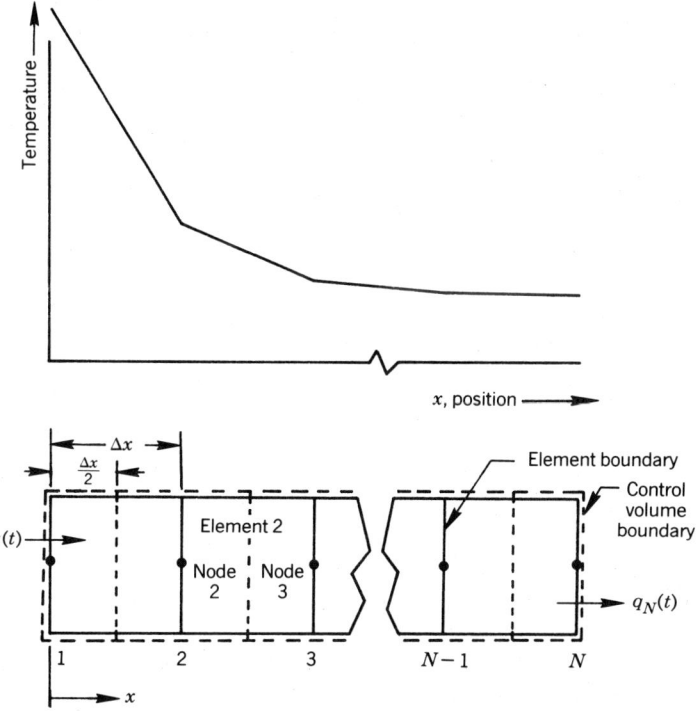

FIGURE 3.3 One-dimensional model for planar geometry showing element and control volume boundaries.

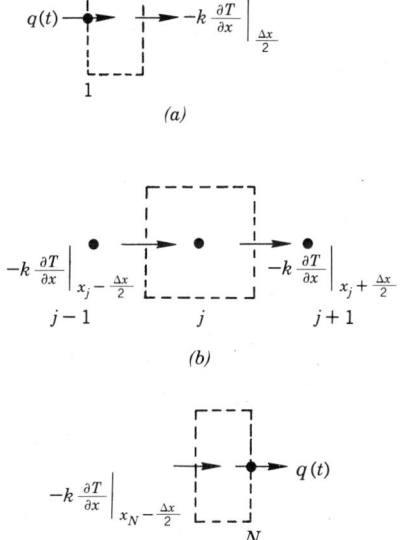

FIGURE 3.4 Expanded view of various control volumes. (*a*), Control volume surrounding surface node 1; (*b*), control volume surrounding arbitrary interior node j; (*c*), control volume surrounding backface surface node N.

into useful computational algorithms, it is necessary to make some assumptions about the temperature profile within each element. Many assumptions can be made concerning the element temperature profile; however, the more physics exercised, the better. As an example, for a steady-state, planar geometry with specified temperatures as boundary conditions and constant k, the temperature profile is linear with position. Thus for the transient case, a linear temperature profile is assumed for the element. For an element bounded by $x_{j-1} \leqslant x \leqslant x_j$, the assumed temperature profile is

$$T(x) = T_{j-1} \frac{(x-x_j)}{(x_{j-1}-x_j)} + T_j \frac{(x-x_{j-1})}{(x_j-x_{j-1})}, \quad x_{j-1} \leqslant x \leqslant x_j \quad (3.3.5)$$

where $T_j = T(x_j)$ is a node point (or nodal) temperature. Equation (3.3.5) is written in the form of a Lagrangian interpolation polynomial. The local heat flux is determined by differentiating Eq. (3.3.5) and using Fourier's law

$$q(x) = -k \frac{\partial T}{\partial x} = -k \frac{(T_j - T_{j-1})}{x_j - x_{j-1}}, \quad x_{j-1} < x < x_j \quad (3.3.6)$$

The linear element temperature profile produces a constant heat flux within elements; however, the heat flux changes from element to element.

Assuming that the volumetric heat capacity (ρc) is a constant for each element, a numerical expression for the energy storage term requires the evaluation of integrals of the form $\int T(x)dx$. It is convenient to perform these integrals over portions of elements and then add the appropriate contribution to the control volume energy balance. The two integrals that occur are evaluated using Eq. (3.3.5) to obtain

$$\int_{x_j}^{x_j + \Delta x/2} T\,dx = \frac{\Delta x}{2} (\tfrac{3}{4} T_j + \tfrac{1}{4} T_{j+1}) \quad (3.3.7)$$

$$\int_{x_j - \Delta x/2}^{x_j} T\,dx = \frac{\Delta x}{2} (\tfrac{1}{4} T_{j-1} + \tfrac{3}{4} T_j) \quad (3.3.8)$$

Note that Eqs. (3.3.7) and (3.3.8) are proportional to the average element temperature over *half* of the element.

From the foregoing linear element temperature profiles, the finite control volume energy balance for nodes 1, j and N, are

$$k \frac{(T_1 - T_2)}{\Delta x} - q(t) + \rho c \frac{\Delta x}{2} \frac{d}{dt} (\tfrac{3}{4} T_1 + \tfrac{1}{4} T_2) = 0 \quad (3.3.9)$$

$$k \frac{(T_j - T_{j+1})}{\Delta x} - k \frac{(T_{j-1} - T_j)}{\Delta x} + \rho c \frac{\Delta x}{2} \frac{d}{dt} (\tfrac{1}{4} T_{j-1} + \tfrac{3}{4} T_j)$$

$$+ \rho c \frac{\Delta x}{2} \frac{d}{dt} (\tfrac{3}{4} T_j + \tfrac{1}{4} T_{j+1}) = 0, \quad j = 2, 3, \ldots, N-1 \quad (3.3.10)$$

$$q_N(t) - k \frac{(T_{N-1} - T_N)}{\Delta x} + \rho c \frac{\Delta x}{2} \frac{d}{dt} (\tfrac{1}{4} T_{N-1} + \tfrac{3}{4} T_N) = 0 \quad (3.3.11)$$

SEC. 3.3 DIFFERENCE METHODS

These three equations represent a system of N first-order ordinary differential equations in time. By assuming an element temperature profile and integrating out the spatial dependence, the FCV procedure yields a system of ordinary differential equations (as opposed to a single partial differential equation). Note that each of the energy storage terms in Eqs. (3.3.9)–(3.3.11) contains a contribution from a node outside the control volume; this occurs because the linear element temperature profile depends on two nodal temperatures, one of which lies outside the control volume of interest. The foregoing results are sometimes referred to as distributed capacitance effects because the thermal capacitance of a control volume may be distributed over as many as three nodes.

The finite difference method is another technique for arriving at a system of ordinary differential equations representing heat conduction within a solid. This procedure is also called lumped capacitance because all of the control volume thermal capacitance is lumped at the center node point for all interior control volumes. This is accomplished by replacing $(\frac{1}{4}T_{j-1}+\frac{3}{4}T_j)$ by T_j and $(\frac{3}{4}T_j+\frac{1}{4}T_{j+1})$ with T_j in Eq. (3.3.10); for surface control volumes, $(\frac{3}{4}T_1+\frac{1}{4}T_2)$ becomes T_1 and $(\frac{1}{4}T_{N-1}+\frac{3}{4}T_N)$ becomes T_N. Additional details on difference methods are available in Myers,[9] Smith,[15] Patankar,[21] Anderson et al.,[22] and Shih.[23]

Another method for the numerical solution of heat conduction problems is the finite element (FE) method.[11-14] Many methods can be used to develop finite element type systems of differential equations. The Galerkin-type method of weighted residuals is one of the simpler approaches and requires that the integral weighted average of the partial differential equation be zero over some domain. For example, for a slab

$$\int_0^L w(x)\left(\frac{\partial T}{\partial t}-\alpha\frac{\partial^2 T}{\partial x^2}\right)dx=0 \qquad (3.3.12)$$

where L is the thickness of the slab and $w(x)$ is a weighting function to be specified. There is an infinite number of possible weighting functions; the trick is to choose a weighting function that gives good results. A popular weighting function is the so-called tent function which is shown in Figure 3.5:

$$w(x)=w_j(x)=\frac{x-x_{j-1}}{x_j-x_{j-1}}, \quad x_{j-1}\leqslant x\leqslant x_j$$

$$=\frac{x_{j+1}-x}{x_{j+1}-x_j}, \quad x_j\leqslant x\leqslant x_{j+1}$$

$$=0, \quad \text{all other } x \qquad (3.3.13)$$

Note the similarity between the weighting function $w_j(x)$ and the Lagrangian interpolation polynomial in Eq. (3.3.5), and that the weighting function is different for each node j. Since the weighting function $w_j(x)$ is zero over a large

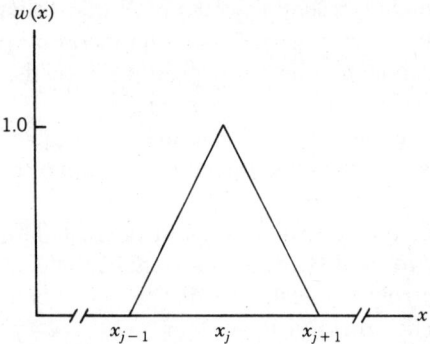

FIGURE 3.5 Weighting function for Galerkin-type method of weighted residuals.

portion of the domain $0 \leq x \leq L$, Eq. (3.3.12) can be written as

$$\int_{x_{j-1}}^{x_{j+1}} w_j(x) \left(\frac{\partial T}{\partial t} - \alpha \frac{\partial^2 T}{\partial x^2} \right) dx = 0 \qquad (3.3.14)$$

Utilizing the linear element temperature profile assumption and Eq. (3.3.13) for $w_j(x)$, the FE equation for an arbitrary interior node j can be written as

$$k \frac{(T_j - T_{j+1})}{\Delta x} - k \frac{(T_{j-1} - T_j)}{\Delta x} + \rho c \frac{\Delta x}{2} \frac{d}{dt} (\tfrac{1}{3} T_{j-1} + \tfrac{2}{3} T_j)$$

$$+ \rho c \frac{\Delta x}{2} \frac{d}{dt} (\tfrac{2}{3} T_j + \tfrac{1}{3} T_{j+1}) = 0, \quad j = 2, 3, \ldots, N-1 \qquad (3.3.15)$$

The only difference between the FCV and FE results is the different weighting coefficients on the thermal capacitance terms; the FCV gives a greater weighting to the center temperature T_j. The FE equations for the surface nodes are

$$k \frac{(T_1 - T_2)}{\Delta x} - q(t) + \rho c \frac{\Delta x}{2} \frac{d}{dt} (\tfrac{2}{3} T_1 + \tfrac{1}{3} T_2) = 0 \qquad (3.3.16)$$

$$q_N(t) - k \frac{(T_{N-1} - T_N)}{\Delta x} + \rho c \frac{\Delta x}{2} \frac{d}{dt} (\tfrac{1}{3} T_{N-1} + \tfrac{2}{3} T_N) = 0 \qquad (3.3.17)$$

Lemmon and Heaton[16] demonstrated that if the weighting function $w(x)$ is chosen to be the Dirac delta function

$$w_j(x) = \delta(x - x_j) = \begin{cases} 1, & x = x_j \\ 0, & \text{all other } x \end{cases} \qquad (3.3.18)$$

then the FE results for interior nodes are identical to the lumped thermal capacitance (LC). This method is also called the collocation method. Additional details of the FE method can be found in Zienkiewicz,[12] Norrie and deVries,[13] Becker et al.,[14] and Baker.[24]

SEC. 3.3 DIFFERENCE METHODS

All of the FCV, LC, and FE results can be put in a similar form:

$$\frac{d}{dt}(\beta T_1 + \gamma T_2) = -\frac{\alpha}{\Delta x^2}(T_1 - T_2) + \frac{q(t)}{\rho c \Delta x} \quad (3.3.19)$$

$$\frac{d}{dt}(\gamma T_{j-1} + 2\beta T_j + \gamma T_{j+1}) = -\frac{\alpha}{\Delta x^2}(T_j - T_{j+1}) + \frac{\alpha}{\Delta x^2}(T_{j-1} - T_j) \quad (3.3.20)$$

$$\frac{d}{dt}(\gamma T_{N-1} + \beta T_N) = \frac{\alpha}{\Delta x^2}(T_{N-1} - T_N) - \frac{q_N(t)}{\rho c \Delta x} \quad (3.3.21)$$

where β and γ have the following values:

	β	γ	$\beta + \gamma$
LC	1/2	0	1/2
FCV	3/8	1/8	1/2
FE	2/6	1/6	1/2

$$(3.3.22)$$

Equations (3.3.19)–(3.3.21) are restricted to temperature-independent thermal properties but the concepts can readily be extended to T-variable cases.

3.3.2 Other Boundary Conditions and Material Interfaces

In Section 3.3.1, a planar body with constant properties subjected to a specified heat flux boundary condition was considered. A heat flux given by a convective heat transfer coefficient and a temperature difference driving potential is also a common boundary condition. For this latter case the energy balance on the control volume surrounding node 1 can be obtained directly from Eq. (3.3.19).

$$\frac{d}{dt}(\beta T_1 + \gamma T_2) = -\frac{\alpha}{\Delta x^2}(T_1 - T_2) + \frac{h}{\rho c \Delta x}(T_\infty - T_1) \quad (3.3.23)$$

Note that convective boundary conditions involve the surface temperature.

The FCV approach can be readily extended to elements of nonuniform size and differing material properties; such a control volume is shown in Figure 3.6.

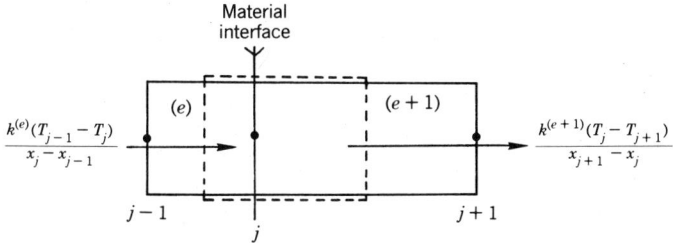

FIGURE 3.6 Control volume for material interface.

The two adjacent elements are denoted (e) and $(e+1)$ with each having its own temperature dependent properties $(\rho c)^{(e)}$ and $k^{(e)}$. Applying an energy balance

$$k^{(e+1)}\frac{T_j - T_{j+1}}{x_{j+1} - x_j} - k^{(e)}\frac{T_{j-1} - T_j}{x_j - x_{j-1}} + (\rho c)^e(x_j - x_{j-1})\frac{d}{dt}(\gamma T_{j-1} + \beta T_j)$$

$$+ (\rho c)^{(e+1)}(x_{j+1} - x_j)\frac{d}{dt}(\beta T_j + \gamma T_{j+1}) = 0 \qquad (3.3.24)$$

The element thermal conductivity is evaluated at the mid-plane of each element; for a linear element temperature profile, the appropriate temperature for element (e) is $\bar{T}^{(e)} = (T_{j-1} + T_j)/2$. There are several choices for the appropriate temperature at which to evaluate the volumetric heat capacity. The simplest procedure is to use the average element temperature to evaluate $(\rho c)^{(e)}$ and assign this value to both halves of the element. A slightly more complicated approach is to compute the average temperature of each half of an element and evaluate the volumetric heat capacity at this temperature. For example, the right half of element (e) contributes to control volume j and the average temperature of this half of element (e) is

$$\bar{T}_R^{(e)} = \tfrac{1}{2}[\bar{T}^{(e)} + T_j] = \tfrac{1}{2}[\tfrac{1}{2}(T_{j-1} + T_j) + T_j]$$

$$= \tfrac{3}{4}T_j + \tfrac{1}{4}T_{j-1} \qquad (3.3.25)$$

where the subscript R indicates right half of element (e). Similarly, for the left half of element $(e+1)$, the average temperature is

$$\bar{T}_L^{(e+1)} = \tfrac{3}{4}T_j + \tfrac{1}{4}T_{j+1} \qquad (3.3.26)$$

At this point, a word of caution is appropriate for the evaluation of temperature-dependent thermal properties. Most thermal properties are known only within $\pm 5\%$ at best. Consequently, it is not particularly fruitful to develop highly sophisticated techniques to precisely handle temperature-dependent thermal properties that are known imprecisely. Many times, the simplest approach is more than adequate.

3.3.3 Numerical Techniques for Solving Systems of First-Order Ordinary Differential Equations

In principle, an analytical solution of the system of energy balance Eqs. (3.3.19)–(3.3.21) is possible. However, numerical techniques offer solutions to more general problems and can be readily extended to the temperature-dependent thermal properties. Numerous numerical techniques are available.

One of the simpler techniques is the Euler (or forward difference, FD) method which can be developed from a Taylor series expansion. Let $u(t)$ be a function of time that satisfies the first-order ordinary differential equation

$$\frac{du}{dt} = f(u, t) \qquad (3.3.27)$$

SEC. 3.3 DIFFERENCE METHODS

Expanding $u(t+\Delta t)$ in a Taylor series about time t, gives

$$u(t+\Delta t) = u(t) + \frac{du(t)}{dt}\Delta t + \cdots \qquad (3.3.28)$$

If the terms $O(\Delta t^2)$ are ignored, then $u(t+\Delta t)$ is given by

$$u(t+\Delta t) \approx u(t) + f[u(t), t]\Delta t \qquad (3.3.29)$$

If $u(t_n)$ is denoted as u_n, then Eq. (3.3.28) becomes

$$u_{n+1} = u_n + f(u_n, t_n)\Delta t = u_n + \left.\frac{du}{dt}\right|_n \Delta t \quad \text{(FD)} \qquad (3.3.30)$$

The algorithm given by Eq. (3.3.30) is shown in Figure 3.7; it is simply the extrapolation of the slope at point (u_n, t_n) to (u_{n+1}, t_{n+1}). As an example of the Euler method, it is applied to Eq. (3.3.19) with $\beta = 1/2$, $\gamma = 0$ (LC). Since the subscript on T represents node number, a superscript is used to indicate time,

$$T_1^{n+1} = T_1^n - 2p(T_1^n - T_2^n) + \frac{2q''\Delta t}{\rho c \Delta x} \qquad (3.3.31)$$

where $p = \alpha\Delta t/\Delta x^2$ is a Fourier number based on the time-space grid. The Euler method for the LC formulation is also known as the explicit method since T_1^{n+1} in Eq. (3.3.31) can be written as an explicit function of known temperatures only at time n; another name is the forward difference method because the time derivative is replaced by a forward time difference. Explicit results are also obtained from Eqs. (3.3.20) and (3.3.21) for the LC formulation.

As can be seen from Figure 3.7, the Euler method underpredicts the temperature if the true solution is concave upward. In an attempt to improve stability, a backward time difference (BD) is often used; analogous to Eq. (3.3.28), the

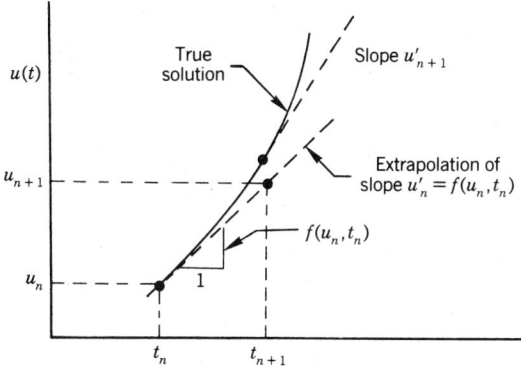

FIGURE 3.7 Schematic of the Euler (forward difference) and implicit (backward difference) methods.

equation is

$$u_n = u_{n+1} - u'_{n+1}\Delta t + u''_{n+1}\frac{\Delta t^2}{2!} - \cdots \qquad (3.3.32)$$

Neglecting terms of $O(\Delta t^2)$, the backward difference integrator gives

$$u_{n+1} \approx u_n + u'_{n+1}\Delta t \quad \text{(BD)} \qquad (3.3.33)$$

The backward time difference integrator is also known as the fully implicit method because the right-hand side contains the unknown u_{n+1}. A central difference (CD) time derivative is known to be more accurate (under certain conditions) than either the forward or backward difference and can be derived from the following Taylor series expansions:

$$u_{n+1/2} = u_n + u'_n \frac{\Delta t}{2} + \frac{u''_n}{2!}\left(\frac{\Delta t}{2}\right)^2 + \cdots \qquad (3.3.34)$$

$$u_{n+1/2} = u_{n+1} - u'_{n+1} \frac{\Delta t}{2} + \frac{u''_{n+1}}{2!}\left(\frac{\Delta t}{2}\right)^2 - \cdots \qquad (3.3.35)$$

Equating these two expressions and solving for u_{n+1} gives the central difference integrator

$$u_{n+1} \approx u_n + (u'_n + u'_{n+1})\frac{\Delta t}{2} \qquad (3.3.36)$$

which is also known as the Crank-Nicolson method. von Rosenberg[18] describes the Crank-Nicolson procedure as successive applications of the forward and backward integrators. The forward, backward, and central difference integrators can all be combined into a *general integrator* (GI) as follows:

$$u_{n+1} = u_n + [\theta u'_{n+1} + (1-\theta)u'_n]\Delta t \qquad (3.3.37)$$

The three special cases are:

$$\begin{aligned}&\theta = 0, \quad \text{forward difference or Euler} \\ &\theta = \frac{1}{2}, \quad \text{central difference or Crank-Nicolson} \\ &\theta = 1, \quad \text{backward difference or fully implicit}\end{aligned} \qquad (3.3.38)$$

3.3.4 General Form of Difference Equations for Heat Conduction in Planar Body

The general integrator given by Eq. (3.3.37) for first-order equations converts an ordinary differential equation into a single difference equation. The next step is to convert the system of ordinary differential equations, Eqs. (3.3.19)–(3.3.21), into a system of algebraic difference equations. The procedure for applying the GI, Eq. (3.3.37), is to evaluate the differential equation at times $n+1$ and n, multiply these equations by θ and $(1-\theta)$, respectively, and add the two equations. Using Eq. (3.3.19) as an example gives

SEC. 3.3 DIFFERENCE METHODS

$$\theta \frac{d}{dt}(\beta T_1 + \gamma T_2)^{n+1} = -\theta \frac{\alpha}{\Delta x^2}(T_1^{n+1} - T_2^{n+1}) + \theta \frac{q^{n+1}}{\rho c \Delta x}$$

$$\theta' \frac{d}{dt}(\beta T_1 + \gamma T_2)^n = -\theta' \frac{\alpha}{\Delta x^2}(T_1^n - T_2^n) + \theta' \frac{q^n}{\rho c \Delta x}$$

where $\theta' = 1 - \theta$ and the superscript n means evaluated at time t_n. Adding the foregoing two equations gives

$$\theta \frac{d}{dt}(\beta T_1 + \gamma T_2)^{n+1} + \theta' \frac{d}{dt}(\beta T_1 + \gamma T_2)^n = -\theta \frac{\alpha}{\Delta x^2}(T_1^{n+1} - T_2^{n+1})$$

$$-\theta' \frac{\alpha}{\Delta x^2}(T_1^n - T_2^n) + \frac{1}{\rho c \Delta x}(\theta q^{n+1} + \theta' q^n) \quad (3.3.39)$$

The general integrator can be applied directly to the left-hand side of Eq. (3.3.39) with the result

$$\beta \frac{(T_1^{n+1} - T_1^n)}{\Delta t} + \gamma \frac{(T_2^{n+1} - T_2^n)}{\Delta t} = -\theta \frac{\alpha}{\Delta x^2}(T_1^{n+1} - T_2^{n+1})$$

$$-\theta' \frac{\alpha}{\Delta x^2}(T_1^n - T_2^n) + \frac{q^{n+\theta}}{\rho c \Delta x} \quad (3.3.40)$$

where for convenience the following definition is employed

$$q^{n+\theta} \equiv \theta q^{n+1} + \theta' q^n \quad (3.3.41)$$

Equation (3.3.40) can be simplified to yield

$$(p\theta + \beta)T_1^{n+1} - (p\theta - \gamma)T_2^{n+1} = -(p\theta' - \beta)T_1^n + (p\theta' + \gamma)T_2^n + \frac{\Delta t}{\rho c \Delta x}q^{n+\theta} \quad (3.3.42)$$

Applying the GI to Eqs. (3.3.20) and (3.3.21) gives

$$-(p\theta - \gamma)T_{j-1}^{n+1} + 2(p\theta + \beta)T_j^{n+1} - (p\theta - \gamma)T_{j+1}^{n+1} = (p\theta' + \gamma)T_{j-1}^n$$

$$+ 2(\beta - p\theta')T_j^n + (p\theta' + \gamma)T_{j+1}^n, \quad j = 2, 3, \ldots, N-1 \quad (3.3.43)$$

$$-(p\theta - \gamma)T_{N-1}^{n+1} + (p\theta + \beta)T_N^{n+1} = (p\theta' + \gamma)T_{N-1}^n + (\beta - p\theta')T_N^n - \frac{\Delta t}{\rho c \Delta x}q_N^{n+\theta} \quad (3.3.44)$$

Equations (3.3.42)–(3.3.44) form a linear system of N algebraic equations with N unknowns. These algebraic equations have a very special structure that is known as tridiagonal and is best visualized by writing them in matrix form for constant thermal properties,

$$\begin{bmatrix} p\theta + \beta & -(p\theta - \gamma) & & & & \\ -(p\theta - \gamma) & 2(p\theta + \beta) & -(p\theta - \gamma) & & & \\ \cdot & \cdot & \cdot & \cdot & & \\ & \cdot & \cdot & \cdot & \cdot & \\ & & \cdot & \cdot & \cdot & \cdot \\ & & & & -(p\theta - \gamma) & p\theta + \beta \end{bmatrix} \begin{bmatrix} T_1^{n+1} \\ T_2^{n+1} \\ \vdots \\ T_{N-1}^{n+1} \\ T_N^{n+1} \end{bmatrix} =$$

$$\begin{bmatrix} \beta-p\theta' & \gamma+p\theta' & & & \\ \gamma+p\theta' & 2(\beta-p\theta') & \gamma+p\theta' & & \\ & \cdot & \cdot & \cdot & \\ & & \cdot & \cdot & \gamma+p\theta' \\ & & & \gamma+p\theta' & \beta-p\theta' \end{bmatrix} \begin{bmatrix} T_1^n \\ T_2^n \\ \vdots \\ T_{N-1}^n \\ T_N^n \end{bmatrix} + \frac{\Delta t}{\rho c \Delta x} \begin{bmatrix} q^{n+\theta} \\ 0 \\ \vdots \\ 0 \\ -q_N^{n+\theta} \end{bmatrix}$$

(3.3.45)

The tridiagonal coefficient matrix has nonzero terms along only three diagonals. This tridiagonal structure allows for a very efficient specialization of the Gauss elimination procedure for solving linear systems of algebraic equations. The details of the algorithm, known as the Thomas algorithm, are presented in Section 6.3.2.

The form of Eq. (3.3.45) shows the explicit dependence of the new temperatures on the old temperatures and the boundary conditions. From a computational point of view, it is more convenient to lump the right-hand side of Eq. (3.3.45) into a constant vector.

The Euler integrator (forward difference, $\theta=0$) for the LC ($\beta=1/2$, $\gamma=0$) is worthy of special note. For this case, the coefficient matrix for the unknown temperatures become diagonal; consequently, each unknown temperature T_i^{n+1} can be expressed explicitly in terms of known temperatures (T_{i-1}^n, T_i^n, T_{i+1}^n). For either the FCV ($\beta=3/8$, $\gamma=1/8$) or FE ($\beta=2/6$, $\gamma=1/6$) method in conjunction with the Euler integrator ($\theta=0$), all three diagonals of the coefficient matrix are nonzero; hence, the term *explicit* integrator is inappropriate for these cases.

EXAMPLE 3.2. Give the difference equations for the Crank-Nicolson approximation for a plate that is heated at $x=0$ and insulated at $x=L$. Let $\Delta x = L/3$ and use a lumped thermal capacitance. The thermal properties are independent of temperature.

Solution. The difference equations can be obtained using Eq. (3.3.45) for $N=4$ (see Figure 3.3); there are two boundary nodes and two interior nodes. From Eq. (3.3.22) for the lumped capacitance approximation, β equals 1/2 and γ equals 0. From Eq. (3.3.38) for the Crank-Nicolson approximation, θ equals 1/2. For the insulated boundary, $q_4^{n+1/2}=0$.

Using these values in Eq. (3.4.45) and multiplying the first and last equations by 2 gives the four difference equations in matrix form as,

$$\begin{bmatrix} p+1 & -p & 0 & 0 \\ -p/2 & p+1 & -p/2 & 0 \\ 0 & -p/2 & p+1 & -p/2 \\ 0 & 0 & -p & p+1 \end{bmatrix} \begin{bmatrix} T_1^{n+1} \\ T_2^{n+1} \\ T_3^{n+1} \\ T_4^{n+1} \end{bmatrix} = \begin{bmatrix} 1-p & p & 0 & 0 \\ p/2 & 1-p & p/2 & 0 \\ 0 & p/2 & 1-p & p/2 \\ 0 & 0 & p & 1-p \end{bmatrix} \begin{bmatrix} T_1^n \\ T_2^n \\ T_3^n \\ T_4^n \end{bmatrix} + \frac{2\Delta t}{\rho c \Delta x} \begin{bmatrix} q^{n+1/2} \\ 0 \\ 0 \\ 0 \end{bmatrix}$$

□

SEC. 3.3 DIFFERENCE METHODS

Comments on Accuracy. In order to obtain the desired accuracy it is necessary to use a sufficient number of nodes N and to take a "small" time step. In many cases a value of N equal to 20 is sufficient. The time steps must be at least as small as the time steps in the measured temperatures but can be considerably smaller; it is convenient to choose the computational time step, Δt, so that the time step between measurements, Δt_{meas}, is equal to

$$\Delta t = \frac{\Delta t_{meas}}{i} \tag{3.3.46}$$

where i is a positive integer. In order to choose the space step, Δx, as well as the time step, Δt, experience is needed for each particular case because their values are affected by many factors including the time period of interest, variation of thermal properties with temperature, and the time variation of the surface heat flux. For more discussion, see books on numerical methods such as References 6, 8, 9 and 15.

3.3.5 Standard Form of Temperature Equations for IHCP

The purpose of this section is to demonstrate that the difference equations produced by the FD, FE, and FCV methods can be used to obtain the same standard equation, Eq. (3.2.19), that was obtained from Duhamel's theorem. The discussion in this section is confined to the linear IHCP but nearly the same approach can be used for the quasilinear analysis which is discussed in Section 6.2.

The analysis can be performed using either algebraic or matrix notation but the latter analysis is preferred because it is more compact and general. If $n+1$ is replaced by the index M, Eq. (3.3.45) can be written as

$$\mathbf{A}\mathbf{T}^M = \mathbf{B}\mathbf{T}^{M-1} + \mathbf{C}\mathbf{q}^M + \mathbf{g}^M \tag{3.3.47}$$

where $\mathbf{A}$ is the square matrix on the left of Eq. (3.3.45), $\mathbf{B}$ is the square matrix on the right of Eq. (3.3.45) and $C = \Delta t / \rho c \Delta x$. The vectors $\mathbf{T}^M$ and $\mathbf{q}^M$ are

$$\mathbf{T}^M = \begin{bmatrix} T_1^M \\ \vdots \\ T_N^M \end{bmatrix}, \quad \mathbf{q}^M = \begin{bmatrix} q^{M+\theta-1} \\ 0 \\ \vdots \\ 0 \end{bmatrix} \tag{3.3.48}$$

where the subscripts relate to the space nodes and the superscripts to time. The $\mathbf{g}^M$ vector represents any *known* energy sources due to an applied heat flux at $x=L$, volumetric heating, or convective heat transfer. If the Crank-Nicolson method of solution is selected, $\theta = 1/2$ and

$$q^{M+\theta-1} = q^{M-1/2}$$

which is the surface heat flux evaluated at time $t_{M-1/2} = (M-1/2)\Delta t$. Observe that $q^{M-1/2}$ means exactly the same heat flux component as denoted q_M in the Duhamel procedure [see Eq. (3.2.2)].

Equation (3.3.47) is solved for $\mathbf{T}^M$ by multiplying by $\mathbf{A}^{-1}$ to obtain

$$\mathbf{T}^M = \mathbf{D}\mathbf{T}^{M-1} + \mathbf{E}\mathbf{q}^M + \mathbf{A}^{-1}\mathbf{g}^M \qquad (3.3.49a)$$

where

$$\mathbf{D} \equiv \mathbf{A}^{-1}\mathbf{B}, \quad \mathbf{E} \equiv \mathbf{C}\mathbf{A}^{-1} \qquad (3.3.49b,c)$$

(Though it is not a good *numerical* procedure to find the inverse of $\mathbf{A}$ when calculating $\mathbf{T}^M$, it is convenient here to do so symbolically.) Notice that $\mathbf{T}^M$ is a linear function of $\mathbf{q}^M$. Replacing M in Eq. (3.3.49a) with $M+1$ gives

$$\mathbf{T}^{M+1} = \mathbf{D}\mathbf{T}^M + \mathbf{E}\mathbf{q}^{M+1} + \mathbf{A}^{-1}\mathbf{g}^{M+1} \qquad (3.3.50)$$

and then using Eq. (3.3.49a) for $\mathbf{T}^M$ yields

$$\mathbf{T}^{M+1} = \mathbf{D}^2\mathbf{T}^{M-1} + \mathbf{D}\mathbf{E}\mathbf{q}^M + \mathbf{E}\mathbf{q}^{M+1} + \mathbf{D}\mathbf{A}^{-1}\mathbf{g}^M + \mathbf{A}^{-1}\mathbf{g}^{M+1} \qquad (3.3.51)$$

This expression is linear in both $\mathbf{q}^M$ and $\mathbf{q}^{M+1}$. Repeating the process for M replaced by $M+2$ in Eq. (3.3.49a) gives

$$\mathbf{T}^{M+2} = \mathbf{D}^3\mathbf{T}^{M-1} + \mathbf{D}^2\mathbf{E}\mathbf{q}^M + \mathbf{D}\mathbf{E}\mathbf{q}^{M+1} + \mathbf{E}\mathbf{q}^{M+2}$$
$$+ \mathbf{D}^2\mathbf{A}^{-1}\mathbf{g}^M + \mathbf{D}\mathbf{A}^{-1}\mathbf{g}^{M+1} + \mathbf{A}^{-1}\mathbf{g}^{M+2} \qquad (3.3.52)$$

which is linear in $\mathbf{q}^M$, $\mathbf{q}^{M+1}$ and $\mathbf{q}^{M+2}$.

Due to the linearity of Eqs. (3.3.49a), (3.3.51), and (3.3.52), they can be written as

$$\mathbf{T}^M = \hat{\mathbf{T}}^M|_{\mathbf{q}^M=0} + \mathbf{E}\mathbf{q}^M \qquad (3.3.53a)$$

$$\mathbf{T}^{M+1} = \hat{\mathbf{T}}^{M+1}|_{\mathbf{q}^M=\mathbf{q}^{M+1}=0} + \mathbf{D}\mathbf{E}\mathbf{q}^M + \mathbf{E}\mathbf{q}^{M+1} \qquad (3.3.53b)$$

$$\mathbf{T}^{M+2} = \hat{\mathbf{T}}^{M+2}|_{\mathbf{q}^M=\mathbf{q}^{M+1}=\mathbf{q}^{M+2}=0} + \mathbf{D}^2\mathbf{E}\mathbf{q}^M + \mathbf{D}\mathbf{E}\mathbf{q}^{M+1} + \mathbf{E}\mathbf{q}^{M+2} \qquad (3.3.53c)$$

where the temperatures evaluated at $\mathbf{q}^M, \ldots,$ equal to $\mathbf{0}$ are defined by

$$\hat{\mathbf{T}}^M|_{\mathbf{q}^M=0} = \mathbf{D}\mathbf{T}^{M-1} + \mathbf{A}^{-1}\mathbf{g}^M \qquad (3.3.53d)$$

$$\hat{\mathbf{T}}^{M+1}|_{\mathbf{q}^M=\mathbf{q}^{M+1}=0} = \mathbf{D}^2\mathbf{T}^{M-1} + \mathbf{D}\mathbf{A}^{-1}\mathbf{g}^M + \mathbf{A}^{-1}\mathbf{g}^{M+1} \qquad (3.3.53e)$$

$$\hat{\mathbf{T}}^{M+2}|_{\mathbf{q}^M=\mathbf{q}^{M+1}=\mathbf{q}^{M+2}=0} = \mathbf{D}^3\mathbf{T}^{M-1} + \mathbf{D}^2\mathbf{A}^{-1}\mathbf{g}^M + \mathbf{D}\mathbf{A}^{-1}\mathbf{g}^{M+1} + \mathbf{A}^{-1}\mathbf{g}^{M+2}$$
$$(3.3.53f)$$

Notice that the temperature components defined by Eqs. (3.3.53d,e,f) are linear functions of only the temperature distribution at time t_{M-1} and the known heat sources at times t_M, t_{M+1}, and t_{M+2}. They are equal to the corresponding temperatures given by Eqs. (3.3.35a,b,c) if the future heat flux vectors are equal to zero. Equations (3.3.35a,b,c) display the important characteristic of linearity in the temperatures defined by Eqs. (3.3.35d,e,f) and the heat fluxes $\mathbf{q}^M$, $\mathbf{q}^{M+1}$, and $\mathbf{q}^{M+2}$.

SEC. 3.3 DIFFERENCE METHODS

Equations (3.3.53a,b,c) are vector equations for the temperatures at each node in the body. Temperature sensors are located at only a few of these nodes, in general. For simplicity, a single sensor is considered and the space dependence is indicated by a subscript K. Then Eqs. (3.3.53a,b,c) can be written for the Kth node as

$$T_K^M = \hat{T}_K^M\big|_{q^M=0} + X_{11}q^M \qquad (3.3.54a)$$

$$T_K^{M+1} = \hat{T}_K^{M+1}\big|_{q^M=q^{M+1}=0} + X_{21}q^M + X_{11}q^{M+1} \qquad (3.3.54b)$$

$$T_K^{M+2} = \hat{T}_K^{M+2}\big|_{q^M=q^{M+1}=q^{M+2}=0} + X_{31}q^M + X_{21}q^{M+1} + X_{11}q^{M+2} \qquad (3.3.54c)$$

where θ in $q^{M+\theta-1}$ is chosen to be 1 to simplify the notation. The symbols X_{11}, X_{21}, and X_{31} denote sensitivity coefficients at location K and are not functions of the time index M as indicated by the corresponding matrix terms of $\mathbf{E}$, $\mathbf{DE}$ and $\mathbf{D^2E}$. The sensitivity components X_{11}, X_{21}, X_{31} are given by

$$E_{K1} = \frac{\partial T_K^M}{\partial q^M} = \frac{\partial T_K^{M+1}}{\partial q^{M+1}} = \frac{\partial T_K^{M+2}}{\partial q^{M+2}} = X_{11} \qquad (3.3.54d)$$

$$\frac{\partial T_K^{M+1}}{\partial q^M} = \frac{\partial T_K^{M+2}}{\partial q^{M+1}} = X_{21} \qquad (3.3.54e)$$

$$\frac{\partial T_K^{M+2}}{\partial q^M} = X_{31} \qquad (3.3.54f)$$

Equations (3.3.54) can be written in the standard matrix form with the K subscript on $\mathbf{T}$ and $\hat{\mathbf{T}}|_{q=0}$ omitted as

$$\mathbf{T} = \mathbf{X}\mathbf{q} + \hat{\mathbf{T}}|_{q=0} \qquad (3.3.55a)$$

where

$$\mathbf{T} = \begin{bmatrix} T^M \\ T^{M+1} \\ T^{M+2} \end{bmatrix}, \qquad \mathbf{X} = \begin{bmatrix} X_{11} & & \\ X_{21} & X_{11} & \\ X_{31} & X_{21} & X_{11} \end{bmatrix} \qquad (3.3.55b,c)$$

$$\hat{\mathbf{T}}|_{q=0} = \begin{bmatrix} \hat{T}^M|_{q^M=0} \\ \hat{T}^{M+1}|_{q^M=q^{M+1}=0} \\ \hat{T}^{M+2}|_{q^M=q^{M+1}=q^{M+2}=0} \end{bmatrix} \qquad (3.3.55d)$$

Whereas the temperature vectors in Eqs. (3.3.47)–(3.3.53) are for the nodes at time t_M, the temperature vectors in Eqs. (3.3.55) are for different times. Equation (3.3.55a) is called the standard form for the temperature because the same equation was derived using Duhamel's theorem [see Eq. (3.2.19)]. In Eqs. (3.3.55b,c,d) the superscript M denotes the *same* time dependence as the M subscripts in Eqs. (3.2.20a,b) and (3.2.22a). The X_{11}, X_{21}, and X_{31} values correspond to

$$X_{11} = \Delta\phi_0, \quad X_{21} = \Delta\phi_1, \quad \text{and} \quad X_{31} = \Delta\phi_2$$

used in the Duhamel theorem method and the numerical values of X_{11} and $\Delta\phi_0$, X_{21} and $\Delta\phi_1$, and so on, will be nearly identical if the finite difference (or FE or FCV) calculations are accurate. The symbol ϕ_i represents the temperature rise at time t_i and node x_K for the numerical solution of Eq. (3.3.47) for $q^{M+\theta-1} = 1$ for $i = M = 1, 2, 3, \ldots$ and $\mathbf{g}^M = \mathbf{0}$.

Since the same standard form, Eq. (3.2.19) or (3.3.55a), is obtained for **T**, IHCP algorithms can be developed that do not depend on the type of numerical approximation of the transient heat conduction equation. Advantage is taken of this fact in Chapter 4 to derive general algorithms. In Chapter 5 the algorithms of Chapter 4 are implemented using Duhamel's integral, and Chapter 6 uses the finite control volume method.

REFERENCES

1. Keltner, N. R. and Beck, J. V., Unsteady Surface Element Method, *J. Heat Transfer* **103**, 759–764 (1981).
2. Beck, J. V. and Keltner, N. R., Transient Thermal Contact of Two Semi-Infinite Bodies Over a Circular Area, in *Spacecraft Radiative Transfer and Temperature Control*, T. E. Horton, ed., Vol. 83 of *Progress in Astronautics and Aeronautics* (1982), AIAA, 61–82.
3. Beck, J. V., Schisler, I. P. and Keltner, N. R., Simplified Laplace Transform Inversion for Unsteady Surface Element Method for Transient Conduction, *AIAA Journal*, **22**, 1328–1333, (1984).
4. Keltner, N. R. and Beck, J. V., Surface Temperature Measurement Errors, *J. Heat Transfer*, **105**, 312–318 (1983).
5. Litkouhi, B. and Beck, J. V., Intrinsic Thermocouple Analysis Using Multinode Unsteady Surface Element Method, AIAA Paper No. 83-1937. (To be published in AIAA Journal.)
6. Ozisik, M. N. *Heat Conduction*, Wiley, New York, 1980.
7. Luikov, A. V., *Analytical Heat Diffusion Theory*, Academic Press, New York, 1968.
8. Arpaci, V. S., *Conduction Heat Transfer*, Addison-Wesley, Reading, MA, 1966.
9. Myers, G. E., *Analytical Methods in Conduction Heat Transfer*, McGraw-Hill, New York, 1971.
10. Carslaw, H. S. and Jaeger, J. C., *Conduction of Heat in Solids*, 2nd ed., Oxford, London, 1959.
11. Bathe, Klaus-Jurgen, *Finite Element Procedures in Engineering Analysis*, Prentice-Hall, Englewood Cliffs, NJ, 1982.
12. Zienkiewicz, O. C., *The Finite Element Method*, 3rd ed., McGraw-Hill, New York, 1977.
13. Norrie, D. H. and deVries, G., *An Introduction to Finite Element Analysis*, Academic Press, New York, 1978.
14. Becker, E. B., Carey, G. F., and Oden, J. T., *Finite Elements an Introduction*, Vol. 1, Prentice-Hall, Englewood Cliffs, NJ, 1981.
15. Smith, G. D., *Numerical Solution of Partial Differential Equations*, Oxford University Press, New York, 1965.
16. Lemmon, E. C. and Heaton, H. S., Accuracy, Stability, and Oscillation Characteristics of Finite Element Method for Solving Heat Conduction Equation, ASME Paper No. 69-WA/HT-35.

17. Myers, G. E., The Critical Time Step for Finite Element Solutions to Two-Dimensional Heat Conduction Transients, *ASME J. Heat Transfer* **100**, 120–127, (1978).
18. von Rosenberg, D. U., *Methods for the Numerical Solution of Partial Differential Equations*, Americal Elsevier, New York, 1969.
19. Richtmyer, R. D. and Morton, K. W., *Difference Methods for Initial Value Problems*, 2nd ed., Interscience Publishers, New York, 1967.
20. Beck, J. V., Green's Function Solution for Transient Heat Conduction Problems, *Int. J. Heat Mass Transfer*, **27**, 1235–1244 (1984).
21. Patankar, S. V., *Numerical Heat Transfer and Fluid Flow*, Hemisphere, Washington, 1980.
22. Anderson, D. A., Tannehill, J. C., and Pletcher, R. H., *Computational Fluid Mechanics and Heat Transfer*, Hemisphere, Washington, 1984.
23. Shih, T. M., *Numerical Heat Transfer*, Hemisphere, Washington, 1984.
24. Baker, A. J., *Finite Element Computational Fluid Mechanics*, Hemisphere, 1983.

PROBLEMS

3.1. Obtain mathematical expressions for the surface temperature of a semi-infinite body that is initially at the uniform temperature of T_0 and then is exposed to the heat flux history given by,

$$q(t) = \begin{cases} q_0, & 0 < t < t_1 \\ -q_0, & t_1 < t < 2t_1 \\ 0, & t < 0 \text{ and } t > 2t_1 \end{cases}$$

A different expression is needed for each time interval. Plot the group of

$$\frac{T - T_0}{q_0}\left(\frac{k\rho c}{t_1}\right)^{1/2}$$

versus t/t_1 for t/t_1 from 0 to 4. Why do the temperatures change after $t = 2t_1$ even though the net energy added is zero?

3.2. The heat flux history at the surface of a body is

$$q(t) = \begin{cases} 10 \text{ W/m}^2, & 2s < t < 4s \\ 0 & \text{otherwise} \end{cases}$$

The temperature at x_1 is 400 K until $t = 2s$, 401 K at $t = 4s$, and 406 K at $t = 6s$. The heat conduction problem is linear. Find the temperature rise at x_1 at times 2 and 4s for a unit step rise at $t = 0$ in the surface heat flux.

3.3. Find exact expressions for the surface temperature of a semi-infinite body that is initially at the uniform temperature of T_0 and that is subjected to a heat flux that varies in a triangular fashion with time as given by

$$q(t) = \begin{cases} \dot{q}_0 t, & 0 < t < t_1 \\ \dot{q}_0(2t_1 - t), & t_1 < t < t_2 = 2t_1 \\ 0 & \text{otherwise} \end{cases}$$

where $\dot{q}_0$ is a constant.

Give expressions for the three time regions of (a) $0 < t < t_1$, (b) $t_1 < t < 2t_1$ and (c) $2t_1 < t$.

Plot the surface temperature in a dimensionless form versus t/t_1 for the range of 0 to 3.

3.4. A heat flux is given by

$$q(t) = \begin{cases} q_a, & 0 < t < t_a \\ q_a + \dfrac{q_b - q_a}{t_b - t_a}(t - t_a), & t_a < t \leq t_b = 2t_a \end{cases}$$

For a semi-infinite body initially at T_0 calculate the surface temperature at time t_b using Eq. (3.2.12).

Use $\Delta t = 0.5$, $t_a = 1$, $q_a = 1$, $q_b = 2$, and $k\rho c = 1$. Compare the answer with the exact value.

3.5. Derive the approximation of the Duhamel theorem given by

$$T_M = T_0 + \sum_{n=1}^{M} q_n a_{M-n+1}$$

where $q_i = q$ at t_i and

$$a_1 = \phi_1^{(1)}/\Delta t; \quad a_i = (\phi_i^{(1)} - 2\phi_{i-1}^{(1)} + \phi_{i-2}^{(1)})/\Delta t, \quad i = 2, 3, \ldots$$

and $\phi_i^{(1)}$ is the temperature rise at time t_i for the heat flux $q = t$. The heat flux is composed of connected linear segments starting with $q_0 = 0$. (The superscript 1 denotes a *linear* heat flux "building block.")

3.6. Write a computer or programmable calculator program for the numerical convolution as given by Eq. (3.2.12). Allow for at least 10 q_n and 10 $\Delta\phi_i$ components. The input is to be T_0 and the q_n's and $\Delta\phi_i$'s. The output is

$$T_1, T_2, \ldots, T_{10}.$$

Verify your program by checking Example 3.1.

3.7. A semi-infinite body initially at the uniform temperature of zero is exposed at $t = 0$ to the heat flux.

$$q = (\pi t)^{-1/2}$$

Let $\alpha = 1$, $k = 1$, and $\Delta t = 1$.

Calculate the temperature at $x = 0$ using Eq. (3.2.12). Use at least 10 time steps. The exact temperature is 1. Comment on the accuracy of the procedure.

Partial Answers: 0.900, 0.784, 0.841, 0.865, and 0.880

3.8. An alternate way to write the convolution integral is

$$T(x, t) = T_0 + \frac{\alpha}{k}\int_0^t q(\lambda)G(x, t - \lambda)d\lambda \qquad (a)$$

PROBLEMS

where $G(x, t) = (k/\alpha)\partial\phi(x, t)/\partial t$. ($G(x, t)$ is a Green's function[20].)

a. Is it possible to use a numerical approximation of Eq. (a) when $G(0, t)$ is needed at $t = 0$? Give a reason for your answer.

b. Derive the numerical approximation of Eq. (a),

$$T_M = T_0 + \sum_{n=1}^{M} q_n H_{M-n+1} \Delta t \qquad (b)$$

where $T_M \equiv T(x, t_M)$, $q_M \equiv q(t_{M-1/2})$ and

$$H_i \equiv G(x, t_{i-1/2}) \frac{\alpha}{k}$$

c. Use Eq. (b) for the linear heat flux case of Example 3.1 and compare the values.

3.9. Show that if ϕ_i is given by

$$\phi_i = Be^{-ci}$$

that the sum

$$T'_M = \sum_{n=1}^{M-m-1} q_n \Delta\phi_{M-n}$$

can be written in the sequential form

$$T'_{M+1} = e^{-1} T'_M + B(e^{-c} - 1) q_{M-m} e^{-(m+1)c}$$

3.10. Assume that the temperature rise for an unknown surface heat flux is $\Delta T(\mathbf{r}, t)$ which is a known function. By taking the Laplace transform of Eq. (3.2.9), derive

$$q(t) = \mathscr{L}^{-1} \frac{\mathscr{L}[\Delta T(\mathbf{r}, t)]}{s\mathscr{L}[\phi(\mathbf{r}, t)]}$$

where s is the Laplace transform parameter. For

$$\phi = 2\left(\frac{t}{\pi k\rho c}\right)^{1/2}, \qquad \Delta T = \frac{4\dot{q}_0(\alpha t)^{3/2}}{3\alpha k(\pi)^{1/2}}$$

find $q(t)$.

3.11. Using the Laplace transform show that the temperature rise in a body exposed to the linearly varying heat flux,

$$q(t) = \dot{q}_0 t$$

is equal to

$$\Delta T(\mathbf{r}, t) = \dot{q}_0 \int_0^t \phi(\mathbf{r}, t') dt'$$

where $\phi(\mathbf{r}, t)$ is for a unit step heat flux starting at $t = 0$.

3.12. Using the divergence theorem of Gauss, demonstrate that Eq. (3.3.1) becomes

$$\nabla \cdot \mathbf{q} + \rho \frac{\partial e}{\partial t} = e'''$$

3.13. For one-dimensional (radial) cylindrical geometries, use the steady-state temperature profile as the element temperature profile and develop the equations analogous to Eqs. (3.3.9)–(3.3.11). Assume the control volume boundaries are at the midpoint of the elements.

3.14. Repeat Problem 3.13 for one-dimensional (radial) spherical geometries.

3.15. Verify the finite element equation, Eq. (3.3.15).

3.16. Give the difference equations in matrix form for a plate with $\Delta x = L/4$ which is exposed to a time-variable heat flux at $x=0$ and a convective boundary condition at $x=L$. The properties are temperature independent. Use the backward difference time approximation and lumped capacitance.

3.17.
   a. For two nodes in a plate, one at $x=0$ and the other at $x=L$, give the two Crank-Nicolson, lumped capacitance equations. The surface at $x=0$ is exposed to a heat flux and the surface at L is insulated.
   b. For the temperature at node 2, derive an expression analogous to Eq. (3.3.54a). Give $T_2^M|_{q^M=0}$ as a function of T_1^{M-1}, T_2^{M-1}, p, and $\Delta t/\rho c \Delta x$.
   c. Give X_{11} as a function of p and $\Delta t/\rho c \Delta x$.

3.18. The one-dimensional Green's function equation analogous to Duhamel's theorem is given in Problem 3.8. Write the equation in the form

$$T(x, t_M) = T_0 + \frac{\alpha}{k} \int_0^{t_M} q(\lambda) G(x, t_M - \lambda) d\lambda$$

$$= T_0 + \frac{\alpha}{k} \sum_{i=1}^{M} \int_{t_{i-1}}^{t_i} q(\lambda) G(x, t_M - \lambda) d\lambda$$

Approximate $q(\lambda)$ for $t_{i-1} < \lambda < t_i$ by

$$q(\lambda) = q_{i-1} \frac{t_i - \lambda}{\Delta t} + q_i \frac{\lambda - t_{i-1}}{\Delta t}$$

where $q_i = q(t_i)$, $t_i = i\Delta t$.
Derive the approximate relation,

$$T(x, t_M) = T_0 + \sum_{i=1}^{M} [iq_{i-1} I_1(x, M, i)$$

$$- q_{i-1} I_2(x, M, i) + q_i I_2(x, M, i) - (i-1) q_i I_1(x, M, i)] \quad (a)$$

PROBLEMS

where

$$I_1(x, M, i) \equiv \frac{\alpha}{k} \int_{t_{i-1}}^{t_i} G(x, t_M - \lambda) d\lambda$$

$$I_2(x, M, i) \equiv \frac{\alpha}{k} \int_{t_{i-1}}^{t_i} \frac{\lambda}{\Delta t} G(x, t_M - \lambda) d\lambda$$

Show that Eq. (a) can be written as

$$T(x, t_M) = T_0 + \sum_{i=0}^{M} a_{iM}(x) q_{M-i}$$

What are the $a_{iM}(x)$ expressions?

CHAPTER 4

INVERSE HEAT CONDUCTION ESTIMATION PROCEDURES

4.1 INTRODUCTION

The inverse heat conduction problem is one of many ill-posed problems. The notion of a well-posed or "correctly set" mathematical problem made its debut in the 1923 discussions in Hadamard's work.[1] The conditions of being well-posed require sufficiently general properties of *existence*, *uniqueness* and (by implication) *stability* of solutions. In Section 4.2 it is shown that the IHCP is not well posed and hence it is called ill-posed.

Ill-posed problems according to Tikhonov and Arsenin[2] can be divided into two subclasses: those involving estimation using data, and those involving design of automatic controls. The IHCP is in the first subclass. In general, this subclass includes problems of mathematical processing and interpretation of data in varied fields where functions are to be estimated. Some areas are nuclear physics, radiophysics, electronics, interpretation of geophysical observations, mineral and petroleum exploration, rocketry, and nuclear reactor engineering. The second subclass includes many optimization problems such as optimal control and optimal economic planning. In this book only problems involving estimation with data are considered with the emphasis on the inverse heat conduction problem but the techniques can be applied to many other problems.

Ill-posed problems include the mathematical problems of solution of singular or ill-conditioned systems of linear algebraic equations, differentiation of functions known only approximately, solution of partial differential equations using "interior" measurements and solution of integral equations of the first kind utilizing measurements. Even though there are many such problems that are of engineering importance, the ill-posed nature of the problems not only defies easy solution, but has served to discourage the type of massive study that has accompanied direct or well-posed problems.

SEC. 4.1 INTRODUCTION

One of the difficulties of ill-posed problems is in defining what is meant by a "solution" because the solution does not satisfy general conditions of existence, uniqueness, and stability. This difficulty is addressed in this chapter.

The plan of this chapter is first to investigate the concept of ill-posed problems in Section 4.2. Sections 4.3 through 4.7 present a number of IHCP algorithms. Section 4.8 presents the mean squared error concept which can be utilized to compare various algorithms; this section also exhibits two conflicting objectives.

To better relate the various algorithms given in Sections 4.3 through 4.8, the contents are outlined next. These sections are restricted to estimating a single heat flux history and the linear IHCP. Section 4.3 presents algorithms for a single future time step. For a single temperature sensor, the heat flux component at any given time is found by setting the calculated temperature equal to the measured temperature at that time. This is called "exact" matching because the calculated temperature is made equal to the measured temperature. It is called a "sequential" procedure because q_M is estimated after q_{M-1}, for $M = 2, 3, \ldots$. If Duhamel's theorem is used to obtain the calculated temperatures, the exact matching algorithm is called the Stolz[9] algorithm. The paper by Stolz in 1960 was one of the earliest on the inverse heat conduction problem. Section 4.3.3 discusses the case of multiple sensors for a single future time; the least squares method is used to obtain approximate agreement between calculated and measured temperatures.

Section 4.4 develops the function specification method in which a functional form for the unknown heat flux is assumed. The functional form contains a number of unknown parameters that are estimated utilizing the method of least squares. Section 4.4.2 describes the whole domain function specification method, a method in which all the parameters for the complete time history are simultaneously estimated. Early proponents of the whole domain method were Frank in 1963[10] and Davies in 1966.[11] Section 4.4.3 gives the sequential function specification method which was first discussed by Beck in 1961.[12-15] Section 4.4.3.2 gives an alternative interpretation of the sequential function specification method which has been proposed by Blackwell.[16]

Section 4.5 presents the regularization method. A number of authors have championed the whole domain regularization procedure, Section 4.5.3, including Tikhonov and Arsenin,[2] Alifanov,[17-19] Cozdoba and Crykowsky,[20] and Bell and Wardlaw.[21,22] The zeroth, whole domain regularization method is related to ridge regression.[6,24-29] A sequential regularization method is given in Section 4.5.3; a related method is given by Beck and Murio.[23]

A generalization of the function specification and regularization methods, called the trial function method, is discussed in Section 4.6. A whole domain version of the method was proposed by Twomey[33,34] but had not been previously applied to the IHCP.

In Section 4.7 it is shown that the algorithms in the previous sections (4.3–4.6) can be given in the form of a digital filter. This form is important for a number of reasons. One is its computational efficiency and another is its convenient form for obtaining the variance of the estimated heat flux components.

110 CHAP.4 INVERSE HEAT CONDUCTION ESTIMATION PROCEDURES

Section 4.8 discusses the two conflicting criteria of unbiased and minimum variance heat flux estimates. The overall criterion of the mean squared error is introduced and is shown to contain the two components of (1) variance resulting from measurement errors and (2) deterministic error. Methods for calculating these two components are given. The mean squared error provides a means for comparison of IHCP algorithms.

4.2 ILL-POSED PROBLEMS

The inverse heat conduction problem is shown to be ill-posed in this section. This problem can be viewed from the perspective of a partial differential equation, an integral equation or a set of linear algebraic equations which could result from the use of finite differences or elements. Comments are made regarding each perspective.

4.2.1 Partial Differential Equation Perspective

The exact IHCP solution for a plate derived in Section 2.5 from the partial differential equation of heat conduction is considered. The properties are constant and the temperature and temperature gradient are given at $x=0$,

$$T(0, t) = f(t) \tag{4.2.1}$$

$$\left.\frac{\partial T(x, t)}{\partial x}\right|_{x=0} = g(t) \tag{4.2.2}$$

where $f(t)$ and $g(t)$ are known functions.

The solution for the temperature and heat flux ($q = -k\partial T/\partial x$) distributions are needed for $x>0$. It is shown in Section 2.5 that the exact solution for $T(x, t)$, provided f and g are infinitely differentiable, is

$$T(x, t) = \sum_{n=0}^{\infty} \frac{1}{(2n)!} \left(\frac{x^2}{\alpha}\right)^n f^{(n)}(t) - x \sum_{n=0}^{\infty} \frac{1}{(2n+1)!} \left(\frac{x^2}{\alpha}\right)^n g^{(n)}(t) \tag{4.2.3}$$

where the (n) superscript notation implies the nth derivative with respect to time,

$$f^{(n)}(t) \equiv \frac{d^n f(t)}{dt^n} \tag{4.2.4}$$

In order for the problem to be well-posed it is necessary that the solution (1) exist, (2) be unique, and (3) be continuously dependent on the data or equivalently, be stable. Each of these aspects is now considered for the foregoing solution.

The *existence* of this solution was proved by Widder.[3] (See also Weber.[4]) The series given by Eq. (4.2.3) converges uniformly for bounded t, provided f

SEC. 4.2 ILL-POSED PROBLEMS

and g satisfy the absolute value conditions,

$$|f^{(n)}(t)| \leqslant M \frac{(2n)!}{L} \tag{4.2.5}$$

$$|g^{(n)}(t)| \leqslant P \frac{(2n)!}{L}, \quad n = 0, 1, 2, \ldots \tag{4.2.6}$$

for some constants M and P and where $x = L$ is the heated surface. Hence, the solution exists for these conditions for the nth time derivatives of f and g growing with n less than the factorial of $2n$.

The *uniqueness* of the solution given by Eq. (4.2.3) is proved by noting that any two solutions must be equal to the right side of Eq. (4.2.3) and therefore to each other.

The question of *stability* is now considered. This requires measures of the differences between corresponding solutions. For that reason, for the time and space intervals of $0 \leqslant t \leqslant t_f$ and $0 < x < L$, the following norms are defined:

$$\|f\| \equiv \max_{0 \leqslant t \leqslant t_f} |f(t)| \tag{4.2.7a}$$

$$\|g\| \equiv \max_{0 \leqslant t \leqslant t_f} |g(t)| \tag{4.2.7b}$$

$$\|T\| \equiv \max_{\substack{0 < x < L \\ 0 \leqslant t \leqslant t_f}} |T(x, t)| \tag{4.2.7c}$$

[The single bars are for absolute values and the double bars are for the norms which are defined in Eqs. (4.2.7).] Let $T_i(x, t)$ be the solution of Eq. (4.2.3) resulting from $f_i(x, t)$ and $g_i(x, t)$ where $i = 1$ and 2.

A necessary condition for stability is that the norm $\|T_1 - T_2\|$ can be made arbitrarily small by choosing f_1, f_2, g_1, and g_2 so that the norms $\|f_1 - f_2\|$ and $\|g_1 - g_2\|$ are sufficiently small. This is impossible, in general, as shown next.

To prove that the solution is not always stable, we need display only one exceptional case; namely

$$g_1 = g_2 = 0 \tag{4.2.8}$$

$$f_2(t) = f_1(t) + \frac{1}{\beta} \cos(\beta^2 t), \quad \beta > 0 \tag{4.2.9}$$

and $f_1(t)$ is an arbitrary analytic function. The error term, $\beta^{-1} \cos \beta^2 t$, can be made small by making β large. This forces f_1 and f_2 to be arbitrarily close as measured by the norm,

$$\|f_1 - f_2\| = \max_{0 \leqslant t \leqslant t_f} \left| \frac{1}{\beta} \cos \beta^2 t \right| = \frac{1}{\beta}$$

The corresponding solutions for T_1 and T_2 may not be close at all.

The difference between T_1 and T_2 is

$$T_1 - T_2 = \sum_{n=0}^{\infty} \frac{1}{(2n)!} \left(\frac{x^2}{\alpha}\right)^n \frac{d^n}{dt^n}\left(\frac{1}{\beta}\cos\beta^2 t\right) \qquad (4.2.10a)$$

$$= \frac{1}{\beta}\sum_{n=0}^{\infty} \frac{(-1)^n}{(4n)!}\left(\frac{x^2}{\alpha}\right)^{2n} \beta^{4n} \cos\beta^2 t$$

$$+ \frac{1}{\beta}\sum_{n=0}^{\infty} \frac{(-1)^{n+1}}{(4n+2)!}\left(\frac{x^2}{\alpha}\right)^{2n+1} \beta^{4n+2} \sin\beta^2 t \qquad (4.2.10b)$$

For the special location x where $x^2/\alpha = 1$ s and for time $t=0$, Eq. (4.2.10b) gives

$$T_1 - T_2 = \frac{1}{\beta}\sum_{n=0}^{\infty} \frac{(-1)^n}{(4n)!} \beta^{2n} \qquad (4.2.11a)$$

$$= \frac{1}{\beta}\cosh(2^{-1/2}\beta)\cos(2^{-1/2}\beta) \qquad (4.2.11b)$$

The equality between Eqs. (4.2.11a) and (4.2.11b) is proved in Reference 4. Using Eq. (4.2.11b) results in

$$\|T_1 - T_2\| \geq \frac{1}{\beta}|\cosh(2^{-1/2}\beta)\cos(2^{-1/2}\beta)| \qquad (4.2.12)$$

which as $\beta \to \infty$ gives $(2\beta)^{-1}\exp(2^{-1/2}\beta)$ which goes to infinity. In other words, for the special case considered, arbitrarily small differences in the input temperature can result in arbitrarily large differences in the surface temperature.

The conclusion is that the solution given by Eq. (4.2.3) is unstable. This is observed to be true although Eq. (4.2.3) is the *exact* solution for an IHCP with continuous data. Thus, even with existence and uniqueness, stability is not present. Hence, it is proved that the IHCP solution coming from a partial differential equation is ill-posed because it does not satisfy the three conditions for a well-posed problem.

4.2.2 Integral Equation Perspective

Linear transient heat conduction problems can be modeled using a convolution integral, as discussed in Section 3.2. For one-dimensional cases, the temperature rise at a point x can be expressed by

$$T(x,t) = \frac{\alpha}{k}\int_0^t G(x, t-\lambda)q(\lambda)d\lambda, \quad G(x,t) = \frac{\partial \phi(x,t)}{\partial t}\frac{k}{\alpha} \qquad (4.2.13)$$

where $G(x,t)$ is a Green's function and λ is a dummy integration variable. In the IHCP, the objective is to estimate the heat flux function, $q(t)$, given measurements of $T(x,t)$, $0 < x < L$. The following shows that Eq. (4.2.13) does not possess the property of stability under the condition of small changes of $T(x,t)$.

SEC. 4.2 ILL-POSED PROBLEMS

To demonstrate the instability, the function

$$q_2(t) = q_1(t) + N \sin \omega t \qquad (4.2.14)$$

is used where $N = N(\omega)$ is a monotonically increasing function of ω. Introducing Eq. (4.2.14) into Eq. (4.2.13) gives

$$T_2(x, t) = T_1(x, t) + \frac{\alpha}{k} \int_0^t G(x, t-\lambda) N \sin(\omega\lambda) d\lambda \qquad (4.2.15)$$

where $T_1(x, t)$ is the $T(x, t)$ function given by Eq. (4.2.13) for $q = q_1(t)$. A measure of the difference between $T_1(x, t)$ and $T_2(x, t)$ over the interval $0 < x < L$ and at time t is

$$\rho_T(T_1, T_2) = |N| \left\{ \int_0^L \left[\frac{\alpha}{k} \int_0^t G(x, t-\lambda) \sin \omega\lambda d\lambda \right]^2 dx \right\}^{1/2} \qquad (4.2.16)$$

which can be made arbitrarily small by letting ω be sufficiently large (provided G is continuous in x) and by properly selecting $N(\omega)$. (The expression inside the λ integration can be made to approach zero by letting ω be sufficiently large; this is a result of the Riemann-Lebesgue theorem.[5]) A similar measure of the difference between $q_1(t)$ and $q_2(t)$ is

$$\rho_q(q_1, q_2) = \left\{ \int_0^t [q_1(\lambda) - q_2(\lambda)]^2 d\lambda \right\}^{1/2}$$

$$= |N| \left(\int_0^t \sin^2 \omega\lambda d\lambda \right)^{1/2}$$

$$= |N| \left[\frac{t}{2} - \frac{1}{2\omega} \sin(\omega t) \cos(\omega t) \right]^{1/2} \qquad (4.2.17)$$

Hence, for arbitrary small discrepancies between $T_1(x, t)$ and $T_2(x, t)$ the ω and $N(\omega)$ can be chosen in such a way that the discrepancy between $q_1(t)$ and $q_2(t)$, given by Eq. (4.2.17), can be arbitrarily large. The foregoing considerations show that the problem of estimating $q(t)$ in the integral equation, Eq. (4.2.13), is unstable and thus is ill-posed.

4.2.3 Difference Equation Perspective

A solution of the transient heat conduction equation can be obtained by using finite differences or elements. These approaches for a linear case provide sets of linear algebraic equations that can be used to obtain [see Eq. (3.3.55a)]

$$\mathbf{Xq} = \mathbf{Y} \qquad (4.2.18)$$

where $\mathbf{X}$ is an $n \times n$ matrix, $\mathbf{q}$ is an n vector of elements of the heat flux, and $\mathbf{Y}$ is a measurement vector of n elements. For an interior temperature measurement location and "small" time steps, the solution of Eq. (4.2.18) for $\mathbf{q}$ is ill

114   **CHAP.4 INVERSE HEAT CONDUCTION ESTIMATION PROCEDURES**

conditioned. That is, small changes in the **Y** vector would make large changes in the **q** vector.

When the IHCP solution employs difference equations to solve the partial differential equation of heat conduction and also in the numerical solution of the convolution equation, the problem reduces to the solution of a set of equations similar to Eq. (4.2.18) as shown in Section 3.3.5.

In the usual solution of a set of algebraic equations, no constraints are imposed on the unknowns. The values of the unknowns are not forced to increase or decrease in some systematic manner. In the IHCP, however, the heat flux components, $q_1, q_2, \ldots, q_n$, represent the values of a heat flux function, $q(t)$. These functions vary more or less smoothly with time such as in Figure 4.1 or even Figure 4.2. The heat flux components, $q_1, \ldots, q_n$, cannot assume completely arbitrary values relative one to the other. An example of an unacceptable case is that wherein the values oscillate about zero with increasing magnitude.

In some cases a direct solution of Eq. (4.2.18) will produce highly oscillatory estimates. Furthermore, if one component of **Y** is changed slightly, the calculated values of $q_1, \ldots, q_n$ might change greatly. For such cases, a satisfactory solution for $q_1, \ldots, q_n$ must constrain the q values so that a smoothly varying function is approximated. Various procedures estimating the surface heat flux are discussed in Sections 4.3 through 4.7.

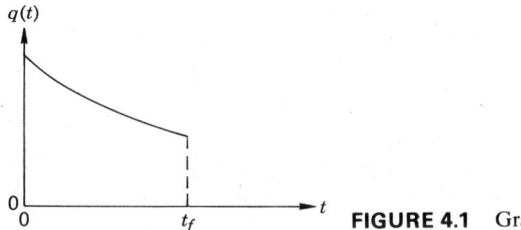

FIGURE 4.1   Gradually changing heat flux.

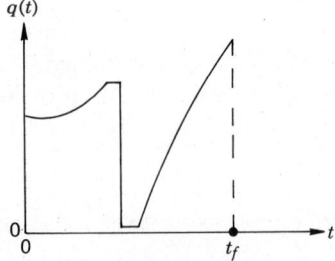

FIGURE 4.2   Abruptly changing heat flux.

4.3 SINGLE FUTURE TIME STEP METHOD

4.3.1 Introduction

The earliest methods proposed for the solution of the IHCP used a single future time step. A single heat flux component was estimated at each time step, producing a *sequential* algorithm. These techniques were restricted to a single temperature sensor.

The plan of this section is to discuss the case of a single future time step and a single temperature sensor in Section 4.3.2 and to discuss the single future time step–multiple sensors case in Section 4.3.3. For simplicity of presentation only the linear IHCP is discussed, leaving a discussion of the nonlinear case to the finite control volume treatment in Section 6.3.

4.3.2 Exact Matching of Measured Temperatures (Single Sensor)

When discrete transient temperatures from a single temperature sensor are used to estimate the surface heat flux as a function of time, the calculated temperatures are made equal to the measured values. (This is called "exact matching" to distinguish it from approximate matching obtained by using least squares.) Exact matching can be obtained by using a numerical form of Duhamel's theorem or a finite control volume (or finite difference or element) method. To be general, a procedure is utilized that includes both Duhamel's theorem and FCV types of solution for the linear IHCP. For both procedures, the temperature at the sensor location and at time t_M can be written as, Eq. (3.2.19) or Eq. (3.3.55a),

$$T_M = T_M|_{q_M=0} + \Delta\phi_0 q_M \qquad (4.3.1)$$

where $\Delta\phi_0$ is the temperature rise at the sensor location at time t_1 for a unit step rise in the surface heat flux at time $t=0$; $\Delta\phi_0$ is also the sensitivity coefficient of T_M for q_M,

$$\Delta\phi_0 = \frac{\partial T_M}{\partial q_M} \qquad (4.3.2)$$

In exact matching T_M is made equal to the measured temperature at time t_M, which is denoted Y_M. Then Eq. (4.3.1) is solved for the estimated heat flux component which is denoted $\hat{q}_M$,

$$\hat{q}_M = \frac{Y_M - \hat{T}_M|_{q_M=0}}{\phi_1}, \quad \text{since } \Delta\phi_0 = \phi_1 - \phi_0 = \phi_1 \qquad (4.3.3)$$

This equation applies for both Duhamel theorem-based and finite control volume-based treatments of the heat conduction equation. It is restricted to a single sensor and a single future temperature at t_M.

An advantage of Eq. (4.3.3) is its simplicity which permits easy understanding of the method and the development of relatively simple and computationally efficient algorithms. The major weakness is its extreme sensitivity to measurement errors particularly as time steps are made *small*, not large, which is the usual difficulty (if any) in finite methods (FCV and FE). Due to the sensitivity of Eq. (4.3.3) to measurement errors and even instability for small Δt values, it is rarely recommended for the IHCP. (For a discussion of the meaning of "small" Δt, see Section 5.3.)

For the Duhamel integral-based form of Eq. (4.3.3), the $\hat{T}_M|_{q_M=0}$ term in Eq. (4.3.3) is the first element of Eq. (3.2.29),

$$\hat{T}_M|_{q_M=0} = \sum_{i=1}^{M-1} \hat{q}_i \Delta \phi_{M-i} + T_0 \tag{4.3.4}$$

and thus Eq. (4.3.3) becomes

$$\hat{q}_M = \frac{Y_M - \sum_{i=1}^{M-1} \hat{q}_i \Delta \phi_{M-i} - T_0}{\phi_1} \tag{4.3.5}$$

This equation is called the Stolz algorithm because Stolz[9] was the first to apply Duhamel's theorem to the IHCP for the single sensor–single future temperature case. The title of Stolz's paper suggests that his method is restricted to simple shapes but the method is much more widely applicable, as mentioned in Chapter 5. An important characteristic of Eq. (4.3.5) is its sequential nature; that is, $\hat{q}_M$ depends on Y_M and the previous $\hat{q}$'s ($\hat{q}_2, \ldots, \hat{q}_{M-1}$) and M is sequentially increased by one at each time step. This is in contrast to estimating all q components simultaneously, which is the whole domain procedure.

Another important characteristic of Eqs. (4.3.3) and (4.3.5) is the linearity in the measurements, $Y_1, Y_2, \ldots$. To demonstrate this, write Eq. (4.3.5) for $\hat{q}_1, \hat{q}_2,$ and $\hat{q}_3$ to obtain

$$\hat{q}_1 = \frac{Y_1 - T_0}{\phi_1} \tag{4.3.6}$$

$$\hat{q}_2 = \frac{Y_2 - \hat{q}_1 \Delta \phi_1 - T_0}{\phi_1} \tag{4.3.7}$$

$$\hat{q}_3 = \frac{Y_3 - \hat{q}_1 \Delta \phi_2 - \hat{q}_2 \Delta \phi_1 - T_0}{\phi_1} \tag{4.3.8}$$

Equation (4.3.6) is a linear function of Y_1. Use Eq. (4.3.6) in Eq. (4.3.7) to get

$$\hat{q}_2 = \frac{Y_2 - (Y_1 - T_0) \Delta \phi_1 / \phi_1 - T_0}{\phi_1} \tag{4.3.9}$$

which shows that $\hat{q}_2$ is a linear function of Y_1 and Y_2. If Eqs. (4.3.6) and (4.3.9) are introduced into Eq. (4.3.8), $\hat{q}_3$ is revealed as a function of Y_1, Y_2 and Y_3,

SEC. 4.3 SINGLE FUTURE TIME STEP METHOD

although the expression is somewhat complicated. This procedure can be extended to show that q_M is a linear function of $Y_1, Y_2, \ldots, Y_M$. This linearity leads to a number of important concepts that are related to insight into the IHCP, stability considerations, and the development of important filter algorithms; see Sections 4.7 and 5.5.

EXAMPLE 4.1 The temperatures at 1 cm inside a semi-infinite steel plate ($k = 40$ W/m-C, $\alpha = 10^{-5}$ m^2/s) initially at 30°C are 37.608°C, 73.226°C, and 136.321°C for $t = 5$, 10, and 15 s, respectively. Calculate the surface heat flux estimates, $\hat{q}_1, \hat{q}_2$, and $\hat{q}_3$ using the Stolz method.

Solution. The $\hat{q}_i$ values are obtained using Eq. (4.3.5) or Eqs. (4.3.6)–(4.3.8). The ϕ_i values are obtained from the solution of a semi-infinite body problem subjected to a step change in surface heat flux. The dimensionless time steps are

$$\Delta t_x^+ = \frac{\alpha \Delta t}{x^2} = (10^{-5}\ \text{m}^2/\text{s})(5\text{s})/(0.01\ \text{m})^2 = 0.5$$

and the times of $t_1^+ = 0.5$, $t_2^+ = 1$, and $t_3^+ = 1.5$ are used in Table 1.2 to get the associated T^+ values by setting $q_0 = 1$ and using

$$\phi_1 = T^+(t_1)\frac{x}{k} = 0.166631\ \frac{0.01}{40} = 4.1658E\text{-}5\ \text{C-m}^2/\text{W}$$

$$\phi_2 = T^+(t_2)\frac{x}{k} = 0.399282\ \frac{0.01}{40} = 9.9821E\text{-}5\ \text{C-m}^2/\text{W}$$

$$\phi_3 = T^+(t_3)\frac{x}{k} = 0.60612\ \frac{0.01}{40} = 1.5153E\text{-}4\ \text{C-m}^2/\text{W}$$

$$\Delta\phi_1 = \phi_2 - \phi_1 = 5.8163E\text{-}5, \quad \Delta\phi_2 = \phi_3 - \phi_2 = 5.1710E\text{-}5$$

For $\hat{q}_1$, Eq. (4.3.6) is used to get

$$\hat{q}_1 = \frac{Y_1 - T_0}{\phi_1} = \frac{37.608 - 30}{4.1658E\text{-}5} = 182{,}630\ \text{W/m}^2$$

For $\hat{q}_2$, Eq. (4.3.7) gives

$$\hat{q}_2 = \frac{Y_2 - \hat{q}_1\Delta\phi_1 - T_0}{\phi_1} = \frac{73.226 - 182630(5.8163E\text{-}5) - 30}{4.1658E\text{-}5}$$

$$= 782{,}650\ \text{W/m}^2$$

and $\hat{q}_3$ is obtained from Eq. (4.3.8),

$$\hat{q}_3 = 1{,}232{,}800\ \text{W/m}^2$$

These $\hat{q}_i$ values can be compared with the exact input values. The exact heat flux input is linearly increasing with time and is given by

$$q(t) = 10^5 t, \quad t\ \text{in s}$$

The exact solution for the temperature is given by Eq. (3.2.13). The calculated heat fluxes are best associated (see Figure 4.3) with $t = 2.5$, 7.5, and 12.5s, respectively, so that the exact heat flux components are $q_1 = 250{,}000$; 750,000; and 1,250,000 W/m^2. The errors

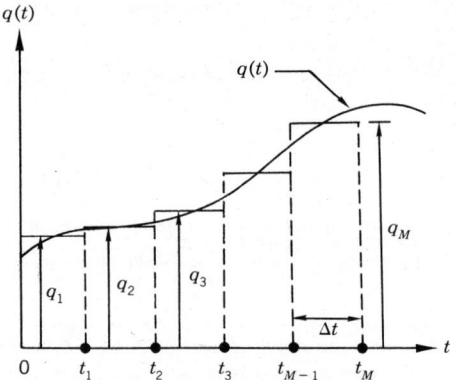

FIGURE 4.3 Constant element approximation for $q(t)$.

in $\hat{q}_i$, $i=1, 2, 3$ are -27%, $+4.4\%$, and -1.4%, respectively. These calculated values are relatively accurate but if the time steps were made smaller, 3 s or less in this example, the Stolz procedure would be unstable. □

4.3.3 Multiple Temperature Sensors

For J temperature sensors, the calculated temperatures can be given by

$$T_{1M} = \hat{T}_{1M}|_{q_M=0} + \phi_{11}q_M$$
$$T_{2M} = \hat{T}_{2M}|_{q_M=0} + \phi_{21}q_M$$
$$\vdots$$
$$T_{JM} = \hat{T}_{JM}|_{q_M=0} + \phi_{J1}q_M \tag{4.3.10}$$

where the first subscript refers to sensor number and the second to time which is the conventional order in the notation $T = T(x, t)$. The corresponding measured temperatures are denoted

$$Y_{jM}, \quad j=1, 2, \ldots, J$$

For J greater than 1, the single heat flux component, q_M, cannot in general be chosen to make,

$$T_{jM} = Y_{jM}, \quad j=1, 2, \ldots, J$$

Hence, agreement only in some average sense is possible.

One way to estimate q_M is by the use of least squares in which the sum of squares,

$$S = \sum_{j=1}^{J} (Y_{jM} - \hat{T}_{jM}|_{q_M=0} - \phi_{j1}q_M)^2 \tag{4.3.11}$$

is minimized with respect to the heat flux component q_M. This is accomplished by differentiating S with respect to q_M, replacing q_M by the estimate $\hat{q}_M$, and setting the equation equal to zero,

$$\left.\frac{dS}{dq_M}\right|_{q_M} = 0 = -2\sum(Y_{jM} - \hat{T}_{jM}|_{q_M=0} - \phi_{j1}\hat{q}_M)\phi_{j1} = 0 \quad (4.3.12)$$

Solving this equation for $\hat{q}_M$ gives

$$\hat{q}_M = \sum_{j=1}^{J}[(Y_{jM} - T_{jM}|_{q_M=0})\phi_{j1}]\left(\sum_{k=1}^{J}\phi_{k1}^2\right)^{-1} \quad (4.3.13)$$

If $J=1$, Eq. (4.3.13) reduces to Eq. (4.3.3). The use of multiple sensors is much less effective in damping measurement errors than the use of future temperatures which is discussed in subsequent sections. When using multiple sensors, the one nearest the heated surface has the dominant effect because $\phi_{11} > \phi_{21} > \phi_{31},\ldots$ if $x_1 < x_2 < x_3,\ldots$ where x_j is the distance of the jth sensor from the heated surface.

4.4 FUNCTION SPECIFICATION METHOD

4.4.1 Introduction

One way to treat the IHCP is to assume a functional form of the surface heat flux variation with time. This is called the function specification method. The function can be a sequence of constant segments, straightline segments, or it can be one of many other forms such as parabolas, cubics, or exponentials.

Other possible variations in this method are (1) to estimate simultaneously all the parameters for the total time interval and (2) to estimate the parameters sequentially. In the first scheme, called whole domain estimation, the complete functional form is found as a unit. In the sequential procedure, one segment after another is estimated, starting with the earliest times and then moving successively to larger times. The following discussion considers various functional forms and both the whole domain (Section 4.4.2) and the sequential procedures (Section 4.4.3). The sequential method is more computationally efficient than the whole domain procedure for the IHCP.

4.4.2 Whole Domain Estimation

One of the first to propose the whole domain estimation procedure for the linear inverse heat conduction problem was Frank.[10] He suggested approximating the heat flux with a polynomial expression and using the method of least squares to estimate the coefficients. Another early paper used whole domain heat fluxes with temperatures measured inside simulated tissue;[11] interior temperatures (nonlinear functions of some parameters) were used. These two papers proceeded from two different basic premises. In the first, the surface heat flux was approximated by polynomials irrespective of the actual history; the polynomials were used for *all* conditions and no knowledge of the surface conditions was used. In the second, the surface flux was considered to have certain known characteristics. In this section ideas related to both premises are given but the emphasis is on methods that do not presuppose knowledge of surface conditions.

4.4.2.1 Smoothly Changing Heat Flux. Consider the IHCP, wherein the surface heat flux function, $q(t)$, is required over a finite time interval, $0 < t < t_f$. The true heat flux is assumed, as an example, to be as shown in Figure 4.1. The heat flux $q(t)$ starts at a some (unknown) value and decreases with a decreasing $|dq/dt|$. If this information regarding the features of the curve is available, then an equation of the functional form can be selected that fits the gradually changing character of the function. For a $q(t)$ similar to Figure 4.1, consider the following forms:

$$q(t) = \beta_1 \tag{4.4.1}$$

$$q(t) = \beta_1 + \beta_2 t \tag{4.4.2}$$

$$q(t) = \beta_1 + \beta_2 t + \beta_3 t^2 \tag{4.4.3}$$

$$q(t) = \beta_1 (1 - e^{-\beta_2 t}) \tag{4.4.4}$$

$$q(t) = \beta_1 [1 - \beta_2 \sin(\beta_3 t)], \quad 0 < \beta_3 t < \frac{\pi}{2} \tag{4.4.5}$$

Only the last three of these contain all the features mentioned, but Eq. (4.4.4) is nonlinear in β_2 and Eq. (4.4.5) is nonlinear in β_3. These expressions are used for the complete time interval. The unknown parameters $\beta_1, \ldots$ are estimated using a mathematical model that relates the temperature and $q(t)$ and which uses all the measured interior temperatures from $t=0$ to t_f.

To clarify the estimation ideas for the whole domain procedure, the case of a body subjected to a heat flux described by the linear expression given in Eq. (4.4.2) is considered. The initial temperature is uniform; namely, T_0. The thermal properties are independent of temperature and a heat flux is known to be applied starting at $t=0$. The partial differential equation can be

$$\frac{k}{r^n} \frac{\partial}{\partial r} \left(r^n \frac{\partial T}{\partial r} \right) = \rho c \frac{\partial T}{\partial t} \tag{4.4.6}$$

where $n=0$ for rectangular coordinates, $n=1$ for cylindrical radial coordinates and $n=2$ for spherical radial coordinates. The boundary condition at $r=r_1$ is

$$-k \left. \frac{\partial T}{\partial r} \right|_{r_1} = \beta_1 + \beta_2 t = q(t) \tag{4.4.7}$$

and the boundary condition at $r=r_2$ is "inactive" such as one of the conditions,

$$T(r_2, t) = T_0, \quad \frac{\partial T(r_2, t)}{\partial r} = 0, \quad \text{or} \quad -k \frac{\partial T(r_2, t)}{\partial t} = h(T(r_2, t) - T_0) \tag{4.4.8}$$

SEC. 4.4 FUNCTION SPECIFICATION METHOD

For this linear problem, direct substitution demonstrates that the temperature distribution given by

$$T(r, t) = T_0 + \beta_1 \phi(r, t) + \beta_2 \phi^{(1)}(r, t) \quad (4.4.9)$$

satisfies Eqs. (4.4.6) and (4.4.7) where $\phi(r, t)$ is the $T(r, t)$ solution of Eq. (4.4.6) with the appropriate condition Eq. (4.4.8), $\beta_1 + \beta_2 t$ in Eq. (4.4.7) replaced by 1, and the T_0 value set equal to zero. The "1" superscript denotes a linear time variation of $q(t)$. The $\phi^{(1)}(r, t)$ solution is a $T(r, t)$ solution similar to $\phi(r, t)$ but now the right side of Eq. (4.4.7) is simply t.

The simplest solution of Eq. (4.4.9) uses exact matching in which two measured temperatures at different times or locations are employed. This gives two simultaneous equations for the two unknowns β_1 and β_2.

Another procedure uses least squares. Suppose that temperatures at times $t_1, t_2, \ldots, t_n$ are measured at $r = r_E$ and are denoted $Y_1, \ldots, Y_n$. Then the least squares method minimizes

$$S = \sum_{i=1}^{n} [Y_i - T_0 - \beta_1 \phi_i - \beta_2 \phi_i^{(1)}]^2 \quad (4.4.10)$$

with respect to β_1 and β_2 and where the notation, $\phi_i \equiv \phi(r_E, t_i)$ and $\phi_i^{(1)} \equiv \phi^{(1)}(r_E, t_i)$, is used. Taking the first derivative of S with respect to β_1 and β_2, replacing β_1 by $\hat{\beta}_1$ and β_2 by $\hat{\beta}_2$, and setting both equations equal to zero gives

$$\sum_{i=1}^{n} [Y_i - T_0 - \hat{\beta}_1 \phi_i - \hat{\beta}_2 \phi_i^{(1)}] \phi_i = 0$$

$$\sum_{i=1}^{n} [Y_i - T_0 - \hat{\beta}_1 \phi_i - \hat{\beta}_2 \phi_i^{(1)}] \phi_i^{(1)} = 0 \quad (4.4.11a,b)$$

or in matrix form

$$\begin{bmatrix} C_{11} & C_{12} \\ C_{12} & C_{22} \end{bmatrix} \begin{bmatrix} \hat{\beta}_1 \\ \hat{\beta}_2 \end{bmatrix} = \begin{bmatrix} d_1 \\ d_2 \end{bmatrix} \quad (4.4.12)$$

where

$$C_{11} = \sum \phi_i^2, \quad C_{12} = \sum \phi_i \phi_i^{(1)}, \quad C_{22} = \sum [\phi_i^{(1)}]^2$$

$$d_1 = \sum (Y_i - T_0) \phi_i, \quad d_2 = \sum (Y_i - T_0) \phi_i^{(1)} \quad (4.4.13)$$

A solution of Eq. (4.4.12) is

$$\hat{\beta}_1 = \frac{d_1 C_{22} - d_2 C_{12}}{C_{11} C_{22} - C_{12}^2} \quad (4.4.14a)$$

$$\hat{\beta}_2 = \frac{d_2 C_{11} - d_1 C_{12}}{C_{11} C_{22} - C_{12}^2} \quad (4.4.14b)$$

A generalization of the foregoing procedure for a polynomial of the kth degree is given in Reference 10.

A case that has convenient expressions for ϕ_i and $\phi_i^{(1)}$ is that of a semi-infinite body, $r=x>0$ and $n=0$ in Eq. (4.4.6). For this case ϕ_i is obtained from Eq. (1.6.18a) with $q_c=1$ and $T_0=0$ and $\phi_i^{(1)}$ is obtained from Eq. (3.2.13) with $\dot{q}_0=1$ and $T_0=0$.

4.4.2.2 Abruptly Changing Heat Flux Histories.

If the $q(t)$ curve is that in Figure 4.2, then the use of the functional forms given by Eqs. (4.4.1)–(4.4.5) is not appropriate for the complete curve. The jump discontinuity portion of the curve followed by a constant part and then an increasing part are very difficult to fit with expressions such as Eqs. (4.4.1)–(4.4.5). Because the heat flux can vary as illustrated by Figure 4.2, a procedure is needed that can accommodate such behavior.

One of the simplest ways to approximate an arbitrary $q(t)$ curve is to divide the $q(t)$ curve into a number of equally spaced time intervals, Δt, and to substitute a constant q during each of these small time intervals for the real $q(t)$. See Figure 4.3 where q_M is an approximation for $q(t)$ between t_{M-1} and t_M. The best value for q_M is the average $q(t)$ between t_{M-1} and t_M,

$$\bar{q}_M = \frac{1}{\Delta t} \int_{t_{M-1}}^{t_M} q(t) dt \qquad (4.4.15)$$

Usually the estimated value, denoted $\hat{q}_M$, is not equal to this true average, $\bar{q}_M$. If a single point for $\hat{q}_M$ is plotted rather than the horizontal line shown in Fig. 4.3, $\hat{q}_M$ is best identified with the time $t_{M-1/2}=(t_{M-1}+t_M)/2$.

Another type of approximation of $q(t)$ is the linear elements depicted in Figure 4.4 which can represent a $q(t)$ curve more accurately than the constant segments of Figure 4.3. This accuracy is gained, however, at the expense of a more complex treatment.

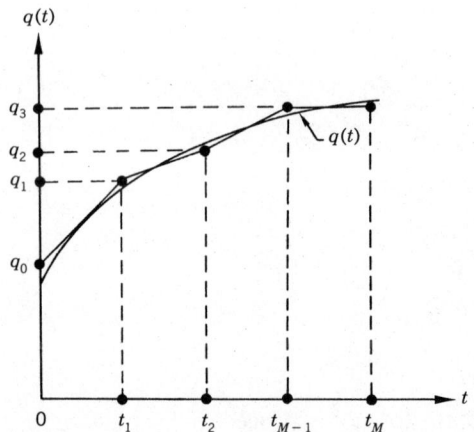

FIGURE 4.4 Linear heat flux approximation for $q(t)$.

SEC. 4.4 FUNCTION SPECIFICATION METHOD

Other possible approximations are the use of parabolic elements between t_{M-1} and t_M such as

$$q(t) = q_{M-1} + \beta_1(t - t_{M-1}) + (q_M - q_{M-1} - \beta_1 \Delta t)\left(\frac{t - t_{M-1}}{\Delta t}\right)^2 \quad (4.4.16)$$

where β_1 is an additional parameter, or the use of cubic splines which are cubic polynomials such that the $q(t)$, $dq(t)/dt$, and $d^2q(t)/dt^2$ values are continuous at the edge of each element.[8]

However, the main case to be considered is the simplest one; namely, the constant element approximation (Figure 4.3). This is consistent with finite difference and finite element solutions which use such an approximation over each calculational time step.

In the function specification method there must be fewer unknown constants to approximate $q(t)$ than times at which the temperatures are measured. This usually results in smoothing of the $q(t)$ function particularly as the ratio of the number of measured temperatures to the number of unknown constants increases. And, in turn, artificially high frequency fluctuations are avoided in $q(t)$. To this end, let there be r measured temperatures and p parameters describing $q(t)$ where $r > p$. With the latter condition the problem of estimating the p parameters is overdetermined because there are more equations than unknowns. Estimates can be made by using the method of least squares. Other estimation procedures are given in Reference 6.

For a uniform initial temperature distribution, T_0, a single interior sensor, a single unknown $q(t)$, and a linear IHCP, a mathematical model for an interior temperature can be obtained from Eq. (3.2.15). [As pointed out in Section 3.3.5, Eq. (3.2.15) can be developed from either Duhamel's integral or finite differences.]

Equation (3.2.15) applies for a constant $q(t)$ during each time interval between measured temperatures. The $q(t)$ can change with each new measurement. One way to reduce fluctuations of q_M is to restrict the q_i's to constant values over certain time intervals; such as, over r time steps where r is any integer greater than one. As a simple example, the values of $n=4$ and $r=2$ are chosen so that:

$$q_1 = q_2 = \beta_1 \quad (4.4.17a)$$

$$q_3 = q_4 = \beta_2 \quad (4.4.17b)$$

where the β's are the heat flux parameters to be estimated. See Figure 4.5. Then Eq. (3.2.15) can be written as

$$\begin{bmatrix} T_1 \\ T_2 \\ T_3 \\ T_4 \end{bmatrix} = \begin{bmatrix} \phi_1 & 0 \\ \phi_2 & 0 \\ \phi_3 - \phi_1 & \phi_1 \\ \phi_4 - \phi_2 & \phi_2 \end{bmatrix} \begin{bmatrix} \beta_1 \\ \beta_2 \end{bmatrix} + \begin{bmatrix} T_0 \\ T_0 \\ T_0 \\ T_0 \end{bmatrix} \quad (4.4.18)$$

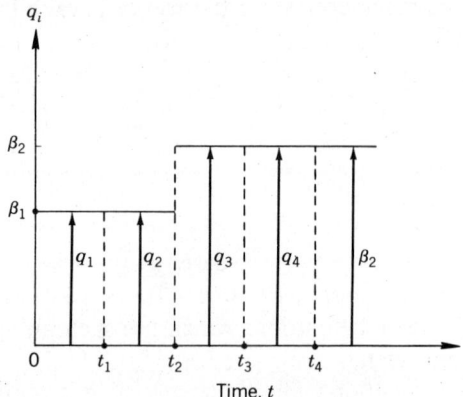

FIGURE 4.5 Functional specification of $q(t)$ with $q_1 = q_2 = \beta_1$ and $q_3 = q_4 = \beta_2$.

since $\Delta\phi_0 = \phi_1$, $\Delta\phi_0 + \Delta\phi_1 = \phi_2$, $\Delta\phi_1 + \Delta\phi_2 = \phi_3 - \phi_1$, and $\Delta\phi_2 + \Delta\phi_3 = \phi_4 - \phi_2$. Notice that there are four equations in Eq. (4.4.18) but only two unknowns, β_1 and β_2. One way to obtain a solution for Eq. (4.4.18) is to minimize the sum of squares function, S,

$$S = \sum_{i=1}^{4} (Y_i - T_i)^2 \qquad (4.4.19)$$

with respect to both β_1 and β_2. The Y_i values are the measured temperatures at times $i = 1, 2, 3$, and 4, and T_i are the corresponding calculated temperatures that come from Eq. (4.4.18).

An alternative to the constant heat flux segments of Figure 4.5 is a sequence of connected linear segments shown in Figure 4.6.

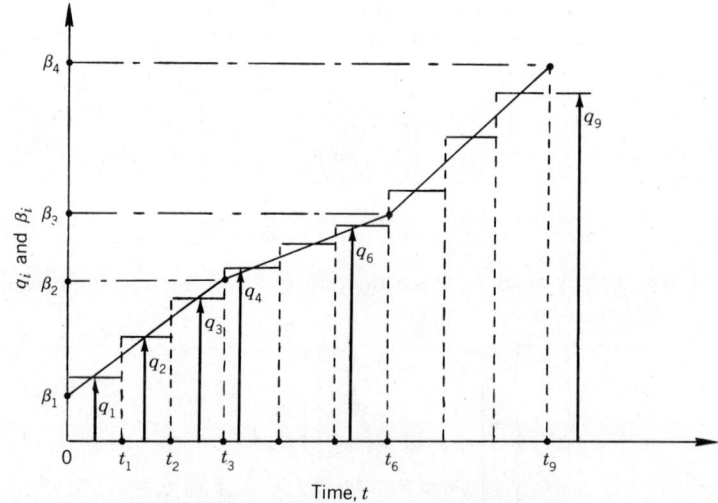

FIGURE 4.6 Linear function specification for $r = 3$.

4.4.3 Sequential Estimation

Several observations based on the characteristics of heat flux sensitivity coefficients are given in Chapter 1. One is that the nature of transient heat conduction is "diffusive"; that is, the effect of heat input at time zero on an interior location is both lagged and damped with respect to the surface. Another aspect of the diffusive character of transient heat conduction is that for two adjacent heat fluxes of similar magnitude the long-time effects on the temperature (see Figure 4.3) are virtually indistinguishable; see Figures 1.10b, 1.12, 1.13, and 1.14. Because of these characteristics, the most computationally efficient estimation procedure for the inverse heat conduction problem does not estimate simultaneously all the heat flux components (q_M, $M = 1, \ldots, n$). Instead, a sequential procedure such as given in this subsection is recommended for the IHCP.

The purpose of this subsection is to present sequential methods for the function specification procedure. The basic principles are developed in a form that can utilize integral models such as Duhamel's integral and also difference equation procedures such as finite differences or elements.

The basic concepts in the function specification sequential procedure are:

1. A functional form for $q(t)$ is assumed for times $t_M, t_{M+1}, \ldots, t_{M+r-1}$ (the heat flux is *known* for $t < t_{M-1}$).
2. A sum of squares function is used for these times and involves squares of differences between the measured and the corresponding calculated temperatures.
3. The heat flux components are estimated for the assumed functional form.
4. Only the first heat flux component, q_M, is retained.
5. M is increased by one and the procedure is repeated.

This is the general procedure for the *linear* IHCP. Some iteration may be necessary for the nonlinear case. The function specification methods for both the linear and nonlinear cases were originally proposed by Beck.[12-15]

4.4.3.1 Constant Heat Flux Functional Form.

The simplest sequential procedure is to assume *temporarily* that several future heat fluxes are constant with time as shown in Figure 4.7. The estimated heat flux components $\hat{q}_1, \hat{q}_2, \ldots, \hat{q}_{M-1}$ are assumed to be known and the objective is to estimate q_M. In order to add stability to the IHCP algorithms, the heat flux components $q_M, q_{M+1}, \ldots, q_{M+r-1}$ are assumed to be equal:

$$q_{M+1} = q_{M+2} = \cdots = q_{M+r-1} = q_M \qquad (4.4.20)$$

Hence r "future" heat flux components are *temporarily* made equal.

If $r = 1$, no "additional" information is introduced for a single interior temperature sensor and the calculated temperature exactly matches the measured value at each time step (for a single sensor). See Section 4.3.

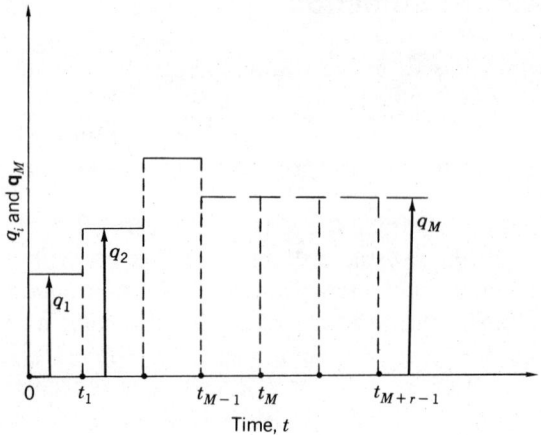

FIGURE 4.7 Constant heat flux functional form for the sequential procedure.

For the sequential estimation of q_M using the assumption given in Eq. (4.4.20), models for $T_M, T_{M+1}, \ldots, T_{M+r-1}$ are needed. The expressions for these temperatures are given by the standard temperature form, Eq. (3.2.19), where the components are given by Eqs. (3.2.20)–(3.2.22). For the constant heat flux assumption given by Eq. (4.4.20), the components of Eq. (3.2.19) are given by

$$T_M = \hat{T}_M|_{q_M = 0} + \phi_1 q_M \qquad (4.4.21\text{a})$$

$$T_{M+1} = \hat{T}_M|_{q_M = q_{M+1} = 0} + \phi_2 q_M \qquad (4.4.21\text{b})$$

$$\vdots$$

$$T_{M+r-1} = \hat{T}_M|_{q_M = \cdots q_{M+r-1} = 0} + \phi_r q_M \qquad (4.4.21\text{c})$$

where the relation

$$\phi_j = \sum_{i=0}^{j-1} \Delta\phi_i \qquad (4.4.22)$$

is used.

The least squares procedure for estimating q_M with the temperature measurements, $Y_M, Y_{M+1}, \ldots, Y_{M+r-1}$, minimizes

$$S = \sum_{i=1}^{r} (Y_{M+i-1} - T_{M+i-1})^2$$

$$= \sum_{i=1}^{r} (Y_{M+i-1} - T_{M+i-1}|_{q=0} - \phi_i q_M)^2 \qquad (4.4.23)$$

with respect to q_M. Equation (4.4.23) is differentiated with respect to q_M, set equal to zero and q_M is replaced by the estimate; the result is the function

SEC. 4.4 FUNCTION SPECIFICATION METHOD

specification equation for q temporarily assumed constant,

$$\hat{q}_M = \frac{\sum_{i=1}^{r} (Y_{M+i-1} - \hat{T}_{M+i-1}|_{q_M = \cdots = 0})\phi_i}{\sum_{i=1}^{r} \phi_i^2} \quad (4.4.24)$$

This equation provides an algorithm that is used in a sequential manner by increasing M by one for each time step. The value of r is commonly chosen to be about 3 or 4; if the r is set equal to 1, the Stolz equation is found from the Duhamel convolution method. (See also Section 4.3.2 and Section 5.3.2.)

A few comments are made about Eq. (4.4.24). The use of several future temperatures, $r > 1$, greatly improves the stability of Eq. (4.4.24) and substantially reduces the sensitivity to measurement errors. This is discussed more completely in Chapters 5 and 6. Another point is that Eq. (4.4.24) is a *linear* function of the temperature measurements, as is the Stolz method. Finally Eq. (4.4.24) is sometimes written as

$$\hat{q}_M = \sum_{i=1}^{r} K_i (Y_{M+i-1} - \hat{T}_{M+i-1}|_{q_M = \cdots = 0}) \quad (4.4.25a)$$

where K_i is called a gain coefficient and is defined by

$$K_i \equiv \frac{\phi_i}{\sum_{j=1}^{r} \phi_j^2} \quad (4.4.25b)$$

Note that K_i has units of reciprocal ϕ. For $r = 2$, $\hat{q}_M$ is given by

$$\hat{q}_M = K_1(Y_M - \hat{T}_M|_{q_M=0}) + K_2(Y_{M+1} - \hat{T}_{M+1}|_{q_M = q_{M+1} = 0}) \quad (4.4.25c)$$

$$K_1 \equiv \frac{\phi_1}{\phi_1^2 + \phi_2^2} ; \quad K_2 \equiv \frac{\phi_2}{\phi_1^2 + \phi_2^2} \quad (4.4.25d)$$

By writing Eq. (4.4.24) in this form, comparisons can be made with a number of other sequential procedures.

Table 4.1 displays the dimensionless gain coefficients, K_i^+, for a sensor at $x = L$ in a finite insulated plate. The ϕ_i's are numerically equal to the dimensionless temperatures taken from Table 1.1 for dimensionless time steps of 0.05, 0.2 and 0.5. Notice that the K_i^+ values* decrease with increasing Δt^+ values, indicating decreased sensitivity to measurement errors with increasing Δt^+. A large reduction in sensitivity to measurement errors is also obtained for a given Δt^+ by increasing r, the number of future time steps.

In some experiments more than one temperature sensor is used because the additional information can aid in more accurately estimating the surface

---

*The dimensionless values of K_i can be denoted K_i^+ and are related to ϕ_i (with dimensions) by $K_i^+ = (q_c L/k)\phi_i(\sum \phi_j^2)^{-1}$ where $q_c = 1$ W/m^2.

TABLE 4.1 Dimensionless Gain Coefficients for Eq. (4.4.25) for Sensor at $x=L$ in a Finite Insulated Plate. Function Specification Method, q_M Assumed Constant

Gain Coefficient	$r=1$	$r=2$	$r=3$	$r=4$	$r=5$
		$\Delta t^+ = 0.05$			
K_1^+	3720	4.33	0.292	0.057	0.018
K_2^+		127	8.56	1.68	0.533
K_3^+			31.8	6.24	1.98
K_4^+				13.08	4.15
K_5^+					6.79
		$\Delta t^+ = 0.2$			
K_1^+	16.3	1.02	0.248	0.095	0.046
K_2^+		3.95	0.955	0.365	0.176
K_3^+			1.75	0.668	0.323
K_4^+				0.975	0.471
K_5^+					0.620
		$\Delta t^+ = 0.5$			
K_1^+	2.9869	0.41509	0.129	0.056	0.029
K_2^+		1.0332	0.322	0.140	0.073
K_3^+			0.516	0.224	0.117
K_4^+				0.308	0.161
K_5^+					0.205

conditions. In this case the sum of squares function,

$$S = \sum_{i=1}^{r} \sum_{j=1}^{J} (Y_{j,M+i-1} - T_{j,M+i-1})^2 \qquad (4.4.26)$$

is minimized with respect to q_M. Again, the first subscript refers to space (sensor number) and the second to time. There are r future times and J temperature sensors. For the temporary assumption of $q_M = \text{const}$, $T_{j,M+i-1}$ can be written as

$$T_{j,M+i-1} = \hat{T}_{j,M+i-1}|_{q_M = \cdots = q_{M+i-1} = 0} + \phi_{ji} q_M \qquad (4.4.27)$$

Introducing Eq. (4.4.27) into Eq. (4.4.26) and performing the usual operations of differentiating with respect to q_M, setting equal to zero and replacing q_M by $\hat{q}_M$ gives

$$\hat{q}_M = \frac{\sum_{i=1}^{r} \sum_{j=1}^{J} (Y_{j,M+i-1} - \hat{T}_{j,M+i-1}|_{q_M = \cdots = 0}) \phi_{ji}}{\sum_{i=1}^{r} \sum_{j=1}^{J} \phi_{ji}^2} \qquad (4.4.28)$$

SEC. 4.4 FUNCTION SPECIFICATION METHOD

Using the gain coefficient concept for this equation yields

$$\hat{q}_M = \sum_{i=1}^{r} \sum_{j=1}^{J} K_{ji}(Y_{j,M+i-1} - \hat{T}_{j,M+i-1}|_{q_M = \cdots = 0}) \quad (4.4.29a)$$

$$K_{ji} \equiv \frac{\phi_{ji}}{\sum_{i=1}^{r}\sum_{j=1}^{J} \phi_{ji}^2} \quad (4.4.29b)$$

The foregoing expressions reduce to the single-sensor and r-future times equations, Eqs. (4.4.24) and (4.4.25), for $J=1$ and they reduce to the multiple-sensor and one future time equation, Eq. (4.3.13), for $r=1$. As in the equations just mentioned, Eqs. (4.4.28) and (4.4.29) are also linear functions of the measured temperatures.

Some gain coefficients are displayed in Table 4.2 for the case of two future times ($r=2$) and two sensors ($J=2$). Dimensionless time steps of 0.05 and 0.5 are used. The geometry is a flat plate heated at $x=0$ and insulated at $x=L$, for which the ϕ values are given in Table 1.1. Three sets of sensor locations are considered: $x=0$ and L, $x=0.5L$ and L, and $x=L$ and L. The last set of x values is similar to the case of single x (which is $x=L$) used in Table 4.1; the K_{ji}^+ values in Table 4.2 for $x=L$ and L are one-half of the corresponding values in Table 4.1. The most striking difference between Tables 4.1 and 4.2 is the great reduction in the gain coefficients for small time steps when a sensor is located at or near the heated surface. This means that the heat flux component, q_M, is much less sensitive to measurement errors when sensors are used near the heated surface. (See Section 4.8.) Also note for $\Delta t^+ = 0.05$ that the gain coefficients are quite small for $x=L$ (for the sensors at $x=0$ and L); as a consequence, the sensor at $x=L$ makes little contribution to the estimation of the surface heat flux for this small Δt^+. For the relatively large time step of $\Delta t^+ = 0.5$,

TABLE 4.2 Dimensionless Gain Coefficients for Eq. (4.4.29) for Two Sensors in a Finite Plate Insulated at $x=L$. Function Specification Method, q_M Assumed Constant. Two Future Times, $r=2$, ($\Delta t^+ \equiv \alpha \Delta t/L^2$)

i	$x=0$ $(J=1)$	$x=L$ $(J=2)$	$x=0.5L$ $(J=1)$	$x=L$ $(J=2)$	$x=L$ $(J=1)$	$x=L$ $(J=2)$
			K_{ji}^+ Values for $\Delta t^+ = 0.05$			
1	1.32	0.001	4.03	0.071	2.16	2.16
2	1.87	0.041	15.5	2.07	63.3	63.3
			K_{ji}^+ Values for $\Delta t^+ = 0.5$			
1	0.254	0.102	0.237	0.173	0.208	0.208
2	0.407	0.254	0.495	0.431	0.517	0.517

the locations of the temperature sensors are much less important. The gain coefficients in Table 4.2 illustrate that, whenever practical, sensors should be located as near the heated surface as possible for small dimensionless times. (As pointed out in Chapter 5, these "small dimensionless times" should be based on the distance E from the heated surface to the temperature sensor nearest the heated surface. Note, however, that $\alpha \Delta t/E^2$ is always large if E approaches zero.)

EXAMPLE 4.2 For the same plate and data as in Example 4.1 calculate the surface heat flux estimates, $\hat{q}_1$ and $\hat{q}_2$, using the function specification method with the q=constant temporary assumption and two future temperatures.

Solution. Equation (4.4.25) is used with $r=2$. The ϕ_i values are the same as for Example 4.1 and

$$K_1 = \frac{\phi_1}{\phi_1^2 + \phi_2^2} = 3560.6 \text{ W/m}^2\text{-C}$$

$$K_2 = \frac{\phi_2}{\phi_1^2 + \phi_2^2} = 8552.0 \text{ W/m}^2\text{-C}$$

Using Eq. (4.4.25) with $M=1$ gives

$$\hat{q}_1 = (Y_1 - T_0)K_1 + (Y_2 - T_0)K_2 = 395{,}900 \text{ W/m}^2$$

and repeating with $M=2$ gives

$$\hat{q}_2 = (Y_2 - \hat{q}_1 \Delta \phi_1 - T_0)K_1 + (Y_3 - \hat{q}_1 \Delta \phi_2 - T_0)K_2 = 804{,}400 \text{ W/m}^2$$

These $\hat{q}$ values are larger than the exact values of 250,000 and 750,000 W/m², with errors of 58% and 7%, respectively. The errors are somewhat larger than those given in Example 4.1 for the Stolz method but the present algorithm is much less sensitive to measurement errors and can permit considerably smaller time steps. □

EXAMPLE 4.3 A finite plate with $\alpha = 1$ m²/s, $k = 1$ W/m-C, and $L = 1$ m has measured temperatures at the insulated surface ($x = L$) of 16, 45, 99, and 179°C for times 0.5, 1, 1.5, and 2 s, respectively. The initial temperature is 10°C. Calculate $\hat{q}_1$, $\hat{q}_2$ and $\hat{q}_3$ using Eq. (4.4.25) with $r=1$ and 2.

Solution. The dimensionless time step is 0.5 and the ϕ_i values are given in Table 1.1, but in this case the K_i values are given in Table 4.1 for $\Delta t^+ = 0.5$. The results are:

t_i(s)	$\hat{q}_i$, $r=1$	$\hat{q}_i$, $r=2$	q_i, exact
0.5	17.92	38.65	25
1	77.85	78.52	75
1.5	123.1	126.8	125

The units are W/m². After the first time step both the $r=1$ and $r=2$ results are acceptable but only the $r=2$ method can produce acceptable results if Δt is decreased to 0.25 s in this example. The errors in the calculated values are caused by inaccuracies of the algorithms and by inaccurate Y_i values. □

SEC. 4.4 FUNCTION SPECIFICATION METHOD

4.4.3.2 Linear Heat Flux Functional Form Another algorithm for the sequential function specification method uses the temporary assumption of a linear heat flux. Figure 4.8 shows linearly connected segments, starting with the known value of $\hat{q}_{M-1}$. The heat flux over each Δt is considered to be constant, as previously done.

For the connected linear segments shown in Figure 4.8, the single unknown is q_M since $\hat{q}_1, \ldots, \hat{q}_{M-1}$ were previously calculated and $q_{M+1}, \ldots, q_{M+r-1}$ can be written in terms of q_M; hence one can write

$$q_{M+1} = q_M + (q_M - \hat{q}_{M-1}) = 2q_M - \hat{q}_{M-1} \quad (4.4.30a)$$

$$q_{M+2} = q_M + 2(q_M - \hat{q}_{M-1}) = 3q_M - 2\hat{q}_{M-1} \quad (4.4.30b)$$

$$q_{M+3} = 4q_M - 3\hat{q}_{M-1} \quad (4.4.30c)$$

$$\vdots$$

$$q_{M+j-1} = jq_M - (j-1)\hat{q}_{M-1} \quad (4.4.30d)$$

The temperatures for times $t_M, \ldots$ are given by Eq. (3.2.19) which yields

$$T_M = \hat{T}_M|_{q_M=0} + q_M \Delta\phi_0 = T_M|_{q_M=0} + q_M \phi_1 \quad (4.4.31a)$$

$$T_{M+1} = \hat{T}_{M+1}|_{q_M=q_{M+1}=0} + q_M \Delta\phi_1 + q_{M+1}\Delta\phi_0$$
$$= \hat{T}_{M+1}|_{q_M=q_{M+1}=0} - \hat{q}_{M-1}\phi_1 + q_M(\phi_1 + \phi_2) \quad (4.4.31b)$$

$$T_{M+2} = \hat{T}_{M+2}|_{q_M=\cdots=0} + q_M\Delta\phi_2 + q_{M+1}\Delta\phi_1 + q_{M+2}\Delta\phi_0$$
$$= \hat{T}_{M+2}|_{q_M=\cdots=0} - \hat{q}_{M-1}(\phi_1+\phi_2) + q_M(\phi_1+\phi_2+\phi_3) \quad (4.4.31c)$$

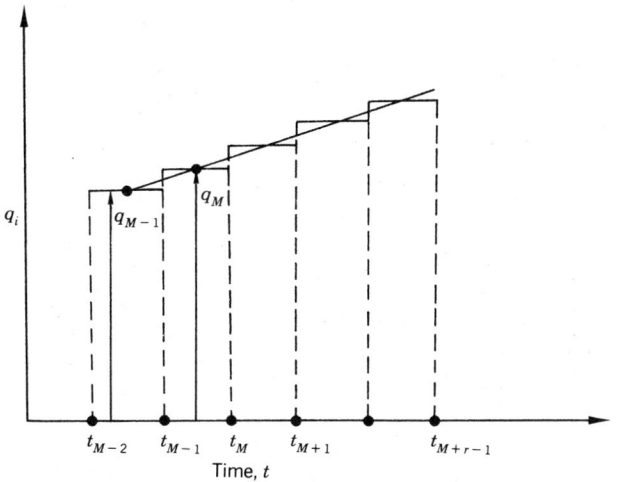

FIGURE 4.8 Connected linear heat flux functional form, $q_{M+j-1} = (1-j)\hat{q}_{M-1} + j\hat{q}_M$.

and the general expression,

$$T_{M+j} = \hat{T}'_{M+j}|_{q_M = \cdots = 0} + q_M \sum_{k=1}^{j+1} \phi_k \qquad (4.4.32a)$$

$$\hat{T}'_{M+j}|_{q_M = \cdots = 0} \equiv \hat{T}_{M+j}|_{q_M = \cdots = 0} - \hat{q}_{M-1} \sum_{k=1}^{j} \phi_k \qquad (4.4.32b)$$

The algorithm for estimating q_M involves minimizing

$$S = \sum_{i=1}^{r} \left(Y_{M+i-1} - \hat{T}'_{M+i-1}|_{q_M = \cdots = 0} - q_M \sum_{k=1}^{i} \phi_k \right)^2 \qquad (4.4.33)$$

Following the usual procedure of differentiating S with respect to q_M and so on yields

$$\hat{q}_M = \frac{\sum_{i=1}^{r} (Y_{M+i-1} - \hat{T}'_{M+i-1}|_{q_M = \cdots = 0}) \sum_{k=1}^{i} \phi_k}{\sum_{i=1}^{r} \left(\sum_{k=1}^{i} \phi_k \right)^2} \qquad (4.4.34)$$

or equivalently

$$\hat{q}_M = \sum_{i=1}^{r} K_i (Y_{M+i-1} - \hat{T}'_{M+i-1}|_{q_M = \cdots = 0}) \qquad (4.4.35a)$$

$$K_i \equiv \left(\sum_{k=1}^{i} \phi_k \right) \left[\sum_{k=1}^{r} \left(\sum_{l=1}^{k} \phi_l \right)^2 \right]^{-1} \qquad (4.4.35b)$$

As for the case of q_M temporarily assumed constant, Eq. (4.4.35) reduces to the Stolz algorithm ($r = 1$),

$$\hat{q}_M = K_1 (Y_M - \hat{T}_M|_{q_M = 0}); \quad K_1 \equiv \frac{1}{\phi_1} \qquad (4.4.36)$$

For $r = 2$, Eq. (4.4.35) becomes

$$\hat{q}_M = K_1 (Y_M - \hat{T}_M|_{q_M = 0}) + K_2 (Y_{M+1} - \hat{T}_{M+1}|_{q_M = q_{M+1} = 0} + \hat{q}_{M-1}\phi_1) \qquad (4.4.37a)$$

$$K_1 \equiv \frac{\phi_1}{\Delta}, \quad K_2 \equiv \frac{\phi_1 + \phi_2}{\Delta}, \quad \Delta \equiv \phi_1^2 + (\phi_1 + \phi_2)^2 \qquad (4.4.37b)$$

Gain coefficients for Eq. (4.4.35) for a plate, insulated at $x = L$ and heated at $x = 0$ are given in Table 4.3. The sensor is at $x = L$. The values can be compared with those in Table 4.1. For $r = 1$ the values are identical with those in Table 4.1 since both the $q = C$ and linear q algorithms reduce to the Stolz method. As r increases, the gain coefficients of the linear q values decrease more rapidly than for the $q = C$ coefficients. Numerical experiments may be necessary to determine the conditions for which the linear variation in heat flux is superior to the constant heat flux case.

SEC. 4.4 FUNCTION SPECIFICATION METHOD

TABLE 4.3 Dimensionless Gain Coefficients for Eq. (4.4.35) for Sensor at $x=L$ in a Finite Insulated Plate. Function Specification Method, Linear q Assumption

Gain Coefficient	$r=1$	$r=2$	$r=3$	$r=4$	$r=5$
		$\Delta t^+ = 0.05$			
K_1^+	3710	4.05	0.18	0.024	0.005
K_2^+		122	5.54	0.724	0.160
K_3^+			25.5	3.33	0.734
K_4^+				8.79	1.84
K_5^+					3.91
		$\Delta t^+ = 0.2$			
K_1^+	16.3	0.66	0.098	0.025	0.008
K_2^+		3.21	0.47	0.120	0.041
K_3^+			1.16	0.294	0.100
K_4^+				0.547	0.186
K_5^+					0.300
		$\Delta t^+ = 0.5$			
K_1^+	2.99	0.227	0.043	0.013	0.005
K_2^+		0.791	0.151	0.044	0.017
K_3^+			0.323	0.094	0.035
K_4^+				0.163	0.061
K_5^+					0.094

4.4.3.3 Alternative Interpretation.

There are alternative interpretations to the IHCP algorithm given by Eq. (4.4.24). A useful one that yields some additional insight into linear inverse problems is presented here. A typical term of Eq. (3.2.19) can be written as

$$T_{M+j-1} = \hat{T}_{M+j-1}|_{q_M = \cdots = q_{M+j-1}=0} + \sum_{i=1}^{j} q_{M+i-1} \Delta \phi_{j-i+1}, \quad j=1,2,\ldots,r \quad (4.4.38)$$

Suppose that during the next r future time steps the temperature data are matched exactly. This implies that $T_{M+j-1} = Y_{M+j-1}$ for $j=1, 2, \ldots, r$; the corresponding heat flux is denoted by $\tilde{q}_{M+i-1}$. For the preceding conditions, Eq. (4.4.38) becomes

$$Y_{M+j-1} = \hat{T}_{M+j-1}|_{q_M = \cdots = q_{M+j-1}=0} + \sum_{i=1}^{j} \tilde{q}_{M+i-1} \Delta \phi_{j-i+1} \quad (4.4.39)$$

Substituting Eq. (4.4.39) into Eq. (4.4.24) yields

$$\hat{q}_M = \frac{\sum_{j=1}^{r} \phi_j \sum_{i=1}^{j} \tilde{q}_{M+i-1} \Delta \phi_{j-i+1}}{\sum_{j=1}^{r} \phi_j^2} \quad (4.4.40a)$$

Expanding the double summation in Eq. (4.4.40a) and collecting coefficients of $\tilde{q}_{M+i-1}$, Eq. (4.4.40a) can be written as

$$\hat{q}_M = \sum_{i=1}^{r} w_i \tilde{q}_{M+i-1} \qquad (4.4.40b)$$

where the heat flux weighting factors are given by

$$w_i = \frac{\sum_{j=1}^{r} \phi_j \Delta\phi_{j-i+1}}{\sum_{j=1}^{r} \phi_j^2} \quad \text{and} \quad \sum_{i=1}^{r} w_i = 1 \qquad (4.4.40c)$$

Equation (4.4.40b) indicates that the solution of the IHCP can be interpreted as a weighted average of the r heat flux values $\tilde{q}_{M+i-1}$, $i = 1, 2, \ldots, r$ determined by exactly matching the r future temperatures. Note that the temperature profile at the beginning of this procedure corresponds to $\hat{q}_{M-1}$. The heat flux weighting factors w_i can be precalculated for a given problem.

In applying Eq. (4.4.40b), a procedure for calculating $\tilde{q}_{M+i-1}$ can be obtained by explicitly writing out the last ($i = j$) term of Eq. (4.4.39).

$$\tilde{q}_{M+j-1} = \frac{Y_{M+j-1} - \hat{T}_{M+j-1}|_{q_M = \cdots = q_{M+j-1} = 0}}{\phi_1} - \frac{1}{\phi_1} \sum_{i=1}^{j-1} \tilde{q}_{M+i-1} \Delta\phi_{j-i+1}$$

(4.4.41)

Additional details of this alternative interpretation of the IHCP algorithm given by (4.4.24) can be found in Blackwell.[35]

EXAMPLE 4.4   Repeat Example 4.3 with $r = 2$ (two future temperatures) using the heat flux weighting factors defined by Eq. (4.4.40c).

Solution.   First Eq. (4.4.40c) is used to get $w_1 = 0.65409$ and $w_2 = 0.34591$. Then Eq. (4.4.41) is used to get $\tilde{q}_1$ and $\tilde{q}_2$ after which Eq. (4.4.40b) is used for $\hat{q}_1$. A summary of results follows:

M	$\tilde{q}_M$	$\tilde{q}_{M+1}$	$\hat{q}_M$
1	17.9216	77.8549	38.6534
2	46.9823	138.148	78.5176
3	91.8682	194.013	126.7559

□

4.5   REGULARIZATION METHOD

4.5.1   Introduction

The regularization method is a procedure which modifies the least squares approach by adding factors that are intended to reduce excursions in the un-

SEC. 4.5 REGULARIZATION METHOD

known function, such as the surface heat flux. These excursions or fluctuations are not of physical origin but are inherent in ill-posed problems unless special treatment of the fluctuations is introduced.

The regularization method has a number of forms and has been studied by many researchers. The Russians, in particular, have pursued various regularization schemes for the IHCP; those who have been active include Tikhonov and Arsenin,[2] Alifanov,[17-19] and Cozdoba and Crykowsky.[20] American authors who have used a regularization method include Bell and Wardlaw[21,22] and Beck and Murio.[23] In most cases the regularization method for the IHCP has been applied in its whole domain form with the only known exception being Reference 23 which covers both whole domain and sequential schemes; it was found that the sequential method can give nearly the same results as the whole domain procedure but with greatly reduced computation.

The regularization method is related to some approximate least squares procedures. One of these is called *ridge regression* or *damped least squares* which has been studied by a number of authors.[24,25] This method is also related to procedures that have been used for solving nonlinear least squares problems; two of these authors are Levenberg,[26] whose work dates back to 1944, and Marquardt,[27,28] who published his frequently cited paper in 1963. Further discussion of ridge regression and nonlinear least squares procedures are in Reference 6. An excellent book on least squares is authored by Lawson and Hanson[29] and includes a discussion of ridge regression (Reference 29, p. 188).

The plan of this section is to first discuss the physical significance of the regularization terms, Section 4.5.2, then the whole domain regularization is developed in Section 4.5.3, and finally Section 4.5.4 covers briefly the sequential regularization method.

4.5.2 Physical Significance of Regularization Terms

In the regularization procedure for the inverse heat conduction problem, an augmented sum of squares function, S, is minimized. In algebraic form S can be written for J temperature sensors in the form

$$S = \sum_{i=1}^{r} \sum_{j=1}^{J} (Y_{j,M+i-1} - T_{j,M+i-1})^2$$

$$+ \alpha \left[W_0 \sum_{i=1}^{r} q_{M+i-1}^2 + W_1 \sum_{i=1}^{r-1} (q_{M+i} - q_{M+i-1})^2 \right.$$

$$\left. + W_2 \sum_{i=1}^{r-2} (q_{M+i+1} - 2q_{M+i} + q_{M+i-1})^2 \right] \quad (4.5.1)$$

If $M=1$ and $r=n$, then Eq. (4.5.1) can be used for the whole domain estimation procedure, whereas in its present form it can be used for the sequential estimation of q_M. Equation (4.5.1) implies that the heat flux components $q_1, q_2, \ldots, q_n$ are used to describe the heat flux history, $q(t)$. The components $q_i, i=1, \ldots, n$

can be constant elements as in Figure 4.3, or related to linear segments as in Figure 4.4, or have other interpretations. The symbol α in Eq. (4.5.1) is called the regularization parameter. Various methods can be used to select α and, fortunately, a relatively large range of values can be employed. The units of α are $(\text{C-m}^2/\text{W})^2$.

The summation in Eq. (4.5.1) that is preceded by W_0 is called the zeroth-order regularization term. If $W_1 = W_2 = 0$ and $W_0 \neq 0$, then minimization of Eq. (4.5.1) yields what is called the zeroth-order regularization procedure. The term preceded by W_1 is called the first-order regularization term and if $W_0 = W_2 = 0$ and $W_1 \neq 0$, then the first-order regularization method involves minimizing the associated S. Similarly the term preceded by W_2 is called the second-order regularization term and so on. Notice that the zeroth-order term involves no differences of the q components, the first-order term involves first differences of the q's, and the second-order term involves second differences of the q's. Higher-order terms than the second can be included in Eq. (4.5.1) but the most commonly used order is the zeroth, with the first order occasionally used and the others infrequently used. The zeroth-order regularization whole domain procedure is similar to the ridge regression procedure.[6,24,28,29] Alliney and Sgallari[36] in a paper on the reconstruction of images from projections use a combination of zeroth-, first- and second-order regularization.

The whole domain zeroth-order regularization procedure for a single sensor involves minimizing

$$S = \sum_{i=1}^{n} (Y_i - T_i)^2 + \alpha \sum_{i=1}^{n} q_i^2 \qquad (4.5.2)$$

with respect to q_i, $i = 1, 2, \ldots, n$. If $\alpha \to 0$, then exact matching of Y_i and T_i is approached but the sum of the q_i^2 terms becomes large for small time steps. For the opposite case of very large α, the magnitudes of the q_i's are reduced with the limit being

$$q_i = 0, \quad i = 1, 2, \ldots, n \qquad (4.5.3)$$

The effect of a nonzero α is to reduce the magnitude of the q_i values. For small time steps for an interior sensor, the case of $\alpha \approx 0$ can result in an unstable procedure with the q_i values oscillating with changing signs and ever-increasing magnitudes. However, by properly selecting α, instability can be eliminated because the effect of the regularization term in Eq. (4.5.2) is to reduce the maximum magnitudes of estimated values of q_i.

The whole domain *first*-order regularization procedure for a single sensor involves minimizing

$$S = \sum_{i=1}^{n} (Y_i - T_i)^2 + \alpha \sum_{i=1}^{n-1} (q_{i+1} - q_i)^2 \qquad (4.5.4)$$

Again if α goes to zero, exact matching is obtained. The extreme value of large α causes q_i to be

$$q_i = \text{constant}, \quad i = 1, 2, \ldots, n \qquad (4.5.5)$$

where the constant can be any positive or negative value. The magnitude of q_i is not affected. For moderate values of α, *differences* in the q_i values are reduced. Reducing the changes in the q_i values from one time step to the next during the minimizing of Eq. (4.5.4) has the effect of improving the stability of the $\hat{q}_i$, $i = 1, 2, \ldots, n$ values. Note that the second summation term of Eq. (4.5.4) is analogous to the integral,

$$\alpha' \int_0^{t_n} \left(\frac{dq}{dt}\right)^2 dt \qquad (4.5.6)$$

which involves the first time derivative of $q(t)$ with $\alpha' = \alpha \Delta t$.

Similarly to Eqs. (4.5.2) and (4.5.4), the second-order regularization procedure minimizes

$$S = \sum_{i=1}^{n} (Y_i - T_i)^2 + \alpha \sum_{i=1}^{n-2} (q_{i+2} - 2q_{i+1} + q_i)^2 \qquad (4.5.7)$$

where the second summation is analogous to

$$\alpha'' \int_0^{t_n} \left(\frac{d^2 q}{dt^2}\right)^2 dt \qquad (4.5.8)$$

The effect of increasing α is to cause q_i to be a straight line or

$$q_i = q_0 + i q' \Delta t, \quad i = 1, 2, \ldots, n \qquad (4.5.9)$$

where q_0 and q' are arbitrary; that is, large α results in approximating $q(t)$ as a straight line with two unknowns, the intercept and slope. Moderate values of α tend to reduce second differences of the estimated q_i's; in other words, the rate of change of $\hat{q}(t)$ is reduced.

The foregoing discussion points out the different effects of the regularization terms of various orders. The zeroth order reduces the magnitude of $\hat{q}_i$, the first order reduces the magnitude of changes in the $\hat{q}_i$ values from one i to the next, and the second-order regularization term tends to reduce rapid oscillations in the estimated heat flux.

4.5.3 Whole Domain Regularization Method

4.5.3.1 Algebraic Formulation. As an example of whole domain estimation, consider the simple case of zeroth-order regularization with only two measurement times. Then Eq. (4.5.2) is used with $n = 2$. Use of the first two equations of Eq. (3.2.15) gives

$$S = (Y_1 - q_1 \phi_1 - T_0)^2 + (Y_2 - q_1 \Delta \phi_1 - q_2 \phi_1 - T_0)^2 + \alpha(q_1^2 + q_2^2) \qquad (4.5.10)$$

This equation is first differentiated with respect to q_1 and set equal to zero and then S is differentiated with respect to q_2 and set equal to zero. In matrix form the two equations can be written as

$$\begin{bmatrix} \phi_1^2 + \Delta\phi_1^2 + \alpha & \phi_1\Delta\phi_1 \\ \phi_1\Delta_1 & \phi_1^2 + \alpha \end{bmatrix} \begin{bmatrix} \hat{q}_1 \\ \hat{q}_2 \end{bmatrix} = \begin{bmatrix} (Y_1 - T_0)\phi_1 + (Y_2 - T_0)\Delta\phi_1 \\ (Y_2 - T_0)\phi_1 \end{bmatrix} \quad (4.5.11)$$

These equations are solved simultaneously for $\hat{q}_1$ and $\hat{q}_2$ but the details are left for an exercise. Note that if α goes to infinity, the solution of Eq. (4.5.11) is $\hat{q}_1 = \hat{q}_2 = 0$.

The regularization parameter α plays a key role and is now discussed. As pointed out in Section 4.5.2 if $\alpha \to 0$, exact matching is obtained, whereas for $\alpha \to \infty$, $\hat{q}_1$ and $\hat{q}_2$ go to zero. To gain insight into the meanings of small and large α's, the diagonal terms of the square matrix of Eq. (4.5.11) are written as

$$C_{11} = \phi_1^2 \left[1 + \frac{\alpha}{\phi_1^2} + \left(\frac{\Delta\phi_1}{\phi_1}\right)^2 \right] \quad (4.5.12a)$$

$$C_{22} = \phi_1^2 \left(1 + \frac{\alpha}{\phi_1^2} \right) \quad (4.5.12b)$$

One way to select a moderate value of α is to make α/ϕ_1^2 about equal to 1. Then in symbols, the regularization parameter is given by

$$\alpha = \phi_1^2 \quad (4.5.13)$$

It should be pointed out that this relation is not recommended but it is a simple criterion which does have the correct units for α. In actual practice, α might be made much larger, such as $\alpha = 1000\phi_1^2$. For more discussion, see Sections 4.5.3.3 and 5.6.2.

4.5.3.2 Matrix Formulation.

In order to develop the whole domain procedure for n times and n components of the heat flux, the matrix approach is used. The sum of squares function becomes

$$S = (\mathbf{Y} - \mathbf{T})^T(\mathbf{Y} - \mathbf{T}) + \alpha[W_0(\mathbf{H}_0\mathbf{q})^T\mathbf{H}_0\mathbf{q}$$
$$+ W_1(\mathbf{H}_1\mathbf{q})^T\mathbf{H}_1\mathbf{q} + W_2(\mathbf{H}_2\mathbf{q})^T\mathbf{H}_2\mathbf{q}] \quad (4.5.14)$$

where for the whole domain method with a single sensor,

$$\mathbf{Y} = \begin{bmatrix} Y_1 \\ \vdots \\ Y_n \end{bmatrix} \quad \mathbf{T} = \begin{bmatrix} T_1 \\ \vdots \\ T_n \end{bmatrix}, \quad \mathbf{q} = \begin{bmatrix} q_1 \\ \vdots \\ q_n \end{bmatrix} \quad (4.5.15a,b,c)$$

The W_i values are the weighting factors for zeroth-, first- and second-order regularization methods. The $\mathbf{H}_i$ matrices are square and are associated with the zeroth-, first- and second-order regularization procedures; analogous to the algebraic expressions given by Eqs. (4.5.2), (4.5.4), and (4.5.7), the $\mathbf{H}_i$ matrices are

$$\mathbf{H}_0 = \mathbf{I} \quad (4.5.16a)$$

SEC. 4.5 REGULARIZATION METHOD

$$\mathbf{H}_1 = \begin{bmatrix} -1 & 1 & 0 & 0 & 0 \\ 0 & -1 & 1 & 0 & 0 \\ 0 & 0 & -1 & 1 & 0 \\ 0 & 0 & 0 & -1 & 1 \\ 0 & 0 & 0 & 0 & 0 \end{bmatrix} \quad (4.5.16b)$$

$$\mathbf{H}_2 = \begin{bmatrix} 1 & -2 & 1 & 0 & 0 \\ 0 & 1 & -2 & 1 & 0 \\ 0 & 0 & 1 & -2 & 1 \\ 0 & 0 & 0 & 0 & 0 \\ 0 & 0 & 0 & 0 & 0 \end{bmatrix} \quad (4.5.16c)$$

where $\mathbf{H}_1$ and $\mathbf{H}_2$ are written for $n = 5$.

The q components in Eq. (4.5.14) are found by minimizing S which is accomplished by taking the matrix first derivative with respect to the n components of the heat flux, replacing $\mathbf{q}$ by $\hat{\mathbf{q}}$, and setting the resulting equation equal to the zero vector,

$$\frac{\partial S}{\partial \mathbf{q}} = \begin{bmatrix} \dfrac{\partial S}{\partial q_1} \\ \vdots \\ \dfrac{\partial S}{\partial q_n} \end{bmatrix}_{\mathbf{q}=\hat{\mathbf{q}}} = \mathbf{0} \quad (4.5.17)$$

The model for the temperature $\mathbf{T}$ is given by Eq. (3.2.17) which is used in Eq. (4.5.14). Taking the matrix derivative as indicated by Eq. (4.5.17) gives

$$[\mathbf{X}^T\mathbf{X} + \alpha(W_0\mathbf{H}_0^T\mathbf{H}_0 + W_1\mathbf{H}_1^T\mathbf{H}_1 + W_2\mathbf{H}_2^T\mathbf{H}_2)]\hat{\mathbf{q}} = \mathbf{X}^T(\mathbf{Y} - T_0\mathbf{1}) \quad (4.5.18)$$

because (see Reference 6, p. 221)

$$\frac{\partial}{\partial \mathbf{q}}(\mathbf{Xq})^T\mathbf{Xq} = 2\mathbf{X}^T\mathbf{Xq} \quad (4.5.19)$$

Note that $\mathbf{1}$ is a vector of ones.

Equation (4.5.18) is called the matrix normal equation. There are n unknown components for $\mathbf{q}$; for n larger than three or four, a linear algebraic-equation computer subroutine should be used; there are many such program packages available, some of which are in the IMSL library[30] and the book by Lawson and Hanson.[29]

The solution of Eq. (4.5.18) for the $\mathbf{q}$ vector can be symbolically written as

$$\hat{\mathbf{q}} = [\mathbf{X}^T\mathbf{X} + \alpha(W_0\mathbf{I} + W_1\mathbf{H}_1^T\mathbf{H}_1 + W_2\mathbf{H}_2^T\mathbf{H}_2)]^{-1}\mathbf{X}^T(\mathbf{Y} - T_0\mathbf{1}) \quad (4.5.20)$$

The terms introduced by the regularization procedure only appear as additive contributions to $\mathbf{X}^T\mathbf{X}$. These terms reduce or eliminate the ill conditioning of the ill-posed inverse heat conduction problem.

If $\alpha=0$ in Eq. (4.5.18), the "exact matching" equation is found. For n measurements $\mathbf{X}$ is a $n \times n$ matrix and

$$(\mathbf{X}^T\mathbf{X})^{-1} = \mathbf{X}^{-1}(\mathbf{X}^{-1})^T \tag{4.5.21}$$

and thus for $\alpha=0$, Eq. (4.5.20) gives

$$\hat{\mathbf{q}} = \mathbf{X}^{-1}(\mathbf{Y} - T_0\mathbf{1}) \tag{4.5.22}$$

which can be obtained from Eq. (3.2.17) with $\mathbf{T}$ replaced by $\mathbf{Y}$. As pointed out before, exact matching is usually not satisfactory.

The first-order regularization term $\mathbf{H}_1^T\mathbf{H}_1$ in Eqs. (4.5.18) and (4.5.20) is

$$\mathbf{H}_1^T\mathbf{H}_1 = \begin{bmatrix} 1 & -1 & 0 & \cdots & & 0 \\ -1 & 2 & -1 & \cdots & & 0 \\ 0 & -1 & 2 & \cdots & & 0 \\ \vdots & & & \ddots & & \\ 0 & 0 & \cdots & -1 & 2 & -1 \\ 0 & 0 & \cdots & 0 & -1 & 1 \end{bmatrix} \tag{4.5.23}$$

Notice that the coefficients of $-1, 2, -1$ for all rows (except first and last) are the same as those for *second* differences.

4.5.3.3 Selection of Regularization Parameter.

Tikhonov[2] called the scalar, α, the "regularization parameter." He suggested that α can be found by making the residual sum of squares approximately equal to the numerical value expected, based on knowledge of the measurement errors. The residual sum of squares is

$$\mathcal{R} = (\mathbf{Y} - \hat{\mathbf{T}})^T(\mathbf{Y} - \hat{\mathbf{T}}) \tag{4.5.24}$$

where $\hat{\mathbf{T}}$ denotes the estimated vector of temperature.

This expression is zero for α set equal to zero in Eq. (4.5.1) provided there is a single interior temperature sensor. As α is increased, $\mathcal{R}$ also increases. Tikhonov's recommendation is that $\mathcal{R}$ be made approximately equal to the sum of squares of the errors in the measurements. The sum of squares S with $\alpha=0$,

$$S_0 = (\mathbf{Y} - \mathbf{T})^T(\mathbf{Y} - \mathbf{T}) \tag{4.5.25}$$

has an expected value of $n\sigma^2$ if the first four standard assumptions of Section 1.4.2 are valid, where $\mathbf{T}$ is the "true" value, σ^2 is the variance of Y_i and n is the number of measurements. The expected value of $\mathcal{R}$ is less than this value of $n\sigma^2$, however, because $\mathcal{R}$ can approach zero. Graham,[31] following a recommendation by Reinsch,[32] suggested that $\mathcal{R}$ be between $[n-(2n)^{1/2}]\sigma^2$ and $[n+(2n)^{1/2}]\sigma^2$.

By choosing the value of $\mathcal{R}$, equal to $n\sigma^2$, which is designated as $\mathcal{R}_s$, the estimation problem can be written as the minimization of S given by Eq. (4.5.1). The condition that $\mathcal{R} = \mathcal{R}_s$ is a necessary one. Tikhonov[2] suggests a number of

SEC. 4.5 REGULARIZATION METHOD

methods of accomplishing this, one of which is a trial-and-error procedure. In practice when using the regularization procedure in a routine manner for inverse heat conduction problems of the same type, the same regularization parameter α can be used in all of them. Another way to select α is to use the concept of the minimum squared error to be discussed in Sections 4.8 and 5.6.

4.5.4 Sequential Regularization Method

The regularization concepts can be implemented in a sequential manner with the advantage of similar results but greatly reduced computer time compared to the whole domain procedure.

The plan is to consider multiple sensors for only the algebraic formulation; the matrix formulation is considered in connection with the more general procedure called the trial function method which is covered in Section 4.6.

The algebraic sequential regularization procedure can start with Eq. (4.5.1). The $q_1, \ldots, q_{M-1}$ components are considered to be estimated and thus $T_{j,M+i-1}$ is replaced by [see Eqs. (3.2.19)–(3.2.22)]

$$T_{j,M+i-1} = \hat{T}_{j,M+i-1}|_{q_M = \cdots = q_{M+i-1} = 0}$$
$$+ q_M \Delta\phi_{j,i-1} + \cdots + q_{M+i-1} \Delta\phi_{j,0} \quad (4.5.26)$$

To simplify the analysis, only the zeroth- and first-order regularization methods are considered, that is, $W_2 = 0$ in Eq. (4.5.1).

Consider the simplest case of $r = 1$ or one future time and J temperature sensors; the sum of squares is

$$S = \sum_{j=1}^{J} (Y_{jM} - \hat{T}_{jM}|_{q_M = 0} - q_M \Delta\phi_{j0})^2 + \alpha q_M^2 \quad (4.5.27)$$

where W_0 is absorbed in α. Note that for $r = 1$, the first order term does not appear. Following the usual minimization of S with respect to q_M yields the estimator,

$$\hat{q}_M = \sum_{j=1}^{J} K_{1j}(Y_{jM} - \hat{T}_{jM}|_{q_M = 0}) \quad (4.5.28a)$$

$$K_{1j} \equiv \Delta\phi_{j0} \left(\alpha + \sum_{k=1}^{J} \Delta\phi_{k0}^2 \right)^{-1} \quad (4.5.28b)$$

For α going to zero, Eq. (4.5.28) reduces to Eq. (4.3.13) which is for a single future time. If, in addition, there is only one sensor, that is, $J = 1$, there is an exact matching of the measured and the calculated temperatures. For the regularization parameter to have significant effect it must be on the order of, or larger than, the square of the largest $\Delta\phi_{k0}$.

Now consider two future times ($r = 2$), J sensors, and zeroth- and first-order regularization; the sum of squares function is

$$S = \sum_{j=1}^{J} (Y_{jM} - \hat{T}_{jM}|_{q_M=0} - q_M \Delta\phi_{j0})^2$$

$$+ \sum_{j=1}^{J} (Y_{j,M+1} - \hat{T}_{j,M+1}|_{q_M=q_{M+1}=0} - q_M \Delta\phi_{j1} - q_{M+1}\Delta\phi_{j0})^2$$

$$+ \alpha W_0(q_M^2 + q_{M+1}^2) + \alpha W_1(q_{M+1} - q_M)^2 \quad (4.5.29)$$

The estimator for $\hat{q}_M$ is given by

$$\hat{q}_M = \frac{d_1 C_{22} - d_2 C_{12}}{C_{11}C_{22} - C_{12}^2} \quad (4.5.30)$$

where

$$C_{11} = \alpha(W_0 + W_1) + \sum_{k=0}^{1} \sum_{j=1}^{J} \Delta\phi_{jk}^2 \quad (4.5.31a)$$

$$C_{22} = \alpha(W_0 + W_1) + \sum_{j=1}^{J} \Delta\phi_{j0}^2 \quad (4.5.31b)$$

$$C_{12} = -\alpha W_1 + \sum_{j=1}^{J} \Delta\phi_{j0} \Delta\phi_{j1} \quad (4.5.31c)$$

$$d_1 = \sum_{k=0}^{1} \sum_{j=1}^{J} (Y_{j,M+k} - \hat{T}_{j,M+k}|_{q_M=q_{M+1}=0}) \Delta\phi_{jk} \quad (4.5.32a)$$

$$d_2 = \sum_{j=1}^{J} (Y_{j,M+1} - \hat{T}_{j,M+1}|_{q_M=q_{M+1}=0}) \Delta\phi_{j0} \quad (4.5.32b)$$

For the case of a single temperature sensor, $J=1$, the estimator for $\hat{q}_M$ given by Eq. (4.5.30) can be written as

$$\hat{q}_M = K_1(Y_M - \hat{T}_M|_{q_M=0}) + K_2(Y_{M+1} - \hat{T}_{M+1}|_{q_M=q_{M+1}=0}) \quad (4.5.33a)$$

$$K_1 = \frac{[\alpha(W_0 + W_1) + \Delta\phi_0^2]\Delta\phi_0}{\Delta} \quad (4.5.33b)$$

$$K_2 = \frac{[\alpha(W_0 + W_1)\Delta\phi_1 + \alpha W_1 \Delta\phi_0]}{\Delta} \quad (4.5.33c)$$

$$\Delta = [\alpha(W_0 + W_1) + \Delta\phi_0^2 + \Delta\phi_1^2][\alpha(W_0 + W_1) + \Delta\phi_0^2] - (\alpha W_1 - \Delta\phi_0 \Delta\phi_1)^2 \quad (4.5.33d)$$

where the K_1 and K_2 expressions are gain coefficients. It can be shown that this algorithm reduces to the Stolz algorithm for $\alpha \to 0$ and to the $q = C$ function specification algorithm for $\alpha \to \infty$. (See Prob. 4.16.)

SEC. 4.5 REGULARIZATION METHOD

Equation (4.5.33) is in exactly the same form as the $r=2$ sequential function specification method, Eq. (4.4.25c), and can be used in the same manner. The only difference is in the values of the gain coefficients. See Example 4.2 for an illustration of the sequential manner in which Eq. (4.5.33) can be employed.

Small or large values of α are defined in relation to the $\Delta\phi_0^2$ and $\Delta\phi_1^2$ values. Figures 4.9a and 4.9b display the gain coefficients for Eq. (4.5.33) for the case of a plate heated at $x=0$ and insulated at $x=L$. The sensor is at $x=L$ and the dimensionless time step is $\Delta t^+ = 0.05$. The ϕ_i values are taken from Table 1.1; the $\Delta\phi_0^2$ and $\Delta\phi_1^2$ values are

$$\Delta\phi_0^2 = \phi_1^2 = (0.000269)^2 = 7.2 \times 10^{-8}$$

$$\Delta\phi_1^2 = (\phi_2 - \phi_1)^2 = (0.007885 - 0.000269)^2 = 5.8 \times 10^{-5}$$

which suggests that 10^{-7} to 10^{-4} is an important range for α values. An inspection of Figures 4.9a and 4.9b reveals that this is a range where interesting changes occur in K_1 and K_2.

Figures 4.9a and 4.9b depict the gain coefficients for three different algorithms; Figure 4.9a is for the zeroth-order case and Fig. 4.9b for the first-order case. The function specification results are shown in both figures. The coefficients for the function specification method are independent of α; the K_1 value is 4.33 and K_2 is 127. These values are also given in Table 4.1 for $\Delta t^+ = 0.05$.

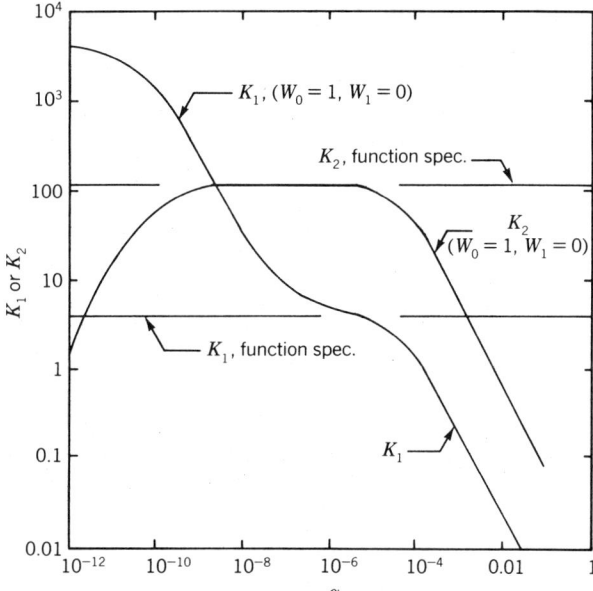

FIGURE 4.9a Gain coefficients for zeroth-order sequential regularization algorithm given by Eq. (4.5.33). $\Delta t^+ = 0.05$ for sensor at $x=L$ in a plate insulated at $x=L$. $r=2$, $W_0=1$, $W_1=0$.

144   CHAP. 4   INVERSE HEAT CONDUCTION ESTIMATION PROCEDURES

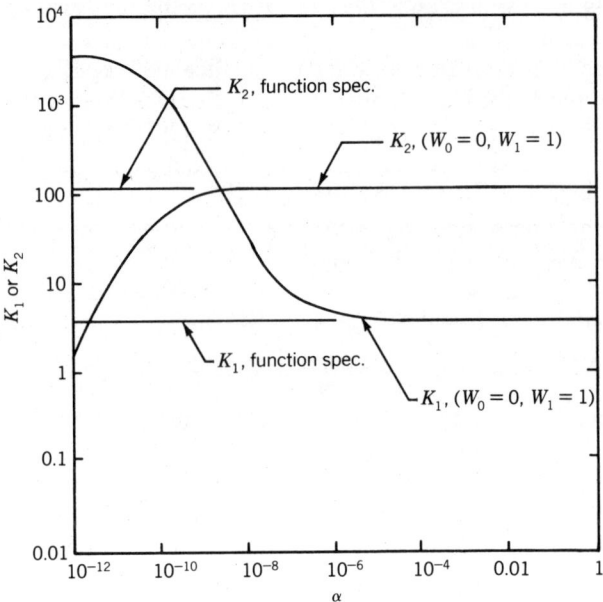

FIGURE 4.9b   Gain coefficients for first-order regularization algorithm given by Eq. (4.5.33). $r=2$, $W_0=0$, $W_1=1$.

The zeroth-order method has $W_0=1$ and $W_1=0$ and the first-order method has $W_0=0$ and $W_1=1$. For the sequential regularization method, the K_1 value for $\alpha \to 0$ starts at 3720, which is the same as the Stolz method value ($r=1$, also see Table 4.1). This is the starting value for the zeroth-order (Figure 4.9a), and the first-order (Figure 4.9b) methods, but K_1 decreases with increasing α and approaches zero for the zeroth-order case. Near $\alpha = 10^{-6}$, the zeroth-order K_1 curve has an inflection but the first-order curve approaches 4.33, the same value as given by the function specification method. The K_2 coefficients for both the zeroth- and first-order cases start near zero for small α's (less than 10^{-12}) and then increase to about 127 with the zeroth-order K_2 decreasing for $\alpha > 10^{-6}$, whereas the first-order K_2 remains constant at 127 for $\alpha > 10^{-9}$. It is remarkable that the first-order sequential regularization procedure (Figure 4.9b) for $r=2$ and $\Delta t^+ = 0.05$ yields almost identical results as the $q=$ constant function specification procedure for the large range of $\alpha > 10^{-7}$. For the zeroth-order regularization method similar results are obtained for α between 10^{-7} and 10^{-5}.

For large α values (α greater than 10^{-4} in this case) both K_1 and K_2 decrease toward zero for the zeroth-order regularization method. This corroborates the previous assertion that the zeroth-order method reduces the $\hat{q}_M$ values for large α's. Since, for α greater than 10^{-6}, the first-order K_1 and K_2 values approach the same values as in the function specification method, the first-order method is seen to be less sensitive to the choice of α than the zeroth-order method.

Based on this limited investigation, the first-order regularization appears to be superior to the zeroth-order procedure but more analysis is needed to verify this conclusion.

4.6 TRIAL FUNCTION METHOD

4.6.1 Introduction

A procedure that combines the function specification and regularization approaches is called the trial solution method by Twomey.[33,34] We shall call it the trial function method. The method as presented here is not identical to that given by Twomey but the basic concept is the same. The basic function to be minimized is a sum of squares criterion plus an additional term that is a generalization of the regularization term in Eq. (4.5.14); the **q** vector is replaced by **q** minus a predetermined "trial" function. The trial function can incorporate prior information regarding the shape of the expected heat flux or it can be a simple function, such as a constant.

The trial function method is potentially very powerful but a choice of α and of the functional form of q must be made. Finding satisfactory combinations may require numerical experimentation.

The derivation is given in matrix notation and includes many interpretations. Sequential and whole domain (Twomey's interest) estimations are included as is the possibility of multiple temperature sensors. A generalized least squares formulation is given to permit nonconstant variance and/or correlated errors in the measured temperatures.

The plan of this section is to present a general matrix approach and then to discuss special cases, some of which are given in algebraic form.

4.6.2 Matrix Analysis

The function to minimize is

$$S = (\mathbf{Y} - \mathbf{T})^T \psi^{-1} (\mathbf{Y} - \mathbf{T})$$
$$+ \alpha \{ [\mathbf{H}_0(\mathbf{q} - \tilde{\mathbf{q}})]^T \mathbf{W}_0 [\mathbf{H}_0(\mathbf{q} - \tilde{\mathbf{q}})] + [\mathbf{H}_1(\mathbf{q} - \tilde{\mathbf{q}})]^T \mathbf{W}_1 [\mathbf{H}_1(\mathbf{q} - \tilde{\mathbf{q}})] \} \quad (4.6.1)$$

which is similar to the S function given by Eq. (4.5.14) for the regularization method. For simplicity in presentation, the term analogous to the second-order regularization term is not included in Eq. (4.6.1). The symmetric square matrix ψ is the covariance matrix of the random measurement errors in **Y**. (See Reference 6, Chapter 6.) It is a known matrix; if the first four standard statistical assumptions are valid, ψ and ψ^{-1} are given by

$$\psi = \sigma^2 \mathbf{I}, \quad \psi^{-1} = \sigma^{-2} \mathbf{I} \quad (4.6.2)$$

where σ^2 is the variance of Y_i.

Equation (4.6.1) can be used for both the whole domain and sequential methods. For the case of a single sensor, the whole domain procedure has $\mathbf{Y}$, $\mathbf{T}$, $\mathbf{q}$, and $\tilde{\mathbf{q}}$ vectors with n elements,

$$\mathbf{Y} = \begin{bmatrix} Y_1 \\ \vdots \\ Y_n \end{bmatrix}, \quad \mathbf{T} = \begin{bmatrix} T_1 \\ \vdots \\ T_n \end{bmatrix}, \quad \mathbf{q} = \begin{bmatrix} q_1 \\ \vdots \\ q_n \end{bmatrix}, \quad \tilde{\mathbf{q}} = \begin{bmatrix} \tilde{q}_1 \\ \vdots \\ \tilde{q}_n \end{bmatrix} \quad (4.6.3)$$

where $\tilde{\mathbf{q}}$ is a trial function vector. For the sequential procedure the subscripts in Eq. (4.6.3) go from M to $M+r-1$ as shown in Eq. (3.2.20).

The square matrix $\mathbf{H}_0$ is equal to $\mathbf{I}$ as given by Eq. (4.5.16a) and $\mathbf{H}_1$ simulates first time differences and is displayed in Eq. (4.5.16b).

The heat flux vector $\mathbf{q}$ in Eq. (4.6.1) is unknown and is to be estimated. The symbol $\tilde{\mathbf{q}}$ denotes the trial function vector which is a linear function of $\mathbf{q}$ and another function $\tilde{\mathbf{q}}^f$,

$$\tilde{\mathbf{q}} = \mathbf{B}\mathbf{q} + \tilde{\mathbf{q}}^f \quad (4.6.4)$$

The matrix $\mathbf{B}$ is square and is selected in accordance with what the analyst seeks to accomplish; $\mathbf{B}$ is discussed later. The vector $\tilde{\mathbf{q}}^f$ can be the prior estimate of $\mathbf{q}$ before the algorithm is used to estimate $\mathbf{q}$. If, for example, the heat flux is known to be nearly constant near a certain value, then $\tilde{\mathbf{q}}^f$ would be chosen to be this value. Another example is when $\mathbf{q}$ is known to vary in a certain way with time such as a decaying exponential; $\tilde{\mathbf{q}}^f$ could then be this decaying exponential function. In the regularization procedure both $\mathbf{B}$ and $\tilde{\mathbf{q}}^f$ are zero matrices.

The general trial function estimator for $\mathbf{q}$ with $\tilde{\mathbf{q}}$ replaced by Eq. (4.6.4) requires the matrix derivative of Eq. (4.6.1) with respect to $\mathbf{q}$. In addition, an expression for $\mathbf{T}$ is needed; for the linear IHCP, $\mathbf{T}$ can be written as

$$\mathbf{T} = \hat{\mathbf{T}}|_{\mathbf{q}=0} + \mathbf{X}\mathbf{q} \quad (4.6.5)$$

As pointed out in Chapter 3, this expression is found for both the Duhamel's numerical and finite difference approaches. Equation (4.6.5) can be used for both the whole domain and sequential procedures. For the whole domain method

$$\hat{\mathbf{T}}|_{\mathbf{q}=0} = T_0 \mathbf{1} \quad (4.6.6)$$

where T_0 is the initial temperature. For the sequential method, see Eq. (3.2.22). The matrix derivative of Eq. (4.6.1) modified by the use of Eqs. (4.6.4) and (4.6.5) yields, after collecting terms and setting equal to zero,

$$(\mathbf{X}^T \psi^{-1} \mathbf{X} + \alpha \{[\mathbf{H}_0(\mathbf{I}-\mathbf{B})]^T \mathbf{W}_0 [\mathbf{H}_0(\mathbf{I}-\mathbf{B})] + [\mathbf{H}_1(\mathbf{I}-\mathbf{B})]^T \mathbf{W}_1 [\mathbf{H}_1(\mathbf{I}-\mathbf{B})]\}) \hat{\mathbf{q}}$$
$$= \mathbf{X}^T \psi^{-1} (\mathbf{Y} - \hat{\mathbf{T}}|_{\mathbf{q}=0}) + \alpha \{[\mathbf{H}_0(\mathbf{I}-\mathbf{B})]^T \mathbf{W}_0 \mathbf{H}_0 \tilde{\mathbf{q}}^f + [\mathbf{H}_1(\mathbf{I}-\mathbf{B})]^T \mathbf{W}_1 \mathbf{H}_1 \tilde{\mathbf{q}}^f\}$$
$$(4.6.7)$$

This matrix equation represents a set of n simultaneous, linear algebraic equations for $\hat{q}_1, \ldots, \hat{q}_n$ for the whole domain method. For the sequential method,

SEC. 4.6 TRIAL FUNCTION METHOD

the unknowns are $\hat{q}_M, \ldots, \hat{q}_{M+r-1}$ but only $\hat{q}_M$ is used. Equation (4.6.7) is valid for both single and multiple sensors.

In order to expedite the examination of Eq. (4.6.7), several simplifications are used; ψ is given by $\sigma^2 \mathbf{I}$, $\mathbf{H}_0$ by $\mathbf{I}$, $\mathbf{W}_0$ by $\mathbf{I}$, and the first-order term is dropped by setting $\mathbf{W}_1 = \mathbf{0}$. Then symbolically, $\hat{\mathbf{q}}$ can be given by

$$\hat{\mathbf{q}} = [\sigma^{-2}\mathbf{X}^T\mathbf{X} + \alpha(\mathbf{I}-\mathbf{B})^T(\mathbf{I}-\mathbf{B})]^{-1}[\sigma^{-2}\mathbf{X}^T(\mathbf{Y}-\hat{\mathbf{T}}|_{\mathbf{q}=0}) + \alpha(\mathbf{I}-\mathbf{B})^T\tilde{\mathbf{q}}^f] \quad (4.6.8)$$

Many special cases can be considered in connection with this equation, several of which are briefly discussed in the following.

4.6.3 Zeroth-Order Regularization Method

The zeroth-order regularization method can be obtained from Eq. (4.6.8) by setting

$$\tilde{\mathbf{q}}^f = \mathbf{0} \quad \text{and} \quad \mathbf{B} = \mathbf{0} \quad (4.6.9)$$

For the whole domain method, Eq. (4.6.8) is used with Eq. (4.6.6). See also Eq. (4.5.20). Equation (4.6.8) can also be used for the sequential regularization method; if there are J temperature sensors then $\mathbf{Y}$ can be partitioned so that

$$\mathbf{Y} = \begin{bmatrix} \mathbf{Y}_M \\ \mathbf{Y}_{M+1} \\ \vdots \\ \mathbf{Y}_{M+r-1} \end{bmatrix}, \quad \mathbf{Y}_{M+i} = \begin{bmatrix} Y_{1,M+i} \\ Y_{2,M+i} \\ \vdots \\ Y_{J,M+i} \end{bmatrix} \quad (4.6.10)$$

where $Y_{j,M+i}$ is understood to be an element of the $\mathbf{Y}$ vector and not part of a two-dimensional array; that is, $\mathbf{Y}$ has dimensions of $rJ \times 1$. An example of the resulting equations is given for $r=2$ by Eqs. (4.5.30) and (4.5.31) where α is replaced by $\alpha\sigma^2$ in Eq. (4.6.8) and $W_0 = 1$ and $W_1 = 0$.

4.6.4 Generalized Sequential Function Specification Method

The trial function method can be used to yield a generalization of the sequential function specification method. For the q = constant temporary approximation,

$$\tilde{\mathbf{q}}^f = \mathbf{0} \quad (4.6.11)$$

$$\mathbf{B} = \begin{bmatrix} 1 & 0 & 0 & 0 & \cdots \\ 1 & 0 & 0 & 0 & \cdots \\ 1 & 0 & 0 & 0 & \cdots \\ \vdots & \vdots & \vdots & \vdots & \end{bmatrix} \quad (4.6.12)$$

which results in the trial function $\tilde{\mathbf{q}}$ being the constant vector of [see Eq. (4.6.4)]

$$\tilde{\mathbf{q}} = q_M \mathbf{1} \quad (4.6.13)$$

148   **CHAP.4 INVERSE HEAT CONDUCTION ESTIMATION PROCEDURES**

For the case of $r=2$ and a single sensor, Eq. (4.6.8) is equivalent to solving the equation,

$$\begin{bmatrix} \sigma^{-2}[(\Delta\phi_0)^2+(\Delta\phi_1)^2]+\alpha & \sigma^{-2}\Delta\phi_0\Delta\phi_1-\alpha \\ \sigma^{-2}\Delta\phi_0\Delta\phi_1-\alpha & \sigma^{-2}(\Delta\phi_0)^2+\alpha \end{bmatrix} \begin{bmatrix} \hat{q}_M \\ \hat{q}_{M+1} \end{bmatrix}$$

$$= \begin{bmatrix} \sigma^{-2}(Y_M-T_M|)\Delta\phi_0+\sigma^{-2}(Y_{M+1}-\hat{T}_{M+1}|)\Delta\phi_1 \\ \sigma^{-2}(Y_{M+1}-\hat{T}_{M+1}|)\Delta\phi_0 \end{bmatrix}$$

(4.6.14)

This equation is solved only for $\hat{q}_M$ because $\hat{q}_{M+1}$ is not used in the sequential procedure. Solving this equation for $\hat{q}_M$ and putting the equation in the gain coefficient form yields Eq. (4.5.33) if $W_0=0$, $W_1=1$, and $\alpha \to \alpha\sigma^2$. It is surprising that for $r=2$ this trial function method gives the same result as the first-order regularization method. Notice that the $W_0=0$, $W_1=1$ curves in Figure 4.9b (which is equivalent to the trial function $r=2$ case) approach the $q=$ constant function specification result for $\alpha\sigma^2$ greater than 10^{-7}, which is a large range of values indeed! The trial function method with **B** given by Eq. (4.6.12) is shown by this example to be a generalization of the $q=$ constant function specification method; this conclusion is also valid for r greater than two.

4.7   FILTER FORM OF LINEAR IHCP

4.7.1   Introduction

The inverse heat conduction algorithms given in previous sections of this chapter can be re-formulated as digital filters. The only restriction is that the problem be linear; the sequential and whole domain procedures can be considered and also single and multiple sensors can be included. Moreover, the method of approximating the heat conduction problem (using Duhamel's theorem, finite differences or other) does not affect the validity of the digital filter approach.

The digital filter approach is important because it can be demonstrated that it is much more computationally efficient than other methods. Due to its efficiency, it can be readily implemented in an on-line method of analysis. Heat flux measuring devices can incorporate the digital filter concept and immediate visual digital output can be provided. (By on-line is meant a measurement that is given in real time, but the IHCP algorithm may necessitate a delay of a few time steps.)

4.7.2   Sequential Filter Algorithm

All the sequential algorithms derived in this chapter for the linear IHCP can be given in terms of gain coefficients. See, for example, Eq. (4.4.25) which can be

SEC. 4.7 FILTER FORM OF LINEAR IHCP

written for a single sensor as

$$\hat{q}_M = \sum_{i=1}^{r} K_i(Y_{M+i-1} - \hat{T}_{M+i-1}|_{q_M = \cdots = 0}) \qquad (4.7.1)$$

where the gain coefficients, K_i, depend on the specific algorithm such as those for the function specification, regularization, or trial function procedures. For present purposes the K_i are assumed known.

A filter form of Eq. (4.7.1) is based on each estimated q_M being linear in the Y_i values. This linearity was demonstrated several times in previous sections and in connection with Eqs. (4.3.6)–(4.3.9). Linearity implies that the effects of various components of Y_i can be individually determined and the total effect can be obtained by superposition.

To illustrate the linearity, the case of a finite plate heated at $x=0$ and insulated at $x=L$ is considered. The sensor is at $x=L$ and the dimensionless time step of $\Delta t^+ = 0.05$ is used with $r=3$ future times in the function specification method with q temporarily held constant. Equation (4.4.25) is used and the gain coefficients from Table 4.1 are $K_1=0.292$, $K_2=8.56$, and $K_3=31.8$ (more significant figures were used in the computations). The initial temperature, T_0, is set equal to zero and a series of separate problems is solved. The first of these problems has $Y_1=1$ and the subsequent Y_i values are equal to zero. The second problem has $Y_i=0$ except $Y_2=1$ and the jth problem has $Y_j=1$ and $Y_i=0$ for $i \neq j$. The associated heat fluxes for each of these problems are given in the respective columns of Table 4.4.

There are several important characteristics of Table 4.4. Except for the first two columns, the calculated heat flux components are the same in each column but each succeeding column is shifted down one line. In order to describe

TABLE 4.4 Heat Flux Components for $Y_j=1$ and $Y_i=0$ for $i \neq j$; Values are Given in Columns for $j=1$ to 6. (For $r=3$, $\Delta t^+ = 0.05$)

| M | $Y_1=1$ | All Y_i Values Equal Zero Except: | | | | |
		$Y_2=1$	$Y_3=1$	$Y_4=1$	$Y_5=1$	$Y_6=1$
1	0.29	8.6	31.8	0	0	0
2	−0.35	−10.1	−29.9	31.8	0	0
3	−0.02	−0.9	−12.1	−29.9	31.8	0
4	0.06	1.7	5.4	−12.1	−29.9	31.8
5	0.03	0.9	4.7	5.4	−12.1	−29.9
6	0.0008	0.05	0.97	4.7	5.4	−12.1
7		−0.15	−0.51	0.97	4.7	5.4
8		−0.08	−0.41	−0.51	0.97	4.7
9		−0.003	0.006	−0.41	−0.51	0.97
⋮	⋮	⋮	⋮	⋮	⋮	⋮
∞	0	0	0	0	0	0

the results of Table 4.4, a notation is needed. Let

$$\frac{\delta \hat{q}_M}{\delta Y_j}$$

represent the entries of the jth column of Table 4.4. The notation denotes the change in $\hat{q}_M$ when Y_j is increased one unit. With this notation, observe that for this case with $r=3$,

$$\frac{\delta \hat{q}_M}{\delta Y_j} = 0 \quad \text{for } M < j - r + 1$$

$$\frac{\delta \hat{q}_1}{\delta Y_3} = \frac{\delta \hat{q}_2}{\delta Y_4} = \cdots = \frac{\delta \hat{q}_M}{\delta Y_{M+r-1}} = 31.8$$

$$\frac{\delta \hat{q}_2}{\delta Y_3} = \frac{\delta \hat{q}_3}{\delta Y_4} = \cdots = \frac{\delta \hat{q}_M}{\delta Y_{M+r-2}} = -29.9$$

Thus for $j \geq r$, the same entries in a given column of Table 4.4 appear in the other columns. Consequently if one column is calculated (for $j \geq r$), the entries for the subsequent columns can be inferred from the one calculated.

The first two columns of Table 4.4 are noted to be different from the subsequent ones. This is true in general for the columns $1, 2, \ldots, r-1$. In order to treat each data point in the same manner, the calculations for sequential IHCP algorithms should start at least $r-1$ time steps *before* heating starts. If this is done, the anomalous first few steps have negligible effect on the calculations.

The entries in Table 4.4 for columns $3, 4, \ldots$ can be given as the filter coefficients $f_{-2}, f_{-1}, f_0, f_1, f_2, \ldots$ as shown in Table 4.5. These f values are related to $\delta \hat{q}_M / \delta Y_j$ by

$$f_{1-r} = \frac{\delta \hat{q}_M}{\delta Y_{M+r-1}} = \frac{\delta \hat{q}_1}{\delta Y_r}$$

$$f_{2-r} = \frac{\delta \hat{q}_M}{\delta Y_{M+r-2}} = \frac{\delta \hat{q}_2}{\delta Y_r}$$

$$\vdots$$

$$f_{-1} = \frac{\delta \hat{q}_M}{\delta Y_{M+1}} = \frac{\delta \hat{q}_{r-1}}{\delta Y_r}$$

$$f_0 = \frac{\delta \hat{q}_M}{\delta Y_M} = \frac{\delta \hat{q}_r}{\delta Y_r}$$

$$f_1 = \frac{\delta \hat{q}_M}{\delta Y_{M-1}} = \frac{\delta \hat{q}_{r+1}}{\delta Y_r}$$

$$\vdots$$

$$f_{1-r+i} = \frac{\delta \hat{q}_M}{\delta Y_{M+r-i-1}} = \frac{\delta \hat{q}_{i+1}}{\delta Y_r} \quad (4.7.2)$$

Note that f_{1-r} is not dependent on M but on the difference in the subscripts of $\delta \hat{q}_M$ and δY_{M+r-1}. Notice also that the first equal sign in each equation of

SEC. 4.7 FILTER FORM OF LINEAR IHCP

TABLE 4.5 Pattern for Digital Filter Coefficients for Table 4.4

M	$Y_1=1$	$Y_2=1$	$Y_3=1$	$Y_4=1$	$Y_5=1$	$Y_6=1$
1			f_{-2}	0	0	0
2			f_{-1}	f_{-2}	0	0
3			f_0	f_{-1}	f_{-2}	0
4			f_1	f_0	f_{-1}	f_{-2}
5			f_2	f_1	f_0	f_{-1}
6			f_3	f_2	f_1	f_0
7			f_4	f_3	f_2	f_1

Eq. (4.7.2) is for a row in Table 4.5, that is, fixed M, and the second equal sign is for a column of Table 4.5, i.e., fixed r.

Using these symbols and the principle of superposition, the linear relation for $\hat{q}_M$ can also be written as

$$\hat{q}_M = \frac{\delta \hat{q}_M}{\delta Y_1}(Y_1 - T_0) + \frac{\delta \hat{q}_M}{\delta Y_2}(Y_2 - T_0) + \cdots + \frac{\delta \hat{q}_M}{\delta Y_{M+r-1}}(Y_{M+r-1} - T_0) \quad (4.7.3a)$$

which looks like a Taylor series expansion with the higher-order terms neglected. However, Eq. (4.7.3a) is an exact result for linear problems. Substituting Eq. (4.7.2) into Eq. (4.7.3a) gives

$$\hat{q}_M = \frac{\delta \hat{q}_{M+r-1}}{\delta Y_r}(Y_1 - T_0) + \frac{\delta \hat{q}_{M+r-2}}{\delta Y_r}(Y_2 - T_0) + \cdots + \frac{\delta \hat{q}_1}{\delta Y_r}(Y_{M+r-1} - T_0) \quad (4.7.3b)$$

$$= f_{M-1}(Y_1 - T_0) + f_{M-2}(Y_2 - T_0) + \cdots + f_1(Y_{M-1} - T_0)$$
$$+ f_0(Y_M - T_0) + f_{-1}(Y_{M+1} - T_0) + \cdots + f_{2-r}(Y_{M+r-2} - T_0)$$
$$+ f_{1-r}(Y_{M+r-1} - T_0) \quad (4.7.4)$$

Notice that the subscripts of f plus those of Y in Eq. (4.7.4) always equal M.

Provided the calculations for q_M start at least $r-1$ steps before heating starts, the IHCP algorithm given in the gain coefficient form by Eq. (4.7.1) can be equivalently written in the digital filter forms as

$$\hat{q}_M = \sum_{i=1}^{M+r-1} \frac{\delta \hat{q}_{M+r-i}}{\delta Y_r}(Y_i - T_0) \quad (4.7.5)$$

$$\hat{q}_M = \sum_{i=1}^{M+r-1} f_{M-i}(Y_i - T_0) \quad (4.7.6)$$

Note that these equations indicate that $\hat{q}_M$ depends on the $M-1$ previous temperature measurements plus the r future temperature measurements. However, in practice the lower limit in the summation need not start at 1 because the associated filter coefficients become negligible as demonstrated in Table 4.4.

152   CHAP. 4   INVERSE HEAT CONDUCTION ESTIMATION PROCEDURES

The digital filter coefficients given by f or $\delta q/\delta Y$ can be calculated by using the IHCP algorithm such as Eq. (4.7.1) with

$$Y_r - T_0 = 1,$$
$$Y_i - T_0 = 0 \quad \text{for } i \neq r \tag{4.7.7}$$

The resulting values of $\hat{q}_M$ are

$$\hat{q}_M \bigg|_{\substack{Y_r - T_0 = 1 \\ Y_i - T_0 = 0 \text{ for } i \neq r}} = \frac{\delta \hat{q}_M}{\delta Y_r} = f_{M-r} \tag{4.7.8}$$

This equation can be verified by using the $Y_3 = 1$ columns of Tables 4.4 and 4.5.

The foregoing discussion of a digital filter is in connection with the sequential method. It is demonstrated in Chapter 5 that the whole domain estimation algorithms can be written in exactly the same form as Eq. (4.7.5), provided r is chosen to be a sufficiently large integer.

EXAMPLE 4.5. For the case of a plate insulated at $x = L$ and heated at $x = 0$, the measured temperatures at $x = L$ for times 0.05, 0.1, and 0.15 s are 30.269, 37.885, and 59.306°C. Relevant values are $L = 1$ m, $\alpha = 1$ m^2/s, $k = 1$ W/m-C, and $T_0 = 30$°C. The value of $r = 3$ is used in the function specification method. Use the filter form of the algorithm to calculate the first three values of heat flux.

Solution. The equation used is Eq. (4.7.5) but the calculational starting time must be moved at least 2 times early. Let the data be arranged as:

Time (s)	i	Y_i
−0.5	1	30
0	2	30
0.05	3	30.269
0.1	4	37.885
0.15	5	59.306

where the first calculation is at $t = -0.05$ s rather than at $t = 0.05$ s.

For the first time of $M = 1$, Eq. (4.7.5) gives (using column 3 of Table 4.4)

$$\hat{q}_1 = \frac{\delta \hat{q}_3}{\delta Y_3}(Y_1 - T_0) + \frac{\delta \hat{q}_2}{\delta Y_3}(Y_2 - T_0) + \frac{\delta \hat{q}_1}{\delta Y_3}(Y_3 - T_0)$$

$$= -12.1(30 - 30) - 29.9(30 - 30) + 31.8(30.269 - 30) = 8.55 \text{ W/m}^2$$

which corresponds to time $t = -0.025$ s. The value for $\hat{q}_2$ is obtained using (4.7.5) to get

$$\hat{q}_2 = \frac{\delta \hat{q}_4}{\delta Y_3}(Y_1 - T_0) + \frac{\delta \hat{q}_3}{\delta Y_3}(Y_2 - T_0) + \frac{\delta \hat{q}_2}{\delta Y_3}(Y_3 - T_0) + \frac{\delta \hat{q}_1}{\delta Y_3}(Y_4 - T_0)$$

$$= 5.4(30 - 30) - 12.1(30 - 30) - 29.9(30.269 - 30) + 31.8(37.885 - 30) = 243 \text{ W/m}^2$$

$$\hat{q}_3 = 693 \text{ W/m}^2$$

SEC. 4.8 TWO CONFLICTING OBJECTIVES

These are the same values as given by the gain coefficient form of the original IHCP algorithm, Eq. (4.7.1). ☐

4.7.3 Prefiltering Temperature Measurements

The subject of digital filters[35] is important in the field of signal processing. Some of the digital filter algorithms can be applied to the temperature measurements *before* the IHCP algorithms are employed. There are many such filters possible and their discussion is beyond the scope of this book. If the statistics of the temperature measurement errors are known, then it is possible to design filters to admit only "low" frequency signals. For more discussion of digital filter design, see Reference 37.

One possible prefilter (Reference 37, p. 57) is

$$y_i = \frac{Y_{i-1} + 2Y_i + Y_{i+1}}{4} \tag{4.7.9}$$

That is, each Y_i value is replaced by y_i before the IHCP algorithm is applied. More specifically, y_i of Eq. (4.7.9) replaces Y_i in Eq. (4.7.5) or Eq. (4.7.6). Notice that the coefficients of Y_{i-1}, Y_i, and Y_{i+1} sum to unity. See Problem 4.15.

4.8 TWO CONFLICTING OBJECTIVES

4.8.1 Minimum Deterministic Bias

One of the characteristics of good estimators is that of minimum bias; in fact, in parameter estimation problems, an unbiased estimator is usually sought. Mathematically, it is desirable for the estimator, $\hat{\beta}$, of a parameter, β, to be the expected value of β,

$$E(\hat{\beta}) = \beta \tag{4.8.1}$$

In other words, the expected value of the estimate is the true value. If Eq. (4.8.1) is true, the estimator $\hat{\beta}$ is said to be unbiased and "on the average" the estimate given by $\hat{\beta}$ would be near β. If $\hat{\beta}$ were a biased estimator of β, $\hat{\beta}$ would on the average tend to be either higher or lower than the true value of β.

In ill-posed problems it is preferable for the estimators for the heat flux components, $\hat{q}_1, \ldots, \hat{q}_n$, to have low bias. It is *not* required that the bias be zero, however. In ill-posed problems, the bias and variability (variance) of the estimators for the unknown parameters are related. Consequently the bias is only one aspect of the estimators that must be considered.

The bias of an estimator can be investigated when the random measurement errors are set equal to zero. This bias or error in the estimator can be called the *deterministic bias*. It is advantageous to make it as small as possible and yet not make the variance unacceptably large.

4.8.2 Minimum Sensitivity to Random Errors

Another characteristic of a good estimator in addition to minimum bias is that of minimum variance. If $\hat{\beta}$ is the estimator of the parameter β, then this means that the variance of $\hat{\beta}$,

$$V(\hat{\beta}) = E\{[\hat{\beta} - E(\hat{\beta})]^2\} \qquad (4.8.2)$$

should be a minimum. For linear parameter estimation problems with the first, second, seventh, and eighth standard assumptions satisfied, the estimator that is a linear function of the measurements and has unbiased and minimum variance is called the Gauss-Markov estimator (Reference 6, p. 232). For ill-posed problems, better estimates are obtained if the requirement of unbiased estimators is not imposed. In the latter case, it is necessary to specify an alternative objective which is discussed next.

4.8.3 Mean Squared Error

For ill-posed problems the common requirements of zero bias and minimum variance do not yield satisfactory estimators. For zero bias, the estimators are very sensitive to measurement errors; that is, the variance of $\hat{\beta}$ is large. There are ways to reduce the variance for the IHCP such as requiring that *all* the components of **q** be equal, or

$$\hat{q}_1 = \hat{q}_2 = \cdots = \hat{q}_n = \text{constant} \qquad (4.8.3)$$

If a requirement such as Eq. (4.8.3) is introduced, the variance of $\hat{q}_i$ is relatively small but it usually has a large bias. Hence, the deterministic bias and variance of the estimator are related and an optimal strategy should consider both aspects.

A function that considers both bias and variability is the mean squared error. Let $\hat{q}_M$ be a component of the estimated **q** vector and let q_M be the true value of the component. The mean squared error of $\hat{q}_M$ is

$$\mathscr{S}_M^2 = E[(\hat{q}_M - q_M)^2] \qquad (4.8.4)$$

In general, estimators that minimize $\mathscr{S}_M^2$ for all values of M, $M = 1, 2, \ldots, n$, are sought.

It is important to observe that Eq. (4.8.4) does not give the variance of $\hat{q}_M$, that is, $\mathscr{S}_m^2 \neq V(\hat{q}_M)$, except for the special case when the expected value of $\hat{q}_M$ is the true value, q_M; this only occurs if $\hat{q}_M$ is an unbiased estimator of q_M. Since for ill-posed problems the estimator is usually biased, Eq. (4.8.4) is not, in general, the variance of $\hat{q}_M$.

It can be shown that Eq. (4.8.4) includes two parts: deterministic and stochastic. Adding and subtracting the expected value of $\hat{q}_M$ inside the right

SEC. 4.8 TWO CONFLICTING OBJECTIVES

side of Eq. (4.8.4) gives

$$\mathcal{S}_M^2 = E(\{[\hat{q}_M - E(\hat{q}_M)] - [q_M - E(\hat{q}_M)]\}^2) \tag{4.8.5a}$$

$$= E\{[\hat{q}_M - E(\hat{q}_M)]^2\}$$
$$- 2E\{[\hat{q}_M - E(\hat{q}_M)][q_M - E(\hat{q}_M)]\}$$
$$+ E\{[q_M - E(\hat{q}_M)]^2\} \tag{4.8.5b}$$

The first term on the right side of Eq. (4.8.5b) is the variance of the estimator, $\hat{q}_M$,

$$V(\hat{q}_M) = E\{[\hat{q}_M - E(\hat{q}_M)]^2\} \tag{4.8.6}$$

The second term on the right side of Eq. (4.8.5b) can be shown to be zero because $q_M - E(\hat{q}_M)$ is not a random variable and also

$$E[\hat{q}_M - E(\hat{q}_M)] = E(\hat{q}_M) - E(\hat{q}_M) = 0 \tag{4.8.7}$$

The last term in Eq. (4.8.5b) is the square of a bias and the outer expected value symbol can be dropped; the resulting expression is given the symbol $\mathcal{D}_M^2$,

$$\mathcal{D}_M^2 = [q_M - E(\hat{q}_M)]^2 \tag{4.8.8}$$

Hence the mean squared error of $\hat{q}_M$ is composed of two parts, a variance and the square of a deterministic error, and the relationship between them is

$$\mathcal{S}_M^2 = V(\hat{q}_M) + \mathcal{D}_M^2 \tag{4.8.9}$$

An expression for $V(\hat{q}_M)$ is developed in Section 4.8.4 and the deterministic error is discussed in Section 4.8.5.

The significance of the components of the mean squared error given by Eq. (4.8.9) can be illustrated by using Figure 4.10. The continuous solid line is a

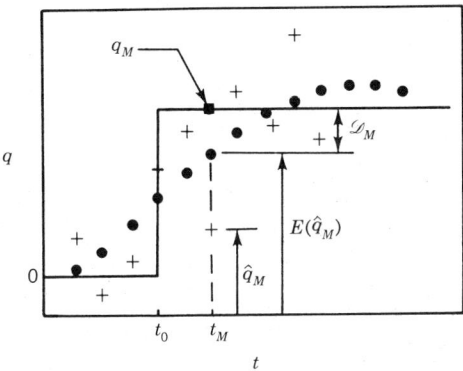

FIGURE 4.10 The true heat flux q_M at time t_M, the estimated value $\hat{q}_M$, and the mean of the estimated value, $E(\hat{q}_M)$.

representation of a "true" heat flux that starts at time zero and at time t_0 jumps to a constant value. The true heat flux at time t_M is shown by a square in Figure 4.10 and is denoted q_M. The dots illustrate estimated values of the heat flux curve for errorless measurements or equivalently the expected (i.e., theoretical average) value of the estimated heat flux, $E(\hat{q}_i)$. Note that the $E(\hat{q}_i)$ values are biased since at most times the values are either greater or less than the true values. For example, at time t_M, the mean estimated value is low. The difference between q_M and $E(\hat{q}_M)$ is the deterministic bias, $\mathscr{D}_M$. The crosses in Figure 4.10 show simulated and calculated heat fluxes with measurement errors; the value at time t_M is denoted $\hat{q}_M$. Data from other sets of measurements would yield a set of estimated heat fluxes at t_M that would center about $E(\hat{q}_M)$, rather than the true value q_M. A measure of the variability of the $\hat{q}_M$ values with respect to $E(\hat{q}_M)$ is called the standard deviation of $\hat{q}_M$ and the square of the standard deviation is the variance of $\hat{q}_M$ which is denoted $V(\hat{q}_M)$. The mean squared error given by Eq. (4.8.9) contains both the effect of measurement errors which is described by the variance, $V(\hat{q}_M)$, and the effect of the deterministic bias, $\mathscr{D}_M$, which is due to a biased IHCP algorithm. In the inverse heat conduction problem some bias is accepted in order to reduce the extreme sensitivity to measurement errors, particularly as the time steps are made small.

4.8.4 Variance of Estimated Heat Flux Component

In Sections 4.3–4.6 two types of estimators are derived: sequential and whole domain. For both types the algorithms can be written in the form of a digital filter. An example is Eq. (4.7.5) which is convenient for determining an expression for the variance of $\hat{q}_M$ because the Y_i components are explicitly given, in contrast to the implicit dependence in the gain coefficient equation, Eq. (4.7.1).

To derive an expression for the variance of $\hat{q}_M$, some statistical assumptions are necessary. The first four standard statistical assumptions are assumed valid; that is, the random errors, ε_i, in Y_i are additive, have zero mean, have constant variance, σ^2, and are uncorrelated,

$$Y_i = T_i + \varepsilon_i \tag{4.8.10a}$$

$$E(\varepsilon_i) = 0 \tag{4.8.10b}$$

$$V(\varepsilon_i) = \sigma^2 \tag{4.8.10c}$$

$$\text{cov}(\varepsilon_i, \varepsilon_j) = 0 \quad \text{for } i \neq j \tag{4.8.10d}$$

For these assumptions, the variance of $\hat{q}_M$ given by Eq. (4.7.5) is

$$V(\hat{q}_M) = \sigma^2 \sum_{i=1}^{M+r-1} \left(\frac{\delta \hat{q}_{M+r-i}}{\delta Y_r} \right)^2 \tag{4.8.11a}$$

SEC. 4.8 TWO CONFLICTING OBJECTIVES

$$= \sigma^2 \sum_{i=1}^{M+r-1} \left(\frac{\delta \hat{q}_i}{\delta Y_r}\right)^2 \qquad (4.8.11b)$$

Notice that the equality of Eq. (4.8.11a) and (4.8.11b) does *not* imply that $\delta \hat{q}_{M+r-i}/\delta Y_r$ is equal to $\delta \hat{q}_i/\delta Y_r$.

EXAMPLE 4.6 Calculate the variance of $\hat{q}_M$ for a flat plate insulated at $x=L$ and heated at $x=0$. The sensor is at $x=L$. Use $\Delta t = 0.05$ s for $L=1$ m, $\alpha = 1$ m^2/s, $k=1$ W/m-C, and $r=3$.

Solution. The $\delta \hat{q}/\delta Y_r$ values are given in the $Y_3 = 1$ column of Table 4.4. For $M=1$, Eq. (4.8.11) gives

$$V(\hat{q}_1) = \sigma^2 \left[\left(\frac{\delta \hat{q}_1}{\delta Y_3}\right)^2 + \left(\frac{\delta \hat{q}_2}{\delta Y_3}\right)^2 + \left(\frac{\delta \hat{q}_3}{\delta Y_3}\right)^2\right]$$

$$= \sigma^2 [(31.8)^2 + (-29.9)^2 + (-12.1)^2] = 2052\sigma^2 \text{ W}^2/\text{m}^4$$

and for $M=2$, one finds

$$V(\hat{q}_2) = \sigma^2 [(31.8)^2 + (-29.9)^2 + (-12.1)^2 + (5.4)^2] = 2080\sigma^2 \text{ W}^2/\text{m}^4$$

For $\hat{q}_3, \hat{q}_4, \hat{q}_5, \hat{q}_6$, and $\hat{q}_7$ the variances are $2103\sigma^2$, $2104\sigma^2$, $2104\sigma^2$, and $2104\sigma^2$, respectively. Notice that the variances approach a constant value as M increases.

The square root of the variance is the standard deviation which can be compared to the estimated $\hat{q}_M$ values. For $\sigma^2 = 2(C)^2$ the standard deviation of $\hat{q}_3$ is 65 W/m^2. This value can be compared to the value of $\hat{q}_3 = 693$ W/m^2 that was calculated for Example 4.5. □

4.8.5 Estimate of Deterministic Error in Surface Heat Flux

There are several ways of estimating the deterministic error in the surface heat flux. Two of these are discussed here. Suppose that the true surface heat fluxes are the values of

$$q_i = 0 \quad \text{for } i \neq r \qquad (4.8.12a)$$

$$q_i = \delta q_r \quad \text{for } i = r \qquad (4.8.12b)$$

The corresponding temperature rises at the interior location x_1 are calculated at times $t_1, t_2, \ldots$ using Eq. (3.2.12) with T_0 equal to zero. An expanded form of Eq. (3.2.12) is

$$T_1 = \Delta \phi_0 q_1 \qquad (4.8.13a)$$

$$T_2 = \Delta \phi_1 q_1 + \Delta \phi_0 q_2 \qquad (4.8.13b)$$

$$T_3 = \Delta \phi_2 q_1 + \Delta \phi_1 q_2 + \Delta \phi_0 q_3 \qquad (4.8.13c)$$

$$\vdots$$

$$T_M = \sum_{i=1}^{M} \Delta \phi_{M-i} q_i \qquad (4.8.13d)$$

For simplicity, consider the case of $r=2$ and use Eq. (4.8.12) in Eq. (4.8.13) to get the simulated temperature measurements of

$$Y_1 = 0 \qquad (4.8.14a)$$

$$Y_2 = \Delta\phi_0 \delta q_r \qquad (4.8.14b)$$

$$Y_3 = \Delta\phi_1 \delta q_r \qquad (4.8.14c)$$

$$\vdots$$

$$Y_M = \Delta\phi_{M-2} \delta q_r \qquad (4.8.14d)$$

These values of Y are then introduced into IHCP algorithms such as Eq. (4.4.25) to calculate the heat flux estimates, δq_M. The ratio of $\delta \hat{q}_M / \delta q_r$ is independent of δq_r; this notation of $\delta \hat{q}_M / \delta q_r$ makes clear that the calculation yields changes in the surface heat flux for temperatures associated with a change in the heat flux at time t_r. Because $\delta \hat{q}_M$ is the change in $\hat{q}_M$ given by an IHCP algorithm and δq_r is the change in the input heat flux, $\delta \hat{q}_M / \delta q_r$ with $M=r$ is not equal to unity (except for the Stolz algorithm). The maximum deviation from the input q_i values given by Eq. (4.8.12) is one estimate of the deterministic error, $\mathscr{D}$,

$$\mathscr{D} = \max_i \left| \frac{\delta \hat{q}_i}{\delta q_r} - \frac{q_i}{\delta q_r} \right| \delta q_r \qquad (4.8.15)$$

For most cases, the i value in Eq. (4.8.15) that yields a maximum for $\mathscr{D}$ corresponds to $i=r$.

Another estimate of the deterministic error $\mathscr{D}$ is obtained from the square root of the sum of squares of the differences between $\delta \hat{q}_M$ and the input values given by Eq. (4.8.12). Mathematically, this estimate of the deterministic error is

$$\mathscr{D} = \left[\sum_{\substack{i=1 \\ i \neq r}}^{\infty} \left(\frac{\delta \hat{q}_i}{\delta q_r} \right)^2 + \left(\frac{\delta \hat{q}_r}{\delta q_r} - 1 \right)^2 \right]^{1/2} \delta q_r \qquad (4.8.16)$$

Surprisingly this definition of $\mathscr{D}$ frequently yields values only 20 or 30% larger than that given by Eq. (4.8.15). Hence the deterministic error $\mathscr{D}$ is usually nearly the same value for both definitions, Eq. (4.8.15) or Eq. (4.8.16). This is important because the two measures emphasize different aspects. The first one, Eq. (4.8.15), obtains its sole contribution at $i=r$ or very near r, whereas the second, Eq. (4.8.16), obtains contributions for an extended range of i's. Even so, the results tend to be close. In later chapters, Eq. (4.8.16) is used rather than Eq. (4.8.15).

EXAMPLE 4.7 For a flat plate insulated at $x=L$, and heated at $x=0$ and the sensor at $x=L$, calculate the deterministic error for the function specification algorithm with $r=3$. Use $\Delta t = 0.05$ s for $L=1$ m, $\alpha = 1$ m^2/s, and $k=1$ W/m-C.

Solution. The deterministic error can be found from both Eq. (4.8.15) and Eq. (4.8.16). In both cases the $\delta \hat{q}_i / \delta q_r$ values are needed which are calculated using the function specification algorithm with the input temperatures of $Y_1 = Y_2 = 0$, $Y_3 = \Delta\phi_0 = 0.000269$; $Y_4 = \Delta\phi_1 = 0.007099, \ldots$. See Table 1.1. The algorithm is Eq. (4.4.25) with $r=3$. The

TABLE 4.6 Results for Example 4.7

M	$\delta\hat{q}_M/\delta q_r$	Sum	$\mathscr{D}$
1	0.00856	0.00856	0.00856
2	0.23427	0.24283	0.23443
3	0.45060	0.69343	0.59733
4	0.29232	0.98575	0.66502
5	0.06487	1.05062	0.66818
6	−0.02623	1.02439	0.66869
7	−0.02382	1.00056	0.66912
8	−0.00513	0.99544	0.66913
9	0.00246	0.99790	0.66914
10	0.00209	0.99998	0.66914
11	0.00042	1.00040	0.66914
12	−0.00020	1.00020	0.66914
13	−0.00021	0.99999	0.66914

gain coefficients are $K_1 = 0.292046$, $K_2 = 8.56054$, and $K_3 = 31.8168$ (see Table 4.1). The $\delta\hat{q}_1/\delta q_3$ value is obtained from Eq. (4.4.25) as

$$\frac{\delta\hat{q}_1}{\delta q_3} = K_3 \Delta\phi_0 = 0.00856$$

and $\delta\hat{q}_2/\delta q_3$ is equal to

$$\frac{\delta\hat{q}_2}{\delta q_3} = K_1(0 - 0.00865 \,\Delta\phi_1)$$
$$+ K_2(\Delta\phi_0 - 0.00856\,\Delta\phi_2) + K_3(\Delta\phi_1 - 0.00856\,\Delta\phi_3)$$
$$= 0.23427$$

Further values are displayed in the second column of Table 4.6. A running sum of the $\delta\hat{q}_M/\delta q_3$ values, given in the third column of Table 4.6, approaches unity, as it should for conservation of energy.

The fourth column gives the deterministic error as defined by Eq. (4.8.16); the entries are running sums of the terms in Eq. (4.8.16). Note that the values quickly converge to the value of 0.66914. This value can be compared to that obtained from Eq. (4.8.15) which is

$$\mathscr{D} = 1 - 0.45060 = 0.54940$$

The value given by Eq. (4.8.16) is only 20% larger than this volume. Since an approximate measure of the deterministic error is satisfactory, both equations can be used; Eq. (4.8.16) is preferred in the remainder of this book. □

REFERENCES

1. Hadamard, J., *Lectures on Cauchy's Problem in Linear Partial Differential Equations*, Yale University Press, New Haven, CT, 1923.

2. Tikhonov, A. N. and Arsenin, V. Y., *Solutions of Ill-Posed Problems*, V. H. Winston & Sons, Washington, D.C., 1977.
3. Widder, D. V., *The Heat Equation*, Academic Press, New York, 1975.
4. Weber, C. F., Analysis and Solution of the Ill-Posed Inverse Heat Conduction Problem, *Int. J. Heat Mass Transfer* **24**, 1783–1792 (1981).
5. Olmsted, J. M. H., *Advanced Calculus*, Prentice-Hall, Englewood Cliffs, NJ, 1961.
6. Beck, J. V. and Arnold, K. J., *Parameter Estimation in Engineering and Science*, Wiley, NY, 1977.
7. Beck, J. V., Criteria for Comparison of Methods of Solution of the Inverse Heat Conduction Problem, *Nucl. Eng. Des.* **53**, 11–22 (1979).
8. deBoor, Carl, *A Practical Guide to Splines*, Springer-Verlag, New York, 1978.
9. Stolz, G., Numerical Solutions to an Inverse Problem of Heat Conduction for Simple Shapes, *J. Heat Transfer* **82**, 20–26 (1960).
10. Frank, I., An Application of Least Square Methods to the Solution of the Inverse Problem of Heat Conduction, *J. Heat Transfer* **85**, 378–379 (1963).
11. Davies, J. M., Input Power Determined From Temperatures in Simulated Skin Protected Against Thermal Radiation, *J. Heat Transfer* **88**, 154–160 (1966).
12. Beck, J. V., Correction of Transient Thermocouple Temperature Measurements in Heat-Conducting Solids, Part II, The Calculation of Transient Heat Fluxes Using the Inverse Convolution, AVCO Corp., Res. and Adv. Dev. Div., Wilmington, MA, Tech. Report RAD-TR-7-60-38 (Part II), March 30, 1961.
13. Beck, J. V., Calculation of Surface Heat Flux from an Internal Temperature History, ASME Paper 62-HT-46 (1962).
14. Beck, J. V., Surface Heat Flux Determination Using an Integral Method, *Nucl. Eng. Des.* **7**, 170–178 (1968).
15. Beck, J. V., Nonlinear Estimation Applied to the Nonlinear Heat Conduction Problem, *Int. J. Heat Mass Transfer* **13**, 703–716 (1970).
16. Blackwell, B. F., An Efficient Technique for the Numerical Solution of the One-Dimensional Inverse Problem of Heat Conduction, *Numer. Heat Transfer* **4**, 229–239 (1981).
17. Alifanov, O. M., Inverse Boundary Value Problems of Heat Conduction, *J. Eng. Phys.* **25** (1975).
18. Alifanov, O. M. and Artyukhin F. A. "Regularized Numerical Solution of Nonlinear Inverse Heat-Conduction Problem," *J. Eng. Phy.* **29**, 934–938, 1975.
19. Alifanov, O. M., *Identification of Processes of Heat Container Apparatus. An Introduction to the Theory of Inverse Problem of Heat Transfer*, Machinery Publisher, Moscow 1979 (in Russian).
20. Cozdoba, L. A. and Crykowsky, P. G., *Methods of Solution to the Inverse Problem*, Scientific Publisher, Kiev, 1982 (in Russian).
21. Bell, J. B. and Wardlaw, A. B., Numerical Solution of an Ill-Posed Problem Arising in Wind Tunnel Heat Transfer Data Reduction, Naval Surface Weapons Center, NSWC TR 82-32, Dec. 1981.
22. Bell, J. B., The Noncharacteristic Cauchy Problem for Class of Equations with Time Dependence. I. Problems in One Space Dimensions, *SIAM J. Math. Anal.* **12**, 759–777 (1981).
23. Beck, J. V. and Murio, D., Combined Function Specification–Regularization Procedure for Solution of Inverse Heat Conduction Problem, AIAA Paper No. AIAA-84-0491, January 1984.
24. Hoerl, A. E. and Kennard, R. W., Ridge Regression Biased Estimation for Nonorthogonal Problems, *Technometrics* **12**, 55–67 (1970).
25. Draper, N. R. and Van Nostrand, R. C., Ridge Regression—Is It Worthwhile, Univ. of Wisconsin Department of Statistics Technical Report 501, March 1978.
26. Levenberg, K., A Method for the Solution of Certain Non-linear Problems in Least Squares, *Q. Appl. Math.* **2**, 164–168 (1944).
27. Marquardt, D. W., An Algorithm for Least Squares Estimation for Nonlinear Parameters, *J. Soc. Ind. Appl. Math.* **11**, 431–441 (1963).
28. Marquardt, D. W., Generalized Inverses, Ridge Regression, Biased Linear Estimation, and Nonlinear Estimation, *Technometrics* **12**, 591–612 (1970).

29. Lawson, C. L. and Hanson, R. J., *Solving Least Squares Problems*, Prentice-Hall, Englewood Cliffs, NJ, 1974.
30. The IMSL Library, International Mathematical and Statistical Libraries, Inc., Houston, TX.
31. Graham, N. Y., Smoothing with Periodic Cubic Splines, *Bell System Tech. J.* **62**, 101–110 (1983).
32. Reinsch, C. H. J., Smoothing by Spline Function, *Numerische Mathematik* **10**, 177–183 (1967).
33. Twomey, S., On the Numerical Solution of Fredholm Integral Equations of the First Kind by the Inversion of the Linear System Produced by Quadrature, *J. Assoc. Comp. Mach.* **10**, 97–101 (1963).
34. Twomey, S., The Application of Numerical Filtering to the Solution of Integral Equations Encountered in Indirect Sensing Measurements, *J. Franklin Inst.* **279**, 95–109 (1965).
35. Blackwell, B. F., Some Comments on Beck's Solution of the Inverse Problem of Heat Conduction Through the Use of Duhamel's Theorem, *Int. J. Heat Mass Transfer* **26**, 302–305 (1983).
36. Alliney, S. and Sgallari, F., An "Ill-Conditioned" Volterra Integral Equation Related to the Reconstruction of Images from Projections, *SIAM J. Appl. Math.* **44**, 627–645 (1984).
37. Hamming, R. W., *Digital Filters*, 2nd ed., Prentice-Hall, Englewood Cliffs, NJ, 1983.

PROBLEMS

4.1. For exact matching of the temperatures given below at $x=0.25L$ in a steel slab, find the heat flux components $\hat{q}_1$, $\hat{q}_2$, and $\hat{q}_3$ in W/m^2.

i	1	2	3
t_i, s	1	2	3
Y_i, °C	26.6	28	29

The initial temperature is 25°C; the slab is 1 cm thick, insulated at $x=L$ and heated at $x=0$; and the thermal properties are $k=40$ W/m-C and $\alpha = 10^{-5}$ m^2/s.

4.2. The surface temperature of a body, T_{1M}, is to be estimated from temperature measurements, Y_{2M} at an interior location of the body. Exact matching is to be used. Derive

$$\hat{T}_{1M} = \hat{T}_{1M}|_{q_M=0} + (Y_{2M} - \hat{T}_{2M}|_{q_M=0})\left(\frac{\phi_{11}}{\phi_{21}}\right)$$

and use this equation to find the heated surface temperature at $t=1, 2$ and 3 s of the slab of Problem 4.1.

4.3. For the Stolz IHCP algorithm and for $Y_1 \neq T_0$ and $Y_i = T_0$, $i = 2, 3, 4, \ldots$, derive

$$\hat{q}_1 = \frac{Y_1 - T_0}{\phi_1}$$

$$\hat{q}_2 = -(Y_1 - T_0)\frac{\Delta\phi_1}{\phi_1^2}$$

$$\hat{q}_3 = -(Y_1 - T_0) \frac{\left[\frac{\Delta\phi_2}{\phi_1} - \left(\frac{\Delta\phi_1}{\phi_1}\right)^2\right]}{\phi_1}$$

$$\hat{q}_4 = -(Y_1 - T_0) \frac{\left[\frac{\Delta\phi_3}{\phi_1} - 2\frac{\Delta\phi_1}{\phi_1}\frac{\Delta\phi_2}{\phi_1} + \left(\frac{\Delta\phi_1}{\phi_1}\right)^3\right]}{\phi_1}$$

$$\hat{q}_5 = -(Y_1 - T_0) \frac{\left[\frac{\Delta\phi_4}{\phi_1} - 2\frac{\Delta\phi_1}{\phi_1}\frac{\Delta\phi_3}{\phi_1} - \left(\frac{\Delta\phi_2}{\phi_1}\right)^2 + 3\left(\frac{\Delta\phi_1}{\phi_1}\right)^2\frac{\Delta\phi_2}{\phi_1} - \left(\frac{\Delta\phi_1}{\phi_1}\right)^4\right]}{\phi_1}$$

4.4. Let the coefficients of $(Y_1 - T_0)$ in Problem 4.3 be denoted f_0, f_1, f_2, f_3, and f_4 so that

$$f_0 = \frac{1}{\phi_1}, \quad f_0 = \frac{-\Delta\phi_1}{\phi_1^2}, \dots$$

a. Show for measurements at $x=0$ for a semi-infinite body that f_0, $f_1, \dots, f_4$ are proportional to $(\Delta t)^{-1/2}$. Calculate and plot $\phi_1 f_i$ values for $i=0, 1, 2, 3,$ and 4.
b. For a lumped body, $\phi_i = i\Delta t/\rho cL$, investigate the f_i values for this case.
c. Compare the results of parts a and b and give conclusions.

4.5. Write a computer or a programmable calculator program for Eq. (4.3.5) and verify values obtained for Example 4.1.

4.6. For a linear problem with the heat flux history for $t>0$ approximated by
$$q(t) = \beta_1 + \beta_2 t + \cdots + \beta_r t^{r-1}$$
give a whole domain estimation algorithm for estimating $\beta_1, \dots, \beta_r$. Use least squares in matrix form for a single temperature sensor with temperatures measured at n discrete times where $n>r$. Show the components of the matrices and let $\phi_i^{(j)}$ be the temperature rise for $q(t) = t^j$.

4.7. Using the least squares method, derive estimators in algebraic form for β_1 and β_2 in Eq. (4.4.18).

4.8. An alternative integral to Duhamel's theorem is obtained by using Green's functions,
$$T(x, t) = T_0 + \frac{\alpha}{k} \int_0^t q(\lambda) G(x, t-\lambda) d\lambda$$
where $G(x, t)$ is the Green's function, α = thermal diffusivity, and k = thermal conductivity. Using the numerical approximation of this equation given in Problem 3.8, derive the following function specifica-

PROBLEMS

tion algorithm for the constant q temporary assumption.

$$\hat{q}_M = \frac{\sum_{i=1}^{r}(Y_{M+i-1}-\hat{T}_{M+i-1}|_{q_M=\cdots=0})\sum_{j=1}^{i}G'_j}{\sum_{i=1}^{r}\left(\sum_{j=1}^{i}G'_j\right)^2}$$

$$G'_j = \frac{\alpha G_j \Delta t}{k}, \quad G_j = G\left(x, \frac{t_j - \Delta t}{2}\right)$$

What expression is used for $\hat{T}_{M+i-1}|_{q_M=\cdots=0}$?

4.9. Calculate the gain coefficients, K_i, for $\Delta t^+ = 0.05$, 0.2, and 0.5 for the algorithm given in Problem 4.8. The Green's function for $x=L$ in a plate heated at $x=0$ and insulated at $x=L$ is

$$G(L,t) = \frac{1}{L}\left[1 + 2\sum_{m=1}^{\infty} e^{-m^2\pi^2\alpha t/L^2}(-1)^m\right]$$

Let α and k be unity.
  Compare the values with those in Table 4.1.
  a. Calculate for $r = 1, 2,$ and 3.
  b. Calculate for $r = 4$.
  c. Calculate for $r = 5$.

4.10. Derive a function specification algorithm for estimating $g(t)$ from two interior temperature measurement histories where $g(t)$ is the time-variable volume-energy generation term in a solid cylinder. The differential equation is

$$k\frac{\partial}{\partial r}\left(r\frac{\partial T}{\partial r}\right) + g(t) = \rho c \frac{\partial T}{\partial t} \qquad (a)$$

Use the temporary assumption of $g(t)$ equal a constant for r future time steps. The solution of (a) is

$$T(r,t) = T_0 + \int_0^t g(\lambda)\frac{\partial \theta(r, t-\lambda)}{\partial t}d\lambda$$

where $\theta(r, t)$ is the temperature rise for a unit step increase in $g(t)$ at time $t=0$.

4.11. Give the matrix elements for the whole domain regularization equation, Eq. (4.5.18), when $W_0 = W_1 = W_2 = 1$.

4.12. The backward heat problem is the estimation of the initial temperature distribution in a body knowing one or more internal temperature histories and the boundary conditions. For the case of a flat plate with $T = T_0$ at $x = 0$ and L and the initial temperature distribution $T(x, 0) =$

$F(x)$, derive a zeroth-order whole domain regularization algorithm for estimating n components of $F(x)$,

$$F_i \equiv F\left(i\Delta x - \frac{\Delta x}{2}\right), \quad \Delta x = \frac{L}{n}$$

Consider the case of three interior sensors and m equally spaced time steps. The describing integral equation is

$$T(x, t) = T_0 + \int_0^L F(x')G(x, t, x')dx'$$

where $G(x, t, x')$ is a Green's function. Modify the notation of Problem 4.8 to permit multiple sensors.

4.13. Modify Eq. (4.6.14) for $r=3$ and calculate the gain coefficients for $\Delta t^+ = 0.05$ for the insulated surface of a hot plate. Vary $\alpha\sigma^2$ over a large range and plot the results. Give conclusions.

4.14. Develop a digital filter procedure for multiple sensors. Give your results in the form of Eqs. (4.7.6)–(4.7.8).

4.15. For the digital prefilter of

$$y_i = aY_{i-1} + bY_i + aY_{i+1}$$

Show that Eq. (4.7.6) can be written as

$$\hat{q}_M = \sum_{i=2}^{M+r-1} F_{M-i}(Y_i - T_0) + a[f_{-r-1}(Y_{M+r} - T_0) + f_M(Y_1 - T_0)$$

$$- f_{2-r}(Y_{M+r-1} - T_0)] + bf_{M-1}(Y_1 - T_0)$$

where

$$F_{M-i} \equiv af_{M-i-1} + bf_{M-i} + af_{M-i+1}$$

4.16. a. Show that Eq. (4.5.33) reduces to the Stolz algorithm for $\alpha \to 0$.
b. Show that Eq. (4.5.33) becomes Eqs. (4.4.25c, d) for $\alpha \to \infty$.

CHAPTER 5

INVERSE CONVOLUTION PROCEDURES FOR A SINGLE SURFACE HEAT FLUX

5.1 INTRODUCTION

In previous chapters, general procedures were given for treating the inverse heat conduction problem (IHCP) and for mathematically modeling the physical problem. The IHCP can be viewed as the estimation of the surface heat flux from transient temperature measurements inside a heat-conducting solid. It is an ill-posed problem which is characterized by extreme sensitivity of the surface heat flux to small variations in the interior temperatures. Methods of reducing this sensitivity were given in Chapter 4. Some of these methods are used in this chapter. Of the various ways of modeling the transient heat conduction in solid bodies, the one used in this chapter employs a convolution integral equation based on Duhamel's theorem. This method requires the problem to be linear; that is, the thermal properties (k, ρ, and c) are not functions of temperature but can be functions of position. Numerical methods of treating the convolution equations were discussed in Chapter 3.

Advantages of the Duhamel's theorem approach are that the body can have an arbitrary shape and the thermal properties can be functions of position (Figure 5.1). The temperature distribution can be one-, two-, or three-dimensional. The only requirement is that the influence function $\phi(\mathbf{x}, t)$ be known. [$\phi(\mathbf{x}, t)$ is the temperature rise at $\mathbf{x}$ due to a unit step increase in the surface heat flux at time zero.] In Figures 5.1a and 5.1b the temperature distributions are functions of only one spatial independent variable; in Figure 5.1a $\mathbf{x}$ becomes x and in Figure 5.1b $\mathbf{x}$ becomes r, the radial coordinate for a cylinder or for a sphere.

The interfaces between dissimilar materials can have perfect or imperfect contact characterized by h_c [see Eq. (1.2.7)].

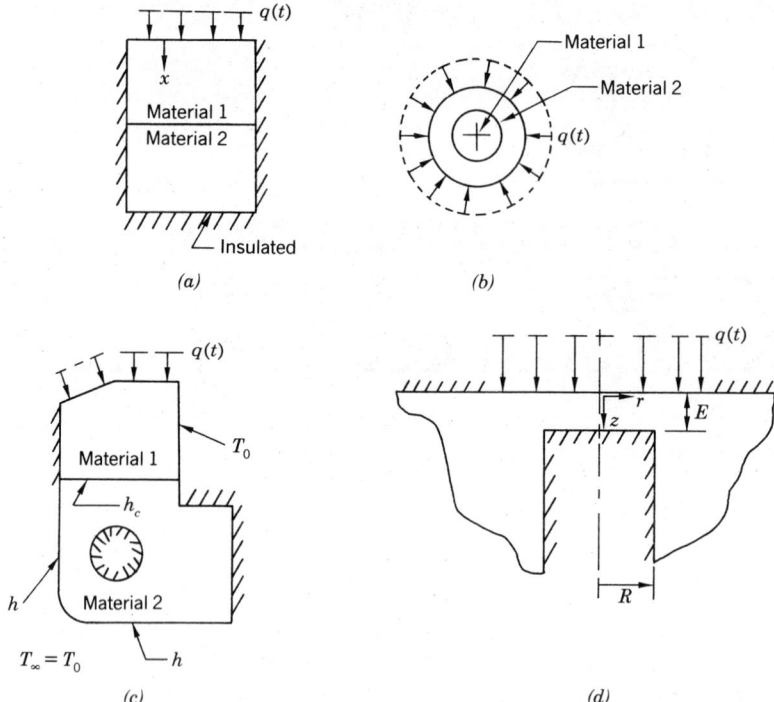

FIGURE 5.1 Various geometries, boundary conditions, and interface conditions that can be treated using Duhamel's theorem. Thermal properties are independent of temperature. (a), Composite plate; (b) composite cylinder or sphere; (c) composite body of irregular shape; T_0 is constant initial temperature; (d) axisymmetric semi-infinite body with cylindrical void and surface partially heated.

The boundary conditions that are permitted include insulation, a constant surface temperature equal to the initial temperature, T_0, or a convective boundary condition provided the ambient temperature, T_∞, is equal to the initial body temperature, T_0 (see Figure 5.1). The conditions of a single prescribed surface heat flux, which is a function of time only, and a uniform initial temperature ensure that there is a single convolution integral in the Duhamel's theorem equation. The only condition that causes the temperature to change is the heat flux, $q(t)$.

The heat flux $q(t)$ is considered to be uniform over the surfaces where it is applied and it can be applied to surfaces that are not single planes as shown by Figure 5.1c and over surfaces that are not completely covered as shown by Figure 5.1d. These two cases are examples of two-dimensional heat flow. Even though the temperature distribution is not one-dimensional, Duhamel's theorem can be used because it is valid for multidimensional linear cases.

It was mentioned previously that the influence function $\phi(\mathbf{x}, t)$ must be known. The $\phi(\mathbf{x}, t)$ solution depends on the geometry, and the interface and boundary conditions; it is the temperature rise for a heat flux of unity at the heated surface and may be solved in many ways. For example, $\phi(\mathbf{x}, t)$ can be determined by some exact method of solution such as separation of variables. It can also be found numerically using finite differences or finite elements. And the surface element method can be used.[1-3] The $\phi(\mathbf{x}, t)$ history is needed only at the sensor locations.

There are many ways to approximate numerically the convolution integral, some of which were discussed in Chapter 3. To illustrate the basic concepts one of these is utilized; in particular, the constant elements of heat flux shown in Figure 3.1. Although linear (or parabolic, cubic, etc.) elements can provide more accurate approximations for the convolution integral, the constant elements are simpler to use. The constant elements are also more analogous to the finite difference and finite element approximations; this is important because improved algorithms can be developed more readily using the convolution approach. In effect, the convolution integral provides a "small-scale laboratory" for the testing of algorithms at much less expense and complexity than the direct use of finite differences/elements.

Even with the restrictions indicated, many algorithms can be applied to the IHCP. (In Chapter 4 more were given than can be effectively explored in this chapter.) Certain function specification and regularization algorithms are used in this chapter. And the emphasis is on sequential rather than whole domain estimation.

Another very important point should be noted regarding the use of a numerical approximation of Duhamel's integral. The numerical values for the heat flux found from a given IHCP algorithm with the $q = C$ approximation in Duhamel's integral are very nearly equal to those found using finite differences or finite elements. The only restrictions are that the IHCP problem is linear and that the finite difference and element methods use sufficiently fine space grids. (The Duhamel's integral approach is exact relative to the space dependence if ϕ_i is exact.) Consequently, for a given IHCP algorithm and test case, the numerical values displayed in this chapter for the Duhamel integral approach can also be considered to be found by using finite differences or elements. Any insight into an IHCP algorithm (such as function specification or regularization) that is gained in this chapter also applies to the same algorithm when the temperatures are approximated using finite differences or elements, with the caveat that the latter must use sufficiently fine space grids.

5.2 TEST CASES

5.2.1 Introduction

In this section several test cases are discussed, some of which were previously used in Chapter 4. One test case is for a step increase in the surface heat flux. Exact values of the simulated temperature history are used. The second case is for a heat flux that varies in time in a triangular fashion; for this case both exact temperatures and temperatures with random errors are used. The third test case is for the temperatures associated with a single heat flux impulse, δq_r, at time t_r; the calculated surface heat flux is called $\delta \hat{q}_M/\delta q_r$. The fourth and last test case is for the input temperatures equal to zero except the temperature at time t_r; the heat fluxes are denoted by $\delta \hat{q}_M/\delta Y_r$. The geometry for each of these test cases is a flat plate which is heated at $x=0$ and insulated at $x=L$; see the inset of Figure 1.7. The measurements are at $x=L$, which is the greatest possible distance from the heated surface and hence poses a greater challenge than any other sensor location for an IHCP algorithm. The dimensionless time is given by $t^+ \equiv \alpha t/L^2$.

For a sensor at a position $x=E$, where $0<E<L$, an IHCP estimation procedure for q_M gives rather similar results to the case for a sensor at $x=L$ if the dimensionless time step, $\Delta t_E^+ = \alpha \Delta t/E^2$, for the $x=E$ location is equal to the dimensionless time step, $\Delta t^+ = \alpha \Delta t/L^2$, for the $x=L$ sensor location. In other words, the test cases in this chapter with the sensor at $x=L$ give insight for cases with a sensor at other locations if the dimensionless time step is based on the distance of the sensor from the heated surface.

5.2.2 Step Change in Surface Heat Flux

A basic test case is for a step change at $t=0$ in the surface heat flux, which is shown in Figure 5.2. This is sometimes called a "constant" heat flux because the heat flux is the constant value of q_c for $t>0$. The use of the words "step change" makes clear, however, that the heat flux is zero for $t<0$. For linear problems neither the sign (positive or negative) nor the magnitude of q_c is important because the estimated heat flux values are linearly proportional to q_c. For this reason and for greater generality most of the examples in this chapter are solved in terms of dimensionless variables. The dimensionless temperatures for a number of locations in the body and the values of dimensionless times are given in Table 1.1 and are plotted in Figure 1.7. The heated surface temperature immediately begins to increase while there is a damped and lagged response at $x^+ = x/L = 1$, the insulated surface. The sensor is located at $x^+ = 1$; the dimensionless time steps of $\Delta t^+ = 0.05$ and 0.5 are chosen. This case can also be related to real time which is discussed in Section 5.2.7.

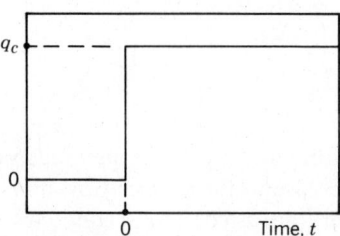

FIGURE 5.2 Heat flux test case for a step change in surface heat flux.

5.2.3 Triangular Heat Flux

The second test case is for heat flux that varies in time in a triangular fashion (Figure 5.3). Before $t^+ = 0$, the heat flux is zero. For t^+ between zero and 0.6, the surface q increases linearly with time, and for $t^+ > 0.6$ the flux decreases linearly to zero at $t^+ = 1.2$ and remains zero thereafter.

The linear portion of the heat flux for $0 < t^+ < 0.6$ is described by

$$q^+ = t^+ \tag{5.2.1}$$

where

$$q^+ = \frac{q}{q_N}, \quad t^+ = \frac{\alpha t}{L^2} \tag{5.2.2a,b}$$

and q_N is a nominal value of heat flux; namely, the value associated with t^+ equal to unity in Eq. (5.2.1). A dimensionless temperature is defined by

$$T^+ \equiv \frac{T - T_0}{(q_N L/k)} \tag{5.2.3}$$

The temperatures at $x^+ = 0$ and 1 for the linear heat flux given by Eq. (5.2.1) are

$$T^+(0, t^+) = \phi^+(0, t^+), \quad T^+(1, t) = \phi^+(1, t^+) \tag{5.2.4}$$

$$\phi^+(0, t^+) = \frac{1}{2}(t^+)^2 + \frac{1}{3}t^+ - \frac{1}{45} + \frac{2}{\pi^4} \sum_{n=1}^{\infty} \frac{1}{n^4} \exp(-\pi^2 n^2 t^+) \tag{5.2.5}$$

$$\phi^+(1, t^+) = \frac{1}{2}(t^+)^2 - \frac{1}{6}t^+ + \frac{7}{360} + \frac{2}{\pi^4} \sum_{n=1}^{\infty} \frac{(-1)^n}{n^4} \exp(-\pi^2 n^2 t^+) \tag{5.2.6}$$

These expressions can be used for t^+ between zero and 0.6. For t^+ between 0.6

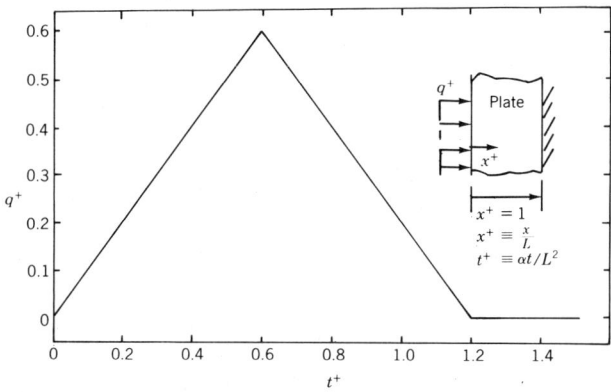

FIGURE 5.3 Triangular heat flux for test case. Finite insulated plate.

and 1.2, the temperature is given by

$$T^+(x^+, t^+) = \phi^+(x^+, t^+) - 2\phi^+(x^+, t^+ - 0.6) \tag{5.2.7}$$

and for $t^+ > 1.2$,

$$T^+(x^+, t^+) = \phi^+(x^+, t^+) - 2\phi^+(x^+, t^+ - 0.6) + \phi^+(x^+, t^+ - 1.2) \tag{5.2.8}$$

These expressions can be derived using simple superposition. The temperature histories for the heated and insulated surfaces are shown in Figure 5.4.

The surface temperature history shown in Figure 5.4 is quite different from that at the insulated surface. It responds immediately both to the onset of heating and to changes in the heating rate; the slope of $T^+(0, t^+)$ changes at both $t^+ = 0.6$ and 1.2. The $x^+ = 0$ curve also has a maximum at an intermediate time, about $t^+ = 0.9$. The insulated surface temperature is negligible until $t^+ = 0.18$ (damped and lagged) and there are no abrupt changes in slope at $t^+ = 0.6$ and 1.2. Both temperatures continue changing until some time after heating ceases. Numerical values of $T^+(1, t^+)$ for the triangular heat flux are given in Table 5.1 for $\Delta t^+ = 0.06$.

5.2.4 Random Errors

To make the previous test cases more realistic, errors can be added to the exact temperatures. Table 5.2 is a short table of normal (also called gaussian) random errors of mean zero and standard deviation of unity. The simulated temperatures are given by additive errors,

$$Y_i = T_i + \varepsilon_i \tag{5.2.9a}$$

$$\varepsilon_i = Cu_i \tag{5.2.9b}$$

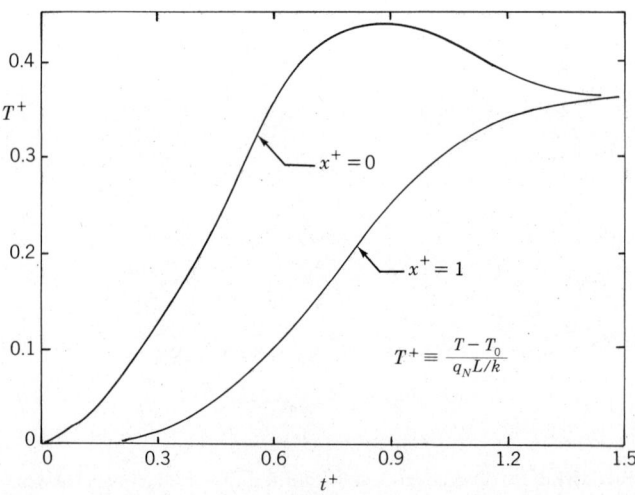

FIGURE 5.4 Heated and insulated surface temperatures for triangular heat flux test case.

SEC. 5.2 TEST CASES

Table 5.1 Temperatures at an Insulated Surface of a Finite Plate Heated with the Triangular Heat Flux shown in Figure 5.3

$\Delta t^+ = 0.06$

t^+	$T^+(1, t^+)$	t^+	$T^+(1, t^+)$	t^+	$T^+(1, t^+)$
0.06	0.000007	0.66	0.127200	1.26	0.348823
0.12	0.000374	0.72	0.157880	1.32	0.353762
0.18	0.002171	0.78	0.189293	1.38	0.356545
0.24	0.006323	0.84	0.219593	1.44	0.358089
0.3	0.013381	0.9	0.247680	1.50	0.358941
0.36	0.023656	0.96	0.272931	1.56	0.359415
0.42	0.037319	1.02	0.295006		
0.48	0.054465	1.08	0.313714		
0.54	0.075145	1.14	0.328954		
0.6	0.099389	1.2	0.340666		

TABLE 5.2 Random Normal Numbers with Zero Mean and Unity Standard Deviation[5]

01	02	03	04	05	06	07	08	09	10
0.464	0.137	2.455	−0.323	−0.068	0.296	−0.288	1.298	0.241	−0.957
0.060	−2.526	−0.531	−0.194	0.543	−1.588	0.187	−1.190	0.022	0.525
1.486	−0.354	−0.634	0.697	0.926	1.375	0.785	−0.963	−0.853	−1.865
1.022	−0.472	1.279	3.521	0.571	−1.851	0.194	1.192	−0.501	−0.273
1.394	−0.555	0.046	0.321	2.945	1.974	−0.258	0.412	0.439	−0.035

where u_i is a random number taken from Table 5.2 or a similar table or produced by a random number generator. (Most large computers have random number generators.) The constant C is chosen to make the standard deviation of ε_i equal the desired value.

In most heat transfer problems, a reasonable simulation of the error is one that has a constant variance with i. This means that the same magnitude of error occurs for low as well as high temperatures. This is usually much better than assuming multiplicative errors, for example,

$$Y_i = (T_i - T_0)(1 + \varepsilon_i) + T_0 \qquad (5.2.10)$$

where again the variance of ε_i is a constant. The assumption in Eq. (5.2.10) is that the *relative* errors are constant; for example, Eq. (5.2.10) implies an error of $0.001°C$ when $T_i - T_0 = 0.1°C$ is as likely as an error of $1°C$ when $T_i - T_0 = 100°C$. This is not reasonable for the same sensor covering the temperature range of $T_i - T_0 = 0$ to $100°C$ in a single transient experiment.

EXAMPLE 5.1. For the triangular heat flux case, give a set of simulated dimensionless temperature measurements with additive, normal, uncorrelated errors with a constant standard deviation of 0.0017. Start the values at time $t^+ = -0.24$.

Solution. The "true" temperatures are given in Table 5.1 for $t^+ \geq 0.06$. For $t^+ \leq 0$, the T values are zero. The simulated temperatures are calculated using Eq. (5.2.9) for which the u_i's values are given in Table 5.2 and with $C = 0.0017$. The equation for the first simulated temperature is

$$Y(1, -0.24) = T(1, -0.24) + 0.0017 u_1$$

$$= 0 + 0.0017(0.464) = 0.000789$$

For the one at $t^+ = 0.3$, which is the tenth time, the simulated temperature measurement is

$$Y(1, 0.3) = T(1, 0.3) + 0.0017 u_{10}$$

$$= 0.013381 + 0.0017(-0.957) = 0.011754$$

The values for the other times are found in a similar manner; see Table 5.3. The first 30 values of random numbers in Table 5.2 were used. Additional sets of simulated temperature measurements for this case can be obtained by using other sets of 30 random numbers taken from Reference 5.

It is instructive to examine the simulated measurements given in Table 5.3. Notice for the earliest times of $t^+ = -0.24$–0.18 that the Y^+'s oscillate and even change sign. Also the Y^+ values for $t^+ = 1.32$–1.5 fluctuate about 0.355. With such variations it is understandable that the IHCP algorithms produce fluctuations in the surface heat flux. (Recall that the Burggraf exact solution, Section 2.5.2, involves derivatives of the measured temperatures.) It is remarkable that the algorithms produce results with the accuracy that they do. □

TABLE 5.3 Simulated Temperature Measurements for Triangular Heat Flux Case (Example 5.1)

t^+	Y^+	t^+	Y^+	t^+	Y^+
−0.24	0.000789	0.36	0.023758	0.96	0.275457
−0.18	0.000233	0.42	0.033025	1.02	0.294404
−0.12	0.004174	0.48	0.053562	1.08	0.312636
−0.06	−0.000549	0.54	0.074815	1.14	0.330139
0	−0.000116	0.6	0.100312	1.20	0.342240
0.06	0.000510	0.66	0.124500	1.26	0.351161
0.12	−0.000116	0.72	0.158198	1.32	0.355097
0.18	0.004378	0.78	0.187270	1.38	0.354908
0.24	0.006733	0.84	0.219630	1.44	0.356639
0.3	0.011754	0.9	0.248573	1.50	0.355771

5.2.5 Heat Flux Impulse Test Case ($\delta\hat{q}_M/\delta q_r$)

One of the most stringent tests of an IHCP algorithm is for a heat impulse that is constant over just one time step and zero at other times. Due to the principle of superposition this case can be considered to be a fundamental case. Any time-variation of the surface heat flux can be approximated by a superposition of such basic elements. If the surface heat flux is formed by steps as illustrated by Figure 4.7, the superposition is exact for linear problems.

This case is important not only for test case purposes but also for investigations of the deterministic error which was discussed in Section 4.8.5.

A heat flux impulse for δq_r is introduced,

$$q_i = \begin{cases} \delta q_r & \text{for } i=r \\ 0 & \text{for } i \neq r \end{cases} \tag{5.2.11}$$

and is illustrated by Figure 5.5a. The associated temperatures are

$$Y_i = 0 \quad \text{for } i \leqslant r-1$$

$$Y_r = (\phi_1)\delta q_r = \Delta\phi_0 \delta q_r$$

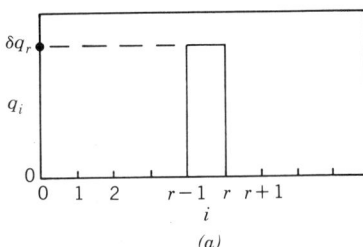

(a)

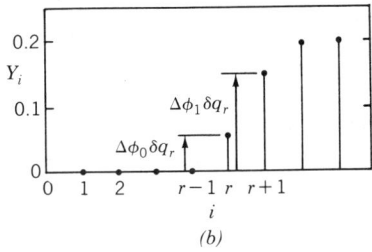

(b)

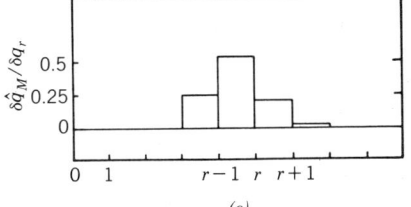

(c)

FIGURE 5.5 Impulse heat flux test case.

174   **CHAP. 5   INVERSE CONVOLUTION FOR A SINGLE SURFACE HEAT FLUX**

$$Y_{r+1} = (\phi_2 - \phi_1)\delta q_r = (\Delta\phi_1)\delta q_r$$
$$\vdots$$
$$Y_{r+j} = (\phi_{j+1} - \phi_j)\delta q_r = \Delta\phi_j \delta q_r \qquad (5.2.12)$$

which are related to the sensitivity coefficients in the **X** matrix; see Eq. (3.2.18a). Some Y_i values for the insulated flat plate case (for $x = L$ and $\Delta t^+ = 0.2$) are shown in Figure 5.5b. The Y_i components go to a constant for "large" i values because $\Delta\phi_j$ approaches a constant; this behavior is also depicted in Figures 1.10 and 1.14.

Using the temperatures given by Eq. (5.2.12), the heat flux values are calculated from an algorithm and are denoted $\delta\hat{q}_M/\delta q_r$, $M = 1, 2, \ldots$. Due to linearity, $\delta\hat{q}_M/\delta q_r$ is independent of the magnitude of δq_r. Results are shown in Figure 5.5c for the function specification algorithm, Eq. (4.4.25), with $r = 2$.

This case can be used to assess the deterministic error in the heat flux. The conservation of energy can also be investigated. These points are discussed in later sections.

5.2.6 Temperature Impulse Test Case ($\delta\hat{q}_M/\delta Y_r$)

Another basic or fundamental case, called the temperature impulse test case, postulates the input temperatures of

$$Y_i = \begin{cases} \delta Y_r & \text{for } i = r \\ 0 & \text{for } i \neq r \end{cases} \qquad (5.2.13)$$

This case was also discussed in Section 4.7.2 in connection with sequential filter algorithms. In addition to the application for digital filters, it can be utilized to investigate the stability of inverse heat conduction algorithms.

The Y_i values are displayed in Figure 5.6a. Some $\delta\hat{q}_M/\delta Y_r$ values for $\Delta t^+ = 0.2$ and the $r = 2$ function specification algorithm are shown in Figure 5.6b. The units of $\delta\hat{q}_M/\delta Y_r$ are W/m^2-C; for the scale shown in Figure 5.6b, the values of k and L are unity or, alternatively, the plot is for $\delta\hat{q}_M L/k\delta Y_r$.

5.2.7 Test Cases with Units

The test cases in this section were previously given in dimensionless forms. This was done for greater generality since each case represented a great many possible plate materials, plate thicknesses, and surface heat fluxes. Nevertheless, it is helpful to be able to interpret the results in relation to physical cases. A way to relate the dimensionless results to those with units is presented.

To cover a wide range of heat conducting characteristics, three materials with dissimilar properties are selected; namely, brick, steel, and copper. See Table 5.4 in which the thermal diffusivities and thermal conductivities are listed.

SEC. 5.2 TEST CASES

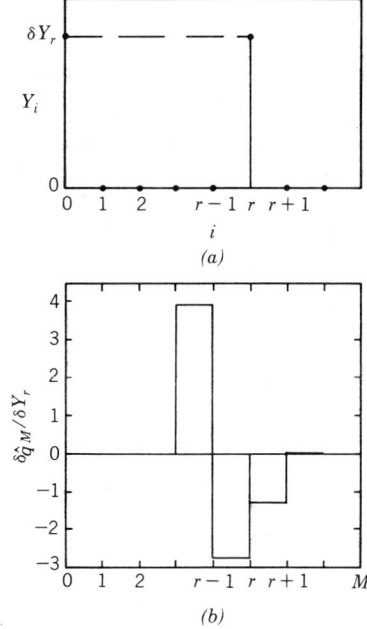

FIGURE 5.6 Impulse temperature test case, $\Delta t^+ = 0.2, r = 2$.

TABLE 5.4 Table of Materials, L, and q_o for Putting the Dimensionless Test Case Results into Dimensional Terms

Material	$\alpha \times 10^5$ (m²/s)	k (W/m-C)	L (cm) for $\Delta t = \Delta t^+$	q_o (W/m²) for $T(°C) = T^+$
Brick	0.04	1	0.0632	1,580
Steel	1.1	40	0.332	12,100
Copper	11	380	1.05	36,200

One way to relate the dimensionless times,

$$\Delta t^+ = \frac{\alpha \Delta t}{L^2} \quad \text{and} \quad t^+ = \frac{\alpha t}{L^2} \qquad (5.2.14a, b)$$

to time steps and times with units of seconds is to establish the following equations

$$\Delta t^+ = \Delta t, s \quad \text{and} \quad t^+ = t, s$$

These relations are valid if L is chosen such that

$$L = \alpha^{1/2} \qquad (5.2.15)$$

The values of L in cm are given in Table 5.4 for the three materials listed. Consequently, for all the test cases for the finite plate, the dimensionless time steps

176 CHAP. 5 INVERSE CONVOLUTION FOR A SINGLE SURFACE HEAT FLUX

(and times) can be interpreted as time steps (and times) in seconds for the cases listed in Table 5.4.

For the step increase in the surface heat flux test case, the temperature rise in Celsius degrees is equal to the dimensionless temperature for the L, k, and q_0 values displayed in Table 5.4. If the temperature rise for this test case (see Table 1.1) seems too small, then the T^+ values, defined by (5.2.3), can be multiplied by a factor, such as 100, if the q_0 values are also multiplied by this factor. For example, the dimensionless calculated heat flux values shown in Figure 5.7 are those for a copper plate 1.05 cm thick with time steps of 0.05 s. The $\hat{q}^+$ values are multiplied by 36,200 W/m² to find the calculated $\hat{q}$ values from the Y_i values in Table 1.1. If the values in Table 1.1 are multiplied by 100 to make the simulated temperature measurements more realistic, then the $\hat{q}^+$ values of Figure 5.7 are multiplied by 3,620,000 W/m² to obtain the $\hat{q}$ values in W/m².

In the case of the triangular heat flux example, the nominal heat flux, q_N, in Eq. (5.2.3) is treated in the same manner as q_0 for the step increase in the surface heat flux test case.

5.3 FUNCTION SPECIFICATION ALGORITHMS

5.3.1 Introduction

In this section some of the sequential function specification algorithms developed in Section 4.4 are used. A numerical form of Duhamel's theorem is employed for the heat conduction model. One of the purposes of this section is to provide greater insight into these IHCP algorithms. This is done by investigating the solutions for the test cases provided in Section 5.2.

The plan of this section is to first investigate the single future temperature algorithm. Then the case of multiple future temperatures is examined.

5.3.2 Single Future Temperature Algorithm (Stolz Method)

The algorithm for a single future temperature is given by Eq. (4.3.5). This equation implies exact matching of the calculated and measured temperatures and is called the Stolz algorithm.[4] For convenience, Eq. (4.3.5) is written in the form

$$\hat{q}_M = \left(Y_M - \sum_{j=1}^{M-1} \hat{q}_j \Delta\phi_{M-j} - T_0 \right) K_1 \tag{5.3.1}$$

where the gain coefficient K_1 is given by

$$K_1 = \frac{1}{\phi_1} \tag{5.3.2}$$

Values of gain coefficients for $\Delta t^+ = 0.05$, 0.2, and 0.5 are given in Table 4.1. A numerical example of the use of Eq. (5.3.1) is provided by Example 4.1.

SEC. 5.3 FUNCTION SPECIFICATION ALGORITHMS

5.3.2.1 Step Heat Flux Test Case. Results for the step heat flux test case are shown by the $r=1$ curves of Figures 5.7 and 5.8. The dimensionless calculated heat flux is denoted $\hat{q}_M^+$ and is defined by $\hat{q}_M^+ \equiv \hat{q}_M/q_c$ where q_c is the constant heat flux actually applied. The dimensionless time step of $\Delta t^+ = 0.05$ is used in Figure 5.7 and the results are excellent until about the eleventh time when rapidly growing oscillations are observable. This case is unstable in the bounded-input, bounded-output sense because the $\hat{q}_M^+$ components continue to increase without bound as M increases. The calculations were performed on a CDC computer using about 15 significant figures. The effect of round-off errors in the fifteenth significant figure is present even at the smallest M values but is not discernable in Figure 5.7 until about $M = 10$. Instability is investigated further in Section 5.3.2.4.

Results for the Stolz method for $\Delta t' = 0.5$ are shown by the $r=1$ curve in Figure 5.8. These results are stable since no oscillations are present. This is the result of the Δt^+ being sufficiently large.

5.3.2.2 Triangular Heat Flux Test Case. Results of calculations for the triangular heat flux test case are displayed in Figure 5.9. There are negligible measurement errors since the Y_i's are taken from Table 5.1. Time steps of $\Delta t^+ = 0.05, 0.3, 0.5$, and 1 are used. The smallest time step, $\Delta t^+ = 0.05$, produces unstable results and hence conveys little useful information regarding the surface heat flux. The other time steps permit stable calculations with the smallest of

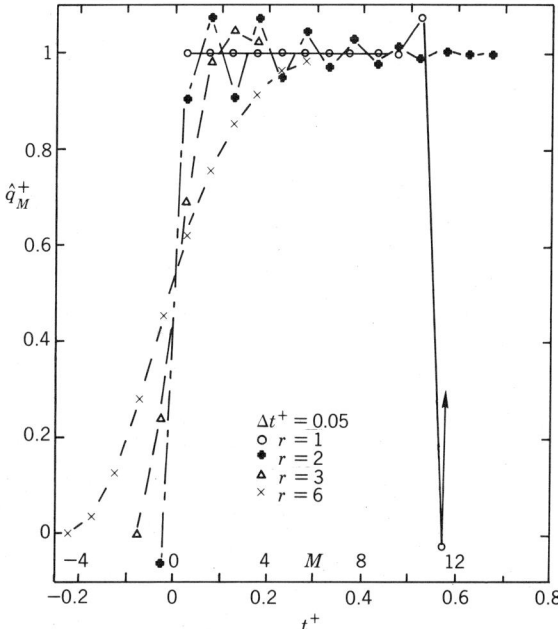

FIGURE 5.7 Calculated surface heat fluxes for constant q input for a plate. Function specification method; $\Delta t^+ = 0.05$. Exact temperature data.

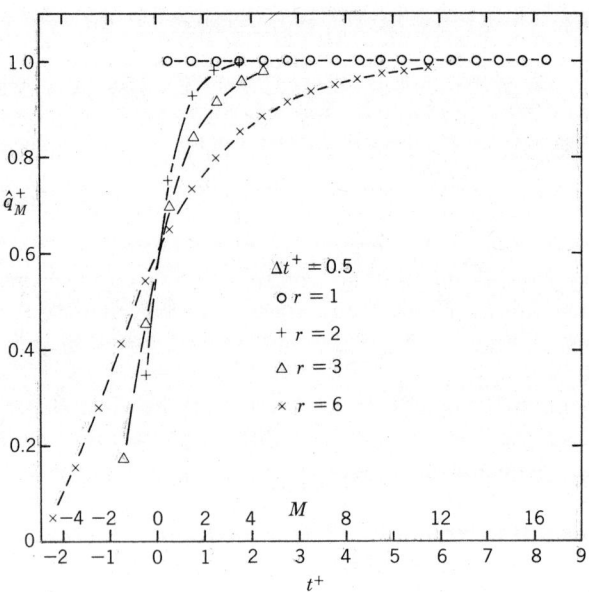

FIGURE 5.8 Calculated surface heat fluxes for constant q input for a plate. Function specification method; $\Delta t^+ = 0.5$. Exact temperature data.

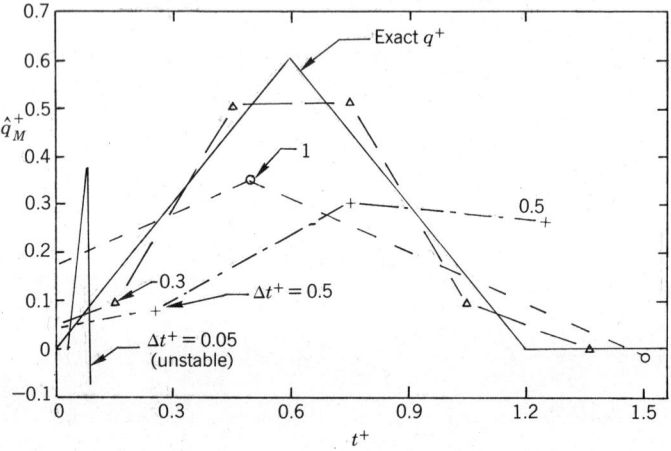

FIGURE 5.9 Calculated surface heat fluxes for a triangular heat flux test case. Exact data used. ($r=1$, Stolz method.)

these, $\Delta t^+ = 0.3$, yielding the most information regarding the applied heat flux. A weakness of the Stolz method is that sometimes the time steps must be so large to insure stability that rapid heat flux variations cannot be followed.

5.3.2.3 Heat Flux Impulse Test Case ($\delta \hat{q}_M / \delta q_r$).
For the heat flux impulse of δq_r the Stolz method ($r=1$) exactly reproduces the impulse, provided

SEC. 5.3 FUNCTION SPECIFICATION ALGORITHMS

the method is stable. Figure 5.10, which is for $\Delta t^+ = 0.05$, shows the "exact" values of $\delta \hat{q}_M / \delta q_r$ for $r = 1$. [The $\delta \hat{q}_M$ values for time t_M are plotted at the time $t_{M-1/2} = (M - \frac{1}{2})\Delta t$.] The results in Figure 5.10 for the Stolz method are somewhat misleading since instability becomes evident for large M values. For dimensionless time steps larger than about 0.3, the results are stable. This is discussed further in connection with the next test case.

5.3.2.4 Temperature Impulse Test Case ($\delta \hat{q}_M / \delta Y_r$).

The temperature impulse test case introduces $Y_i - T_0 = 0$ for all i values except $i = r$ where $Y_r - T_0 = \delta Y_r$. One of the purposes of this test case is to investigate the stability of the Stolz algorithm; that is, the $r = 1$ case. Figure 5.11 shows some $\delta \hat{q}_M / \delta Y_r$ values that are normalized with respect to the first value. For these $r = 1$ cases, the $\delta \hat{q}_1 L / k \delta Y_r$ values are equal to the gain coefficients, K_1, which are 9.94, 6.95, and 2.99 for $\Delta t^+ = 0.25$, 0.3, and 0.5, respectively. The $\Delta t^+ = 0.25$ result is unstable since it continues to oscillate with increasing amplitudes with M; for $M \geq 4$, the amplitude gain factor,

$$\text{amplitude gain factor} = \frac{|\delta \hat{q}_{M+1}|}{|\delta \hat{q}_M|} \qquad (5.3.3)$$

is 1.221 which is greater than one. The amplitude gain factors for $\Delta t^+ = 0.3$ and 0.5 are 0.951 and 0.481, respectively; since these values are less than one,

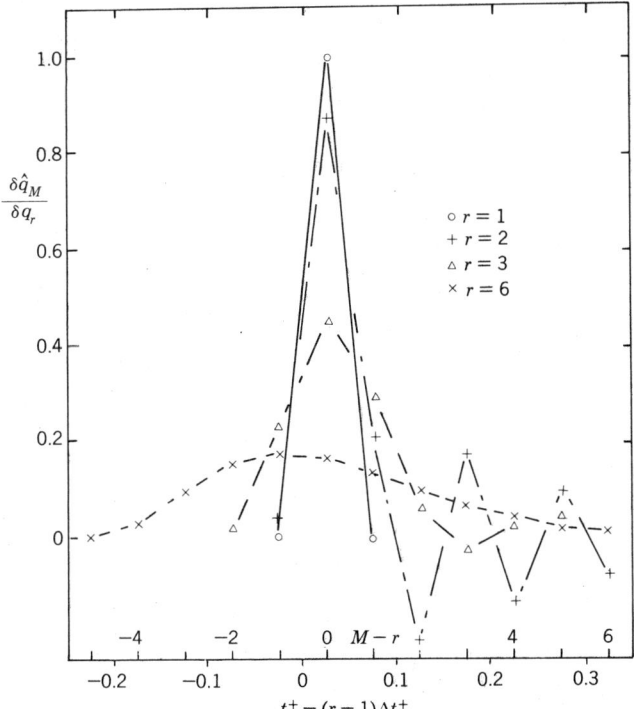

FIGURE 5.10 Calculated surface heat fluxes for impulse q input for a plate; $\Delta t^+ = 0.05$.

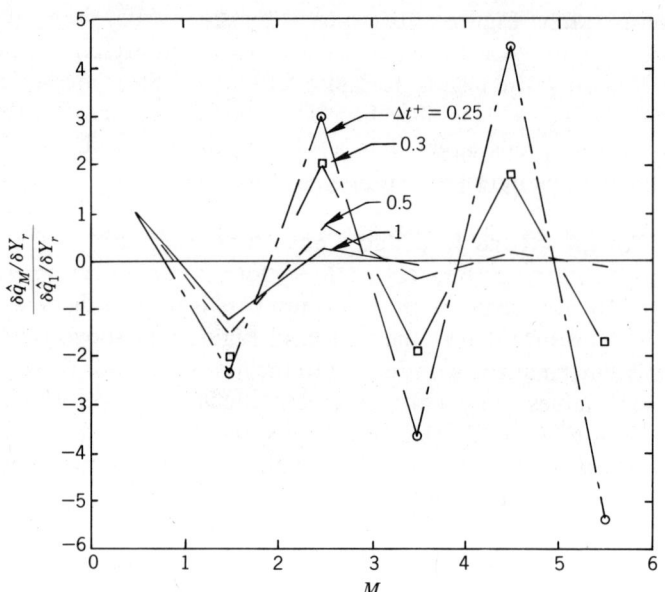

FIGURE 5.11 Curves useful for a stability study of the Stolz method. Finite plate; sensor at insulated surface.

the Stolz algorithm is stable for $\Delta t^+ > 0.3$ (for the flat plate geometry heated at $x=0$ and insulated at $x=L$ and the sensor at $x=L$). (See Problems 5.7 and 5.8 which investigate stability of the Stolz method.)

Even though the Stolz algorithm is stable down to $\Delta t^+ = 0.3$, somewhat larger Δt^+ values are recommended for several reasons. First, the effect of an error at a particular time persists for many subsequent time steps as indicated by Figure 5.11. Second, the disturbance in the surface heat flux for a single measurement error of δY_r has a much larger magnitude for $\Delta t^+ = 0.3$ than for a larger time step such as $\Delta t^+ = 0.5$. For example, at the second time step ($M=2$) after the disturbance, $\delta \hat{q}_M L / k \delta Y_r$ is -4.4 for $\Delta t^+ = 0.5$ but is -14.4 for $\Delta t^+ = 0.3$, a factor of about 3 larger in amplitude.

To gain more insight into the effects of errors in temperature measurement, consider the case of a steel plate 1 cm thick. The thermal conductivity is 40 W/m-C and the thermal diffusivity is 1.1×10^{-5} m^2/s. The Δt^+ value of 0.3 corresponds to a time step of

$$\Delta t = \frac{\Delta t^+ L^2}{\alpha} = \frac{0.3(0.01)^2}{1.1 \times 10^{-5}} = 2.7 \text{ s}$$

For a temperature measurement error of 1°C, the largest disturbance in the calculated surface heat flux is

$$\delta \hat{q}_M = \frac{\delta \hat{q}_M L}{k \delta Y_r} \frac{k \delta Y_r}{L}$$

$$= (-14.4)(40 \times 1/0.01) = -57,600 \text{ W/m}^2$$

SEC. 5.3 FUNCTION SPECIFICATION ALGORITHMS

To put this value in perspective, a steady heat flux of this magnitude applied to a plate with fixed side temperatures causes a temperature difference across the plate of 14°C. Hence, there is a substantial amplification of the 1°C error. Other examples can be constructed using Table 5.4.

5.3.3 Multiple Future Temperatures Algorithm

The use of future temperatures yields much more powerful algorithms than exact matching. The sensitivity to measurement errors is reduced thus permitting smaller calculational time steps.

The sequential function specification algorithm can be obtained from Eq. (4.4.24) which is based on the temporary assumption of a constant heat flux. Using the Duhamel's integral numerical approximation, Eq. (3.2.29) yields

$$\hat{q}_M = \sum_{i=1}^{r} K_i \left(Y_{M+i-1} - \sum_{j=1}^{M-1} \hat{q}_j \Delta\phi_{M-j+i-1} - T_0 \right) \quad (5.3.4a)$$

where the gain coefficient K_i is given by

$$K_i = \frac{\phi_i}{\sum_{j=1}^{r} \phi_j^2} \quad (5.3.4b)$$

and r is the number of future temperatures. If $r=1$, Eq. (5.3.4) reduces to the Stolz algorithm, Eq. (5.3.1). Some K_i values for various time steps and $r=1, 2, \ldots, 5$ are given in Table 4.1. The K_i values tend to decrease with increasing values of r and also increasing values of Δt^+. Two numerical solutions using Eq. (5.3.4) are given in Examples 4.2 and 4.3.

5.3.3.1 Step Heat Flux Test Case.
Figures 5.7 and 5.8 display results for the step heat flux test case with $\Delta t^+ = 0.05$ and 0.5, respectively. Each case of r greater than one *anticipates* the step increase in q with the largest effect associated with the largest r. Moreover, the larger r results tend to predict low q values just *after* the step change in heat flux. These effects are not observed if the algorithms are started at $t=0$, the time of the q step change; such calculations are not rigorous tests of the IHCP algorithms, however, because the instant of heating initiation is usually not known.

For the dimensionless time step of $\Delta t^+ = 0.05$ and $r = 2, 3,$ and 6, the results are stable, unlike the $r=1$ case (Stolz method). The $r=2$ case oscillates slightly about the correct values, the $r=3$ case has a small overshoot, and the $r=6$ case approaches the step change from below. Each case of $r>1$ shown in Figure 5.7 creditably approximates the step change in q.

For the $\Delta t^+ = 0.5$ calculations shown in Figure 5.8, the $r=1$ results are the best. The $r=2$ case is also quite accurate and is much less sensitive to measurement errors than the $r=1$ algorithm.

5.3.3.2 Triangular Heat Flux Test Case.
Calculated heat flux values for the triangular flux test case are displayed in Figures 5.12 and 13. Results for $r=4$ and for errorless temperatures are given in Figure 5.12. The agreement with the exact input heat flux is very good. The only periods with discernible errors are those for abruptly changing heat fluxes.

The $r=4$ case is shown in Figure 5.12 rather than the $r=2$ and $r=3$ values because the agreement with the exact q's is even better for $r=2$ and 3 so the plot is not instructive. To show the differences, Table 5.5 is given. Only the time region near the peak of the triangular heat flux is tabulated since the greatest errors occur there. The maximum errors in Table 5.5 are 2, 2.5, and 6.5% for

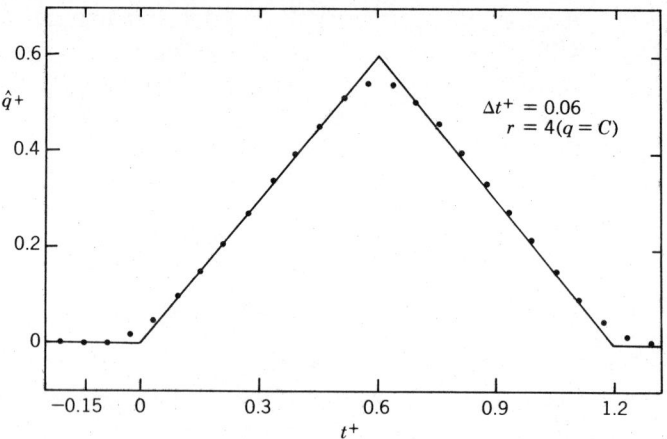

FIGURE 5.12 Calculated surface heat flux for triangular heat flux case with errorless data. Function specification method.

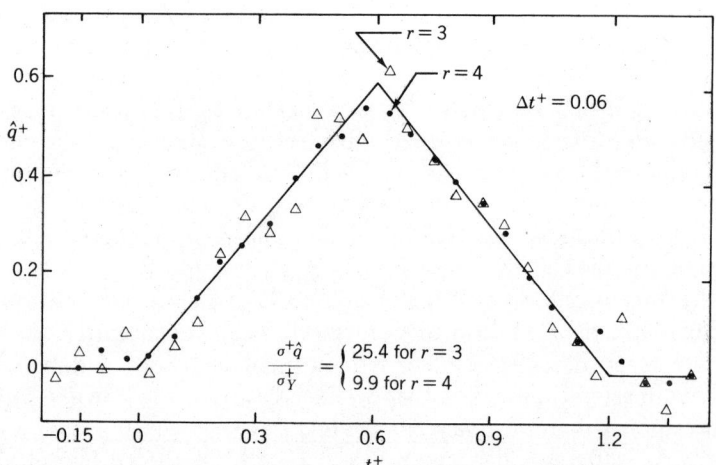

FIGURE 5.13 Calculated surface heat flux for triangular heat flux case with added, normal uncorrelated random errors with $\sigma_Y^+ = 0.0017$. Function specification method.

SEC. 5.3 FUNCTION SPECIFICATION ALGORITHMS

TABLE 5.5 Calculated Heat Fluxes at Certain Times for the Triangular Heat Flux Example Using the Function Specification Method; Errorless Data

		$\hat{q}^+$		
t^+	Exact	$r=2$	$r=3$	$r=4$
0.51	0.51	0.5100	0.5097	0.5038
0.57	0.57	0.5691	0.5559	0.5347
0.63	0.57	0.5818	0.5577	0.5327
0.69	0.51	0.5072	0.5164	0.5018

$r=2$, 3, and 4, respectively. These errors are a result of the bias introduced by the IHCP algorithm.

The heat flux estimates shown in Figure 5.12 clearly represent a triangular heat flux. This is in marked contrast to any of the time step results for $r=1$ that are shown in Figure 5.9. Hence the function specification method with future temperatures is much better for this test case than the exact matching of temperatures.

The effect of a set of random errors is illustrated by Figure 5.13. The errors are additive, zero mean, constant dimensionless standard deviation of 0.0017, uncorrelated, and normal (see Table 5.3). The function specification IHCP algorithm with three and four future temperature measurements is used. Even though the $r=4$ case has the greater *bias* error near the maximum heat flux for errorless data (as indicated by Table 5.5), the $\hat{q}$ values for $r=4$ in Figure 5.13 are much better than those for $r=3$ for this case of random temperature errors.

The dimensionless standard deviation of the estimated heat flux is

$$\sigma_{\hat{q}}^+ \equiv \frac{\sigma_{\hat{q}}}{q_N} \qquad (5.3.5)$$

where $\sigma_{\hat{q}}$ is approximated by

$$\sigma_{\hat{q}} \approx \left[\frac{1}{n} \sum_{i=1}^{n} (\hat{q}_i - \hat{q}_i|_{\sigma_Y=0})^2 \right]^{1/2} \qquad (5.3.6)$$

and q_N is the nominal heat flux for the triangular q test case [see Eq. (5.2.2)], n is the number of measurements, which is about 30, and $\hat{q}_i$ is an estimated heat flux shown in Figure 5.13. (A more accurate way of finding $\sigma_{\hat{q}}$ uses $\delta\hat{q}_M/\delta Y_r$. See Section 4.8.4.)

The σ_Y^+ value is dimensionless and is defined to be

$$\sigma_Y^+ \equiv \frac{\sigma_Y k}{q_N L} \qquad (5.3.7)$$

where σ_Y is the standard deviation of the random temperature error (in °C).

The ratio of

$$\frac{\sigma_{\hat{q}}^+}{\sigma_Y^+}$$

is 25.4 and 9.9 for $r=3$ and 4, respectively. Hence, the $r=3$ case for this example is about three times as sensitive to random errors as the $r=4$ case. This confirms the visual inspection of the variability of $\hat{q}^+$ shown in Figure 5.13. For the value of $\sigma_Y^+ = 0.0017$ the effect of the random errors is considerably larger than that due to bias. Consequently, for this example the $r=4$ algorithm is clearly superior to that for $r=3$. If $r=2$ is used, the effect of the random errors is much larger. The $r=1$ case is unstable.

Due to the linearity of the inverse heat conduction problem, the ratio $\sigma_{\hat{q}}^+/\sigma_Y^+$ for this triangular heat flux case is a constant. Even though the ratio is given for $\sigma_Y^+ = 0.0017$, the ratio is valid for any σ_Y^+ value. The doubling of σ_Y^+ would result in a doubling of the random errors in Figure 5.13.

5.3.3.3 Heat Flux Impulse Test Case ($\delta \hat{q}_M / \delta q_r$).

Results of calculated heat fluxes for the heat flux impulse test case are shown in Figure 5.10 for $\Delta t^+ = 0.05$. The actual input is shown by the $r=1$ values. The effect of increasing r is to spread the estimate of the impulse over more time steps and to decrease its amplitude. For $r=2$ the shape of the impulse is preserved but is followed by some oscillations. For $r=6$, the impulse has been reduced to a relatively flat gaussian curve; notice, however, that the curve is not quite symmetrical and is shifted about one time step early.

An important property of an IHCP algorithm is the conservation of energy. Even though the shape of the impulse is not preserved, the total energy under each curve for $r=2, 3, \ldots$ must equal the input energy. If the case examined allows heat to escape from the body, then the conservation of energy is not satisfied. See Problem 5.16.

For a larger time step such as $\Delta t^+ = 0.5$, the results are similar to Figure 5.10 and the curves are even lower and more spread out.

5.3.3.4 Temperature Impulse Test Case ($\delta \hat{q}_M / \delta Y_r$).

In any realistic experiment there are measurement errors. Some understanding of a set of errors can be obtained by examining the effect of a single error, δY_r. The calculated heat flux for this single error is denoted $\delta \hat{q}_M / \delta Y_r$ with units of W/m²-°C. A dimensionless form of this group is

$$\frac{\delta \hat{q}_M^+}{\delta Y_r^+} \equiv \frac{\delta \hat{q}_M L}{k \delta Y_r} \qquad (5.3.8)$$

Values of this group are given in Table 5.6 for $\Delta t^+ = 0.05$ for $r=1, 2, 3$, and 6 and in Figure 5.14 for $\Delta t^+ = 0.5$ for $r=2, 3$, and 6. The sensitivity to temperature measurement errors is shown to decrease rapidly with both increasing r and Δt^+ values. For $\Delta t^+ = 0.05$, the $r=2$ results given in Table 5.6 oscillate about

TABLE 5.6 Calculated Heat Fluxes, $\delta \hat{q}_M^+ / \delta Y_1^+$, for $Y_1 = 1$ and $Y_i = 0$ otherwise[a]. Plate Geometry and $\Delta t^+ = 0.05$

M	r = 1	r = 2	r = 3	r = 6
−4				4.05
−3				1.18
−2				−0.52
−1			31.8	−1.32
0		126.7	−29.9	−1.44
1	3712	−343.6	−12.1	−1.05
2	−1.05E5	415.8	5.4	−0.59
3	unstable	−354.9	4.7	−0.29
4		273.4	0.97	−0.11
5		−204.7	−0.51	−0.027
6		151.9	−0.41	0.012
10		45.4	0.006	0.016
15		−10.0	0.0003	0.0008
20		2.2	0.0000	−0.0001

[a] The same results are obtained for $Y_r = 1$ and $Y_i = 0$, if $i \neq r$ and if M is replaced by $M - r + 1$.

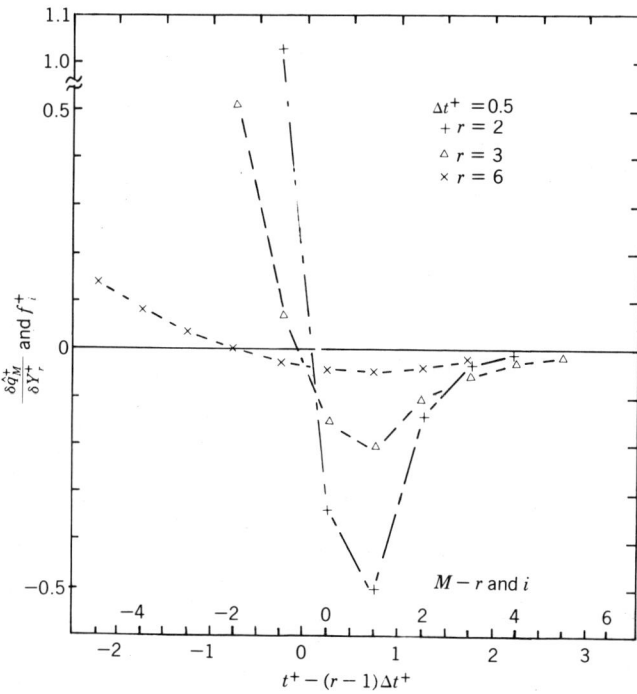

FIGURE 5.14 Calculated heat fluxes for $Y_i = \delta Y_r$ for $i = r$ and $Y_i = 0$ otherwise. Function specification method. Plate geometry and $\Delta t^+ = 0.5$.

zero almost one time step to the next, whereas the $r=6$ results oscillate with a larger period and are more damped.

Provided the algorithm is stable, the sum of the $\delta\hat{q}_M/\delta Y_r$ values is zero. This can be verified, for example, by adding the values of the $r=3$ results in Table 5.6. In a sense this also verifies conservation of energy because the initial and final internal energies for this test case are zero.

EXAMPLE 5.2. A copper calorimeter 1.05 cm thick is heated at $x=0$ by a constant heat flux of 3.62×10^6 W/m^2 and is insulated at $x=L$. From temperature measurements at $x=L$ and with $\Delta t = 0.05$ s time steps, calculate the maximum error in the surface heat flux for a temperature measurement error of 1% of the maximum temperature rise in the calorimeter at $t = 0.5$ s.

Solution. From Table 5.4 it is observed that the dimensionless time steps are numerically equal to the time steps in seconds. Also the dimensionless temperatures are 1/100th of the dimensional temperature rises. Hence Table 1.1 can be used to find the heated surface temperature rise; the maximum dimensionless temperature at $t^+ = 0.5$ is 0.833. Then the maximum surface temperature rise is about 83.3°C at 0.5 s so that the error is 0.83°C. At $t^+ = 0.5$ the insulated surface temperature rise is about 33.3°C.

The error in the surface heat flux can be found using Table 5.6 because the Δt^+ value is 0.05. For two future temperatures ($r=2$) the maximum dimensionless error in the surface heat flux is

$$\frac{\delta\hat{q}_M L}{k\delta Y_r} = 415.8$$

and thus the maximum error in the surface heat flux is

$$\delta\hat{q}_M = 415.8 \frac{k\delta Y_r}{L} = 415.8 \frac{(380)(0.833)}{0.0105}$$

$$= 12.5 \times 10^6 \text{ W/m}^2$$

where the k value of 380 W/m^2-C is taken from Table 5.4. Notice that this error in the surface heat flux is very large, since it is an error of about 346% of the applied heat flux. If the IHCP calculation were performed with six future times steps, then the maximum error would be only 3.4%.

This example demonstrates again the necessity for a compromise between bias and sensitivity to measurement errors. Minimum bias is associated with small r's, but minimum variance is associated with large r's. In most IHCP problems that are solved using the function specification algorithm, an r value of about 3 or 4 is recommended. The optimal r values depend on both the dimensionless time step (based on the distance of the temperature sensor from the nearest heated surface) and the variances of the measurement errors. For further discussion, see Section 5.6. □

5.4 REGULARIZATION ALGORITHMS

5.4.1 Introduction

In this section the whole domain and sequential regularization algorithms are discussed. The basic algorithms were developed in Section 4.5 but are applied

SEC. 5.4 REGULARIZATION ALGORITHMS

here using the Duhamel's theorem to provide numerical values for the temperature. Some of the test cases of Section 5.2 are again used to obtain insight into the algorithms.

The plan of this section is to discuss the whole domain method in Section 5.4.2 and the sequential regularization method in Section 5.4.3. To reduce the scope of this section only the first-order regularization method is investigated; the effect of the first-order regularization term is to reduce fluctuations and eliminate instability in the heat flux estimates.

5.4.2 Whole Domain Regularization Method

In the literature the regularization method is usually presented as an approach that uses the complete time domain of the measurements. All the components of the unknown functions are found simultaneously. In this book this method is called whole domain regularization. An advantage of the method is that stability can be assured. Another advantage is that basically the same procedure can be used whether the origin of the problem is an elliptic, parabolic, or hyperbolic partial differential equation. However, methods necessary for elliptic or hyperbolic equations may not be required for the heat conduction equation (which is parabolic). The main disadvantage is the expense of solving a large set of linear simultaneous algebraic equations. The difficulty is compounded for the nonlinear IHCP (for which finite difference or element methods are required) because a large set of nonlinear algebraic equations must be iteratively solved.

The whole domain first-order regularization equations for the heat flux can be obtained from Eq. (4.5.18). For simplicity let the weighting factor W_1 be equal to unity. Then Eq. (4.5.18) gives the matrix normal equation,

$$[\mathbf{X}^T\mathbf{X} + \alpha \mathbf{H}_1^T\mathbf{H}_1]\hat{\mathbf{q}} = \mathbf{X}^T(\mathbf{Y} - T_0\mathbf{1}) \qquad (5.4.1)$$

This is a set of n equations for the components of $\hat{\mathbf{q}}$ displayed by

$$\hat{\mathbf{q}}^T = [\hat{q}_1 \quad \hat{q}_2 \quad \cdots \quad \hat{q}_n] \qquad (5.4.2)$$

All n components of $\hat{\mathbf{q}}$ are obtained simultaneously. In this method a value of the regularization parameter, α, must be selected.

Equation (5.4.1) applies for one or more temperature sensors. For simplicity, a single sensor is considered. For this case $\mathbf{Y}$ also has n components,

$$\mathbf{Y}^T = [Y_1 \quad Y_2 \quad \cdots \quad Y_n] \qquad (5.4.3)$$

The $\mathbf{X}$ matrix is given by Eq. (3.2.18a) so that $\mathbf{X}^T\mathbf{X}$ is

$$\mathbf{X}^T\mathbf{X} = \begin{bmatrix} \sum_{i=0}^{n-1}\Delta\phi_i^2 & \sum_{i=1}^{n-1}\Delta\phi_i\Delta\phi_{i-1} & \sum_{i=2}^{n-1}\Delta\phi_i\Delta\phi_{i-2} & \cdots & \Delta\phi_{n-1}\Delta\phi_0 \\ & \sum_{i=0}^{n-2}\Delta\phi_i^2 & \sum_{i=1}^{n-2}\Delta\phi_i\Delta\phi_{i-1} & \cdots & \Delta\phi_{n-2}\Delta\phi_0 \\ & & & \vdots & \vdots \\ \text{symmetric} & & & & \Delta\phi_0^2 \end{bmatrix} \qquad (5.4.4)$$

The $\mathbf{H}_1$ matrix, which involves first difference coefficients, is given by Eq. (4.5.16b) and the $\mathbf{H}_1^T\mathbf{H}_1$ matrix is given by Eq. (4.5.23), which surprisingly contains second difference coefficients.

To gain further insight into the effect of the $\alpha\mathbf{H}_1^T\mathbf{H}_1$ matrix added to $\mathbf{X}^T\mathbf{X}$, some of the components of $\mathbf{X}^T\mathbf{X}$ are displayed for $\Delta t^+ = 0.06$. The values of $\Delta\phi_0 = 0.000786$, $\Delta\phi_1 = 0.0141, \ldots, \Delta\phi_6 = 0.0574$ are taken from Table 1.1. (Notice the typical small-time-step characteristic of $\Delta\phi_0$ being much smaller than $\Delta\phi_i$.) With only a few significant figures shown the $\mathbf{X}^T\mathbf{X}$ matrix is

$$\mathbf{X}^T\mathbf{X} = 10^{-3} \begin{bmatrix} \ddots & \vdots & \vdots & \vdots & \vdots & \vdots & \vdots \\ & 12 & 10 & 8 & 5 & 3 & 0.85 & 0.045 \\ & & 9 & 7 & 5 & 3 & 0.82 & 0.043 \\ & & & 6 & 4 & 2 & 0.76 & 0.040 \\ & & & & 3 & 2 & 0.66 & 0.035 \\ & & & & & 1 & 0.47 & 0.026 \\ & \text{symmetric} & & & & & 0.2 & 0.011 \\ & & & & & & & 0.0006 \end{bmatrix} \quad (5.4.5)$$

It is significant that the lower right diagonal element is much smaller than the other diagonal elements. This is because $\Delta\phi_0$ is small compared to $\Delta\phi_1, \ldots$ and causes the $\mathbf{X}^T\mathbf{X}$ matrix to be poorly conditioned. Note that the diagonal elements continually increase from the lower right to the upper left, with the largest relative increase at the lower right. The regularization procedure has the greatest relative effect on the lower right element of $\mathbf{X}^T\mathbf{X}$.

A value of the regularization parameter, α, that has been used is 0.001. For this value of α, the $\alpha\mathbf{H}_1^T\mathbf{H}_1$ matrix is [see Eq. (4.5.23)]

$$\alpha\mathbf{H}_1^T\mathbf{H}_1 = 10^{-3} \begin{bmatrix} 1 & -1 & & & & \\ -1 & 2 & -1 & & & \\ & -1 & \ddots & \ddots & & \\ & & \ddots & 2 & -1 & \\ & & & -1 & 2 & -1 \\ & & & & -1 & 1 \end{bmatrix} \quad (5.4.6)$$

If Eq. (5.4.6) is added to Eq. (5.4.5), the lower right element becomes 1.0006×10^{-3} instead of the Eq. (5.4.5) value of 0.0006×10^{-3}, which is a factor of 1670 larger. The other diagonal elements are also increased but the relative increase is much less. The elements in the $\mathbf{X}^T\mathbf{X}$ matrix that are next to the main diagonal are decreased by unity and again the greatest relative effect is in the lower right portion of $\mathbf{X}^T\mathbf{X} + \alpha\mathbf{H}_1^T\mathbf{H}_1$.

It is interesting to consider the effect on the $\mathbf{X}^T\mathbf{X}$ matrix for a zeroth-order regularization procedure, which is actually the most-used regularization method employed by mathematicians. In this case, $\alpha\mathbf{I}$ is added to $\mathbf{X}^T\mathbf{X}$ and the lower right diagonal term of Eq. (5.4.5) is changed in exactly the same manner as in the first-order regularization method. The total effect is not identical for the

SEC. 5.4 REGULARIZATION ALGORITHMS

zeroth- and first-order regularization procedures but the very important lower diagonal element is changed in exactly the same manner.

5.4.2.1 Triangular Heat Flux Test Case.
The triangular heat flux test case with errorless data is solved using the first-order whole domain regularization method. The regularization parameter, α, is chosen to be 0.001. In this test case $\Delta t^+ = 0.06$ so that $\phi_1^+ = 0.000786$; hence, α is about 1600 times as large as ϕ_1^+ squared. The Tikhonov criterion given in Section 4.5.3.3 cannot be utilized here for selecting α because the "measurements" are errorless. (The α value of 0.001 is consistent with measurement errors having $\sigma_Y^+ = 0.0017$ which is discussed later.)

Results for several values of n are displayed in Figure 5.15. There are 20 time steps of 0.06 in the heating period of $t^+ = 0$-1.2. To adequately test the method, the calculations should start well before the heat is applied; for that reason, the calculations were started at $t^+ = -0.18$. Calculations for $n = 20$, 25, and 30 were performed. The $n = 30$ values cover the complete time period shown and represent the triangular heat flux very well. There is some rounding of the sharp changes of the triangular flux. It is also significant that the $n = 30$ results are symmetric about $t^+ = 0.6$. Though the function specification results for four future temperatures (see Figure 5.12) are also very good, a slight lack of symmetry is noted.

The $n = 20$ and 25 results of Figure 5.15 illustrate a weakness of the whole domain procedure. Though most of the $\hat{q}_i$ values are virtually identical for each n value in Figure 5.15, the last four $\hat{q}_i$ values for $n = 20$ and 25 deviate considerably from the correct values. Consequently the last few elements obtained using this whole domain IHCP algorithm can be inaccurate. Incidently, the

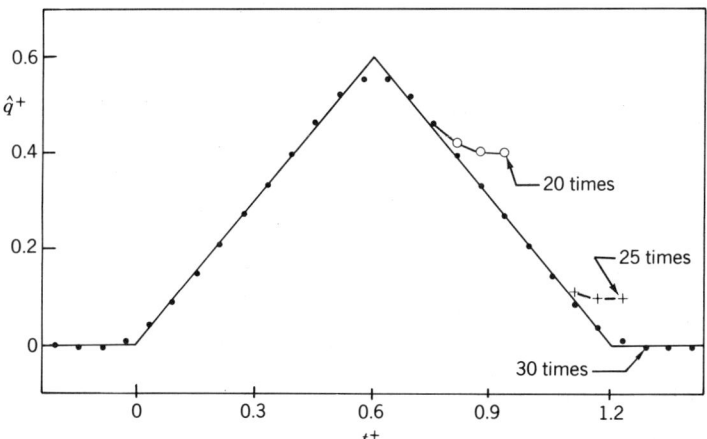

FIGURE 5.15 Calculated surface heat flux for triangular heat flux case with errorless data. Whole domain first-order regularization with $\alpha = 0.001$ and $\Delta t^+ = 0.06$.

number of $\hat{q}_i$ values that must be discarded is about equal to the number of future time steps needed in sequential procedures.

As pointed out in Chapter 1 regarding the sensitivity coefficients (see Figure 1.13, for example), a heat flux element at the surface requires some time before it significantly affects the temperature at an interior location. The last few $\hat{q}_i$ elements in a whole domain analysis may not have sufficient time to receive information from the surface. If so, the dominant effect of the algorithm is due to the regularization term; for the first-order method this means that the slope of $\hat{q}$ with respect to t approaches zero which is illustrated in Figure 5.15.

The effects of measurement errors for this triangular flux test case are almost identical to those for the *sequential* regularization method. For that reason the discussion of errors is deferred to Section 5.4.3 which covers the sequential method.

5.4.2.2 Heat Flux Impulse Test Case.
In the heat flux impulse test case, a single impulse at time t_r is introduced. (Note that in Section 5.4.2, r is not the number of future temperatures as used in the sequential methods.) In the present case the dimensionless time step is 0.06 rather than 0.05 that was used in the test case of Section 5.2.5. For the whole domain regularization method, the time of the heat flux impulse relative to the time domain is important. This is not true for sequential methods provided the calculations start well before the time of the impulse.

Figure 5.16 shows the $\delta\hat{q}_M/\delta q_r$ values for three r values and for a time domain of $n=30$ measurements. The α value is 0.001. One of the r values is $r=1$; this is the dashed curve with crosses. The response is relatively large reaching about 75% of the true δq_r value. The other extreme is $r=n=30$ which has a very small response. Clearly for this $r=n$ value there is not conservation of energy. The most interesting region is the central region of $10 \leqslant r \leqslant n-10$. Notice that the result is independent of r in this region. Also note the symmetrical result. The same results are valid even if n is a much larger value than 30, such as 100. For this region the result is essentially the same for any r value. This is an important observation. It means that the same results are obtained for the central region provided the calculations start early enough and continue long enough. Hence, the results obtained by the whole domain method can also be obtained by other techniques, such as properly designed sequential procedures.

It should be noted that the characteristics shown in Figure 5.16 for the central region are representative of the present problem. These characteristics include the independence of heat fluxes for any r in the central region and also the symmetrical character. They are inherent for transient heat conduction with small time steps. If convection were present, if steady state were approached (i.e., large Δt's), or if another partial differential equation was used, then these characteristics might not be present. The behaviors shown in Figure 5.16 are partly a consequence of the diffusive nature of parabolic partial differential equations. For other equations such as elliptic partial differential equations, each δq_r might have a unique effect for each $r=1, 2, \ldots, n$. In such cases the whole domain estimation would be appropriate.

SEC. 5.4 REGULARIZATION ALGORITHMS

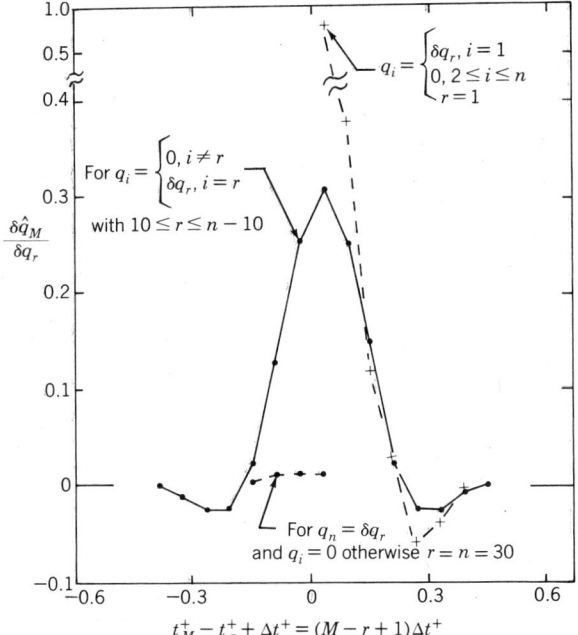

FIGURE 5.16 Calculated heat fluxes for heat flux impulse test case. Whole domain regularization method. $\Delta t^+ = 0.06$ and $\alpha = 0.001$.

5.4.2.3 Temperature Impulse Test Case, $\delta \hat{q}_M / \delta Y_r$.

The temperature impulse test case reinforces some of the observations and conclusions drawn from the heat flux impulse test case. See Figure 5.17, which is also for the first-order whole domain regularization method with $\alpha = 0.001$ and $\Delta t^+ = 0.06$. The case of $Y_i = 0$ except for $i = r = 1$ where $Y_1 = \delta Y_1$ is shown by the dashed line with crosses; the $\hat{q}$ values are quite small. For the middle range of $10 \leq r \leq n - 10$, a single temperature impulse of δY_r gives the $\hat{q}$ results which are independent of r when plotted versus $(M - r + 1)\Delta t^+$. A damped sinusoidal curve is obtained which is not quite antisymmetric about the $(M - r + 1)\Delta t^+ = 0$ axis. The sum of the $\delta \hat{q}_M^+$ values is nearly zero, thus approximating conservation of energy. Results for the extreme value of $r = n$ are also shown; the effect is very large and does not give conservation of energy.

The $\delta \hat{q}_M^+ / \delta Y_r^+$ values can be used to calculate the variances of the $\hat{q}_M$ elements.

5.4.3 Sequential Regularization Method

The sequential implementation of the regularization method applied to the IHCP improves considerably the calculational efficiency. The sequential first-order regularization equation is

$$[\mathbf{X}^T\mathbf{X} + \alpha \mathbf{H}_1^T\mathbf{H}_1]\hat{\mathbf{q}} = \mathbf{X}^T(\mathbf{Y} - \hat{\mathbf{T}}|_{\mathbf{q}=\mathbf{0}}) \qquad (5.4.7)$$

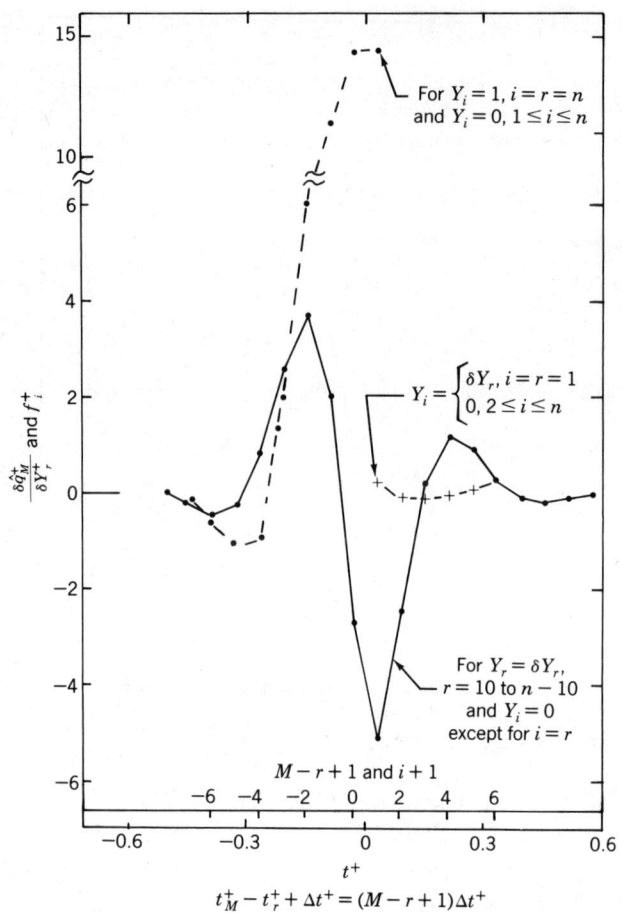

FIGURE 5.17 $\delta\hat{q}_M^+/\delta Y_r^+$ values for whole domain regularization with $\alpha=0.001$ and $\Delta t^+ = 0.06$. Curve with dots and dashed line is for the error in the first measurement. Curve with crosses is for the error in the last measurement. Curve with solid lines is representation of the error in the central time range.

where for r future temperatures,

$$\mathbf{X} = \begin{bmatrix} \Delta\phi_0 & & & \\ \Delta\phi_1 & \Delta\phi_0 & & \\ \vdots & \vdots & & \\ \Delta\phi_{r-1} & \Delta\phi_{r-2} & \cdots & \Delta\phi_0 \end{bmatrix} \qquad (5.4.8)$$

$$\hat{\mathbf{q}} = \begin{bmatrix} \hat{q}_M \\ \hat{q}_{M+1} \\ \vdots \\ \hat{q}_{M+r-1} \end{bmatrix} \quad \mathbf{Y} = \begin{bmatrix} Y_M \\ Y_{M+1} \\ \vdots \\ Y_{M+r-1} \end{bmatrix}, \quad \hat{\mathbf{T}}|_{q=0} = \begin{bmatrix} \hat{T}_M|_{q_M=0} \\ \hat{T}_{M+1}|_{q_M=q_{M+1}=0} \\ \vdots \\ \hat{T}_{M+r-1}|_{q_M=\cdots=0} \end{bmatrix} \qquad (5.4.9a,b,c)$$

SEC. 5.4 REGULARIZATION ALGORITHMS

$$\hat{T}_{M+i}|_{q_M = \cdots q_{M+i} = 0} = \sum_{j=1}^{M-1} \hat{q}_j \Delta\phi_{M-j+i-1} \qquad (5.4.10)$$

Equation (5.4.7) is solved for the $\hat{\mathbf{q}}$ vector but only the q_M component is retained. There are two parameters to select in Eq. (5.4.7), α and r. The regularization parameter is selected in the same manner as for the whole domain procedure. The number of future temperatures is chosen to be sufficiently large such that the $\hat{q}_M$ values are negligibly affected. Actually another choice in the regularization procedure is the order. The zeroth-order procedure is obtained using Eq. (5.4.7) by simply replacing $\mathbf{H}_1^T\mathbf{H}_1$ with the identity matrix, $\mathbf{I}$. An algebraic form of the $r=2$, combined zeroth and first order, sequential regularization algorithm is given by Eq. (4.5.33). If the regularization parameter, α, is set equal to zero, the sequential regularization method reduces to the Stolz method.

5.4.3.1 Triangular Heat Flux Test Case.
Results for the triangular heat flux test case with errorless data are shown in Figure 5.18. The sequential first-order regularization method is used with $\alpha = 0.001$. The results for r equal to 4 and 5 are essentially the same. See also Table 5.7. The results are quite similar to the whole domain method shown in Figure 5.15 and the sequential function specification method results displayed in Figure 5.12 and Table 5.5. The effect of simulated measurement errors (for the standard statistical assumptions) with $\sigma_y^+ = 0.0017$ is illustrated by Figure 5.19. The results are remarkably similar to those for the function specification method that are given in Figure 5.13. The $r=3$ results for both cases seem to be nearly identical and the ratio of $\sigma_{\hat{q}}^+/\sigma_Y^+$ is about 25 in both cases. The $r=4$ results are a little different but the $\sigma_{\hat{q}}^+/\sigma_Y^+$ ratio is about 10 in both cases.

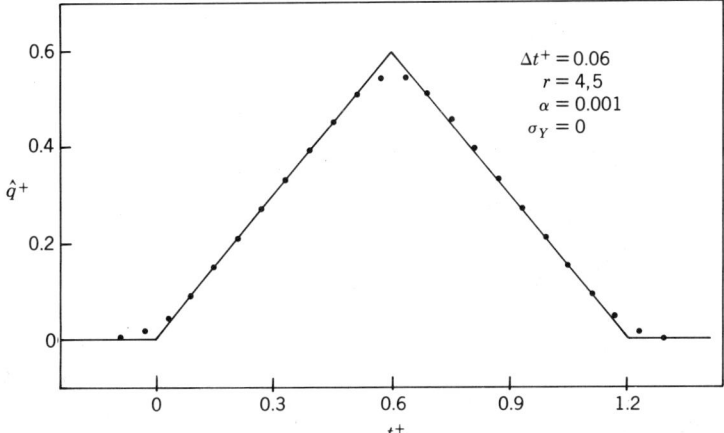

FIGURE 5.18 Calculated surface heat flux for triangular heat flux case with errorless data. Sequential first-order regularization method.

TABLE 5.7 Calculated Heat Fluxes at Certain Times for the Triangular Heat Flux Example Using the First-Order Sequential Regularization Method. Errorless Data and α=0.001

		$\hat{q}^+$		
t^+	Exact	$r=3$	$r=4$	$r=5$
0.51	0.51	0.5097	0.5034	0.5057
0.57	0.57	0.5561	0.5412	0.5415
0.63	0.57	0.5583	0.5428	0.5430
0.69	0.51	0.5168	0.5104	0.5104

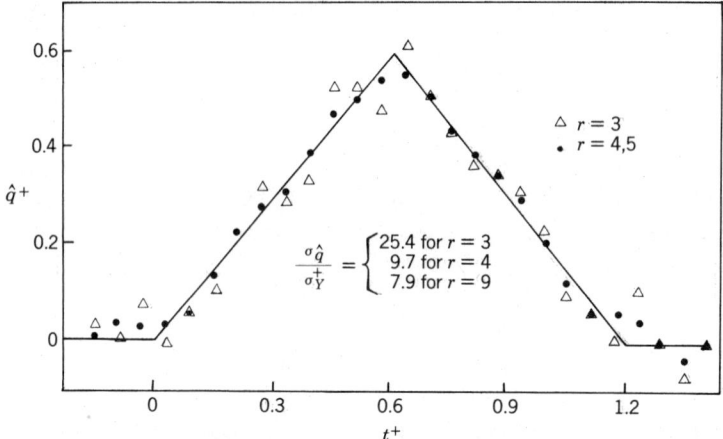

FIGURE 5.19 Calculated surface heat flux for triangular heat flux case with random errors with $\sigma_Y = 0.0017$. Sequential first-order regularization method with $\alpha = 0.001$ and $\Delta t^+ = 0.06$.

5.4.3.2 Heat Flux Impulse Test Case.
The sequential first-order regularization results for the heat flux impulse test case are displayed in Figure 5.20. The cases of $r=4$ and $r=9$ are given; the differences between these two cases are not large. Note that the $r=9$ case of Figure 5.20 is very similar to the midrange results of the whole domain regularization method shown in Figure 5.16. There is a slight loss of symmetry in the sequential regularization method, Figure 5.20.

5.4.3.3 Temperature Impulse Test Case.
Figure 5.21 shows results for the temperature impulse test case for $r=4$ and $r=9$. The $r=9$ curve is almost identical to the midrange whole domain regularization result in Figure 5.17.

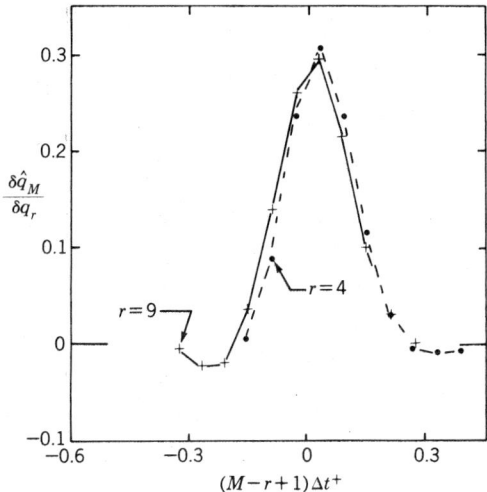

FIGURE 5.20 $\delta\hat{q}_M/\delta q_r$ values for the sequential regularization with $\alpha = 0.001$ and $\Delta t^+ = 0.06$.

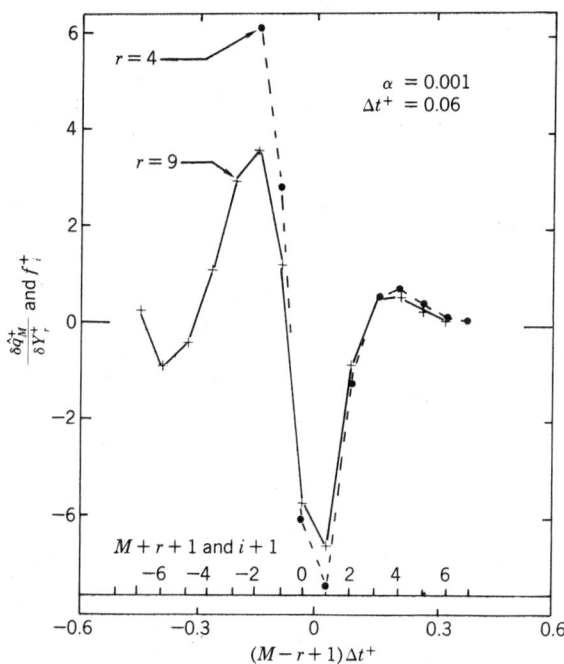

FIGURE 5.21 $\delta\hat{q}_M^+/\delta Y_r^+$ and f_i^+ values for the sequential regularization method with $\alpha = 0.001$ and $\Delta t^+ = 0.06$. (See Sect. 5.5.2 for definition of f_i^+.)

5.4.3.4 Comparison of Whole Domain and Sequential Regularization Methods.
One of the remarkable aspects of the comparison of the sequential and whole domain regularization methods is the similarity in the results. One conclusion is that these methods give about the same results for given heat flux input. The sequential method is much more efficient, however; this is because the computations for a solution of a set of algebraic equations with a full square matrix tend to increase as n^3 where n is the number of equations. If $n=100$, for example, $n^3=10^6$ for the whole domain method, but for $r=6$ future times, the number of computations for the sequential method is proportional to $nr^3=100\cdot 6^3=21{,}600$, which is much smaller than 1,000,000. Since the sequential method is so much more efficient and gives only slightly different results from the whole domain method, the sequential method is preferred. This is even more so for the nonlinear IHCP. For the linear problem, however, the filter concept discussed in the next section makes the computations even more efficient.

5.4.3.5 Comparison of Sequential Regularization and Function Specification Methods.
The function specification method can produce results that are surprisingly close to those of the sequential regularization method. An example is the errorless data case for the triangular heat flux as can be verified by comparing Figures 5.12 and 5.18. Furthermore, Figures 5.13 and 5.19, which are for random temperature errors, give very similar results with the regularization results being very slightly better, showing only about a 2% reduction in the standard deviation.

The sequential regularization method has two parameters to select, α and r; but if r is made sufficiently large, it has little effect on the heat flux estimates. Thus there is only one critical choice to make; namely, the regularization parameter. In the function specification method, there is only one parameter, the number of future temperatures, r. When both r and α are appropriately chosen, there is little difference in the results; the choice of the technique can be made on other grounds such as ease of use or of writing the computer program. The sequential function specification method is the simpler of the two both in understanding and in computer programming; for these reasons it is recommended for the IHCP.

5.5 DIGITAL FILTER ALGORITHM

5.5.1 Introduction

The filter approach for solving the linear inverse heat conduction problem has significant computational advantages over direct application of the IHCP algorithms; the filter algorithm, which was derived and discussed in Section 4.7, can be more computationally efficient.

SEC. 5.5 DIGITAL FILTER ALGORITHM

The filter approach can use filter coefficients, f_i, that are calculated using Duhamel's theorem, finite differences, or another method. The emphasis in this section is on the use of Duhamel's integral but essentially the same results are obtained if the f_i's are obtained using finite differences, provided the same algorithm is used for both.

The filter algorithms are independent of the procedure used to calculate the filter coefficients. For example, if the f_i's are found for the function specification algorithm, the use of these f_i's is the same as if they were found using the whole domain regularization method (if the end conditions are avoided as discussed in Section 5.4.2.2).

The filter algorithm given by Eq. (4.7.6) can be written more generally as

$$\hat{q}_M = \sum_{i=M-m_1}^{M+m_2} f_{M-i}(Y_i - T_0) \qquad (5.5.1)$$

where, if t_{M-1} corresponds to the "present" time, there are $m_2 + 1$ "future" times and $m_1 - 1$ "past" times. The choice of m_1 and m_2 depends on the algorithm ($m_2 = r - 1$ for the function specification algorithm, for example) and the geometry. For the finite body and for "large" dimensionless time steps, m_1 tends to be a small integer; for small dimensionless times and semi-infinite bodies the m_1 value can be large. The filter coefficients are calculated from $\delta\hat{q}_M/\delta Y_r$ as indicated in Section 4.7.2. Representative units of f_i are W/m^2-C.

The plan for this section is to investigate the methods for determining the filter coefficients for first the function specification method and then the whole domain regularization method. Some methods for truncating the summation in Eq. (5.5.1) are discussed.

5.5.2 Function Specification-Based Filter

5.5.2.1 Finite Plate Case. In the use of the digital filter algorithm it is first necessary to determine the filter coefficients which in turn require an IHCP algorithm. In this section the function specification method is used to obtain the filter coefficients. The procedure is described in Section 4.7.2 but greater detail is given here.

The Duhamel's theorem function specification algorithm is given by Eq. (5.3.4). The filter coefficients are found by setting $Y_r - T_0 = 1$ and $Y_i - T_0 = 0$ for $i \neq r$. Then from Eq. (4.7.8)

$$f_{M-r} = \hat{q}_M \Big|_{\substack{Y_r - T_0 = 1 \\ Y_i - T_0 = 0, \ i \neq r}} \qquad (5.5.2)$$

Hence for $M = 1$, use of (5.3.4a) gives

$$f_{1-r} = \hat{q}_1 \Big|_{\substack{Y_r - T_0 = 1 \\ Y_i - T_0 = 0, \ i \neq r}} = K_r \qquad (5.5.3)$$

and for $M=2$,

$$f_{2-r} = \hat{q}_2 \bigg|_{\substack{Y_r - T_0 = 1 \\ Y_i - T_0 = 0, \, i \neq r}} = K_{r-1} - K_r \sum_{i=1}^{r} K_i \Delta \phi_i \qquad (5.5.4)$$

The K's denote gain coefficients which were discussed in Chapter 4.

Since the algebraic expressions become complicated, it is simpler to use a computer program to calculate the $\hat{q}_M$ values with $Y_r - T_0 = 1$ and $Y_i - T_0 = 0$, $i \neq r$. For the standard problem of a plate heated at $x=0$ and insulated at $x=L$, some ϕ_i^+ values from Table 1.1 for $\Delta t^+ = 0.06$ and $x/L = 1$ are $\phi_1^+ = 0.000786$, $\phi_2^+ = 0.014887$ and $\phi_3^+ = 0.047584$. Using these and other values of ϕ_i^+ in Eq. (5.3.4) for $r=3$ gives $K_1 = 0.315$, $K_2 = 5.987$, and $K_3 = 19.137$ and also $\hat{q}_1^+ = 19.137$, $\hat{q}_2^+ = -14.220$, $\hat{q}_3^+ = -8.86$ and $\hat{q}_4^+ = 0.901$. Notice that $\hat{q}_1^+ = K_3$ for $r=3$, as indicated by Eq. (5.5.3).

Dimensionless values for the filter coefficients are given in Table 5.8 for $r=3$, 4, and 5 and for $\Delta t^+ = 0.06$ and $x/L = 1$. After obtaining the $\hat{q}_M^+$ values, Eq. (5.5.2) is used to obtain the f_i^+ values. The f_i^+ values in Table 5.8 illustrate that the i subscript of f_i^+ corresponds to $M = i - r$ of $\hat{q}_M^+$; that is, increasing M values correspond to increasing i values. The $\hat{q}_i^+$ values in the previous paragraph for $r=3$ are given as f_{-2}^+, f_{-1}^+, f_0^+, and f_1^+.

The values of f_i with units can be obtained from ϕ_i values with units or from

TABLE 5.8 Digital Filter Coefficients for Function Specification Algorithm, $\Delta t^+ = 0.06$

i	f_i^+ $r=3$	f_i^+ $r=4$	f_i^+ $r=5$
-4			4.538
-3		8.387	0.716
-2	19.137	-1.301	-1.209
-1	-14.220	-4.060	-1.775
0	-8.861	-2.885	-1.383
1	0.901	-0.849	-0.685
2	2.217	0.045	-0.273
3	0.961	0.263	-0.070
4	0.0849	0.220	0.0127
5	-0.1333	0.126	0.0363
6	-0.0830	0.053	0.0347
7	-0.0182	0.0143	0.0254
8	0.0059	-0.0014	0.0160
9	0.0063	-0.0050	0.0089
10	0.0021	-0.0041	0.0044
11	-0.00005	-0.0023	0.0018

SEC. 5.5 DIGITAL FILTER ALGORITHM

dimensionless f_i^+ values using

$$f_i = k \frac{f_i^+}{L} \tag{5.5.5}$$

where k is the thermal conductivity and L is the plate thickness.

Examination of Table 5.8 reveals that the f_i^+ values are both positive and negative, with the values oscillating with decreasing amplitude as i increases. The frequency of the oscillations tends to decrease with increasing r values. See Figure 5.14 for filter coefficients for $\Delta t^+ = 0.5$. For a given r value, the sum of the f_i^+ values tends to zero; hence, the T_0 term in Eq. (5.5.1) can be dropped in this case.

EXAMPLE 5.3. Use the filter algorithm to obtain heat flux estimates for the exact triangular heat flux data (Table 5.1) for several time steps. Consider the $r = 3$ and 4 cases.

Solution. The case of $r = 3$ is solved first. The temperature data are given in Table 5.1 and the filter coefficients are given in Table 5.8. Since the initial temperature is zero, and r is equal to 3, and m_1 from Table 5.8 is 11, Eq. (5.5.1) can be written as

$$\hat{q}_M^+ = \sum_{i=M-11}^{M+2} f_{M-i}^+ Y_i^+ \tag{5.5.6}$$

where $m_2 = r - 1 = 2$ for function specification with $r = 3$.
The first nonzero value of $\hat{q}_M^+$ is for $M = -1$ and thus

$$\hat{q}_{-1}^+ = \sum_{i=1}^{1} f_{-1-i}^+ Y_i^+ = f_{-2}^+ Y_1^+ = (19.137)(0.000007)$$

$$= 0.0001$$

since $Y_i^+ = 0$ for $i \leq 0$. Similarly for $M = 0$

$$\hat{q}_0^+ = \sum_{i=1}^{2} f_{-i}^+ Y_i^+ = f_{-1}^+ Y_1^+ + f_{-2}^+ Y_2^+$$

$$= (14.220)(0.000007) + (19.137)(0.000374) = 0.0071$$

The $M = 1$ equation is

$$\hat{q}_1^+ = f_0^+ Y_1^+ + f_{-1}^+ Y_2^+ + f_{-2}^+ Y_3^+ = 0.0362$$

Additional values are $\hat{q}_2^+ = 0.0868$ and $\hat{q}_3^+ = 0.1473$.

These $\hat{q}_i^+$ values can be compared with the exact values at times $t^+ = -0.09, -0.03, 0.03, 0.09,$ and 0.15, which are 0, 0, 0.03, 0.09, and 0.15, respectively. The agreement is quite good.

For $r = 4$ the calculational procedure is similar with the upper limit of the summation in Eq. (5.5.6) replaced by $M + 3$. Some values of $\hat{q}_M^+$ are 0.0001, 0.0031, 0.0177, 0.0487, 0.0941, and 0.1487 for $M = -2, -1, 0, 1, 2,$ and 3, respectively.

Some comments on this example are now given. First, the calculated values for the heat flux are exactly the same as those given by the function specification method. Furthermore, though the Duhamel's theorem method was used, finite difference modeling of the transient heat conduction equation would give nearly the same results.

Second, the same procedure can be used if there are errors in the measurements. Third, the computational load does not continue to increase as it does for the Duhamel's theorem algorithm, Eq. (5.3.4), where the number of terms in the inner summation continues to increase with increasing M. This is because the filter coefficients in Table 5.8 tend to zero for large positive values of i. For smaller dimensionless time steps the f_i^+ values decrease in magnitude more slowly for increasing values of i. Fourth, even though the filter coefficients for $r=3$ and 4 of Table 5.8 are quite different, the estimated values of q are relatively close. Finally, there is no accumulation of errors in the filter algorithm, Eq. (5.5.1), since the calculation of $\hat{q}_M$ is independent of $\hat{q}_i$ for $i \neq M$. □

5.5.2.2 Semi-Infinite Body.

The same filter procedure used for the finite body, just described, can also be utilized for a semi-infinite body, which unfortunately has a "long" memory. By "long" memory is meant that the filter coefficients, f_i, decrease very slowly as i becomes large. A method of improving the efficiency of the filter algorithm is suggested below.

For simplicity, the sensor is considered to be at the heated surface and the Stolz method (i.e., $r=1$) is used to generate the filter coefficients. The equation for calculating the ϕ function is [see Eq. (1.6.20)]

$$\phi = 2 \left(\frac{t}{\pi k \rho c} \right)^{1/2} \tag{5.5.7}$$

Also for simplicity the time steps are chosen to be 1 s and $k\rho c$ is equal to unity with proper units. The initial temperature, T_0, is set equal to zero and all Y_i values are zero except $Y_1 = 1$. Using Eq. (5.3.1) yields the $\hat{q}_M$ values, the first few of which are 0.886227, -0.367087, and -0.129623 for $M=1$, 2, and 3. The values decrease slowly in magnitude, and are -0.0035, -0.0012, and 0.0006 for $M=20$, 40, and 60, respectively. The sum of the $\hat{q}_M$'s over all M's tends to zero but the convergence is slow; for $M=10$ and 20 the sums are 0.187 and 0.129, respectively.

One way to speed up the convergence—that is, use fewer terms in the filter algorithm—is to take advantage of the small differences between succeeding f_i values for large i values. Equation (5.5.1) is written again for $M-1$ and the equations are subtracted to obtain

$$\hat{q}_M - \hat{q}_{M-1} = \sum_{i=M-m_1}^{M+m_2} f_{M-i}(Y_i - T_0) - \sum_{i=M-1-m_1}^{M-1+m_2} f_{M-i-1}(Y_i - T_0) \tag{5.5.8}$$

which can be rearranged to the recursive form of

$$\hat{q}_M = \hat{q}_{M-1} + \sum_{i=M-m_1-1}^{M-1+m_2} (f_{M-i} - f_{M-i-1})(Y_i - T_0)$$
$$+ f_{-m_2}(Y_{M+m_2} - T_0) - f_{m_1}(Y_{M-1-m_1} - T_0) \tag{5.5.9}$$

Due to the recursive nature of Eq. (5.5.9), information regarding the Y_i values for small i's is contained in $\hat{q}_{M-1}$. [Incidentally, since Eq. (5.5.9) is recursive, more care in the calculations is needed because the errors can accumulate, unlike the result from Eq. (5.5.1).] The $f_{M-i} - f_{M-i-1}$ values decrease in magni-

SEC. 5.5 DIGITAL FILTER ALGORITHM

tude much more rapidly than f_i for large positive i's. This can be verified by examining Table 5.9. Consequently, for cases of large memory the recursive form of Eq. (5.5.9) requires fewer terms in the summation than Eq. (5.5.1) and thus is more computationally efficient.

5.5.3 Whole Domain Regularization Filter

The whole domain regularization filter also uses Eq. (5.5.1) but now m_2 in the upper limit is not equal to $r-1$ as it was for the function specification method. An example is given for the standard case of the flat plate with $\alpha = 0.001$, $\Delta t^+ = 0.06$, and $n = 30$ times. The case of $Y_{15} - T_0 = 1$ and other $Y_i - T_0$ values equal to zero is considered. The f_i^+ values are given in Table 5.10 and are plotted as the solid line in Figure 5.17. The value of m_1 in Eq. (5.5.1) for the f_i^+ values of Table 5.10 is 14 and the m_2 value is 15.

If the f_i^+ values are calculated simultaneously for a larger number of times such as $n=40$, the f_i^+ values for small absolute values of i are nearly identical to those in Table 5.10. For the extreme i values there are small differences but the associated f_i^+ values have small magnitudes and affect the $\hat{q}_M$ estimates only slightly.

A comparison of the f_i^+ values for the function specification method given by Table 5.8 and the regularization method values of Table 5.10 reveals similar orders of magnitude and periodic behavior. At any given i value, however, the f_i^+ values are quite different.

For this case of a plate heated at $x=0$ and insulated at $x=L$, the sum of the filter coefficients should be equal to zero. A running sum of the coefficients is

TABLE 5.9 Filter Coefficients for Semi-Infinite Body with $r=1$, $\Delta t=1$ s, $k\rho c=1$, and $x=0$

j	f_j	$f_j - f_{j-1}$
0	0.886227	
1	−0.367087	−1.253314
2	−0.129623	0.237464
3	−0.06710	0.062525
4	−0.04186	0.02524
5	−0.02909	0.01277
6	−0.02166	0.00743
10	−0.00961	0.00174
15	−0.00510	0.00058
20	−0.00327	0.00027
40	−0.00114	0.00004
80	−0.00040	0.00001

TABLE 5.10 Filter Coefficients for Whole Domain Regularization Method. Flat Plate Case. $\alpha = 0.001$ and $\Delta t^+ = 0.06$. $f_i^+ \equiv \delta q_{i\ +r}^+ / \delta Y_r^+$

i	f_i^+	i	f_i^+
−14	−0.0191	1	−2.4851
−13	−0.0021	2	0.1889
−12	0.0335	3	1.1485
−11	0.0713	4	0.8765
−10	0.0623	5	0.2898
−9	−0.0574	6	−0.0854
−8	−0.2938	7	−0.1729
−7	−0.4815	8	−0.1066
−6	−0.2379	9	−0.0222
−5	0.8328	10	0.0210
−4	2.5795	11	0.0249
−3	3.6912	12	0.0131
−2	2.0263	13	0.0032
−1	−2.7314	14	−0.0002
0	−5.1572	15	−0.0004

TABLE 5.11 Running Sums of Filter Coefficients of Table 5.10

Summation Limits	Sum
0	−5.1572
−1 to 1	−10.3737
−2 to 2	−8.1585
−3 to 3	−3.3188
−4 to 4	0.1372
−5 to 5	1.2598
−6 to 6	0.9365
−7 to 7	0.2821
−8 to 8	−0.1183
−9 to 9	−0.1979
−10 to 10	−0.1146
−11 to 11	−0.0184
−12 to 12	0.0282
−13 to 13	0.0293
−14 to 14	0.0100
−14 to 15	0.0096

SEC. 5.6 OPTIMAL CONSIDERATIONS

TABLE 5.12 Heat Flux Values for Triangular Heat Flux Test Case. Filter Regularization Algorithm (Whole Domain) $n=30$, $\Delta t^+ = 0.06$, and $\alpha = .001$

M	Time	Exact q_M	Uncorrected $\hat{q}_M$	Corrected $\hat{q}_M$
−3	−0.15	0.0	−0.0054	−0.0062
−2	−0.09	0.0	−0.0043	−0.0052
−1	−0.03	0.0	0.0083	0.0072
1	0.03	0.03	0.0383	0.0372
2	0.09	0.09	0.0857	0.0844
3	0.15	0.15	0.1440	0.1426
9	0.51	0.51	0.5187	0.5170
10	0.57	0.57	0.5538	0.5521
11	0.63	0.63	0.5540	0.5523
12	0.69	0.51	0.5191	0.5174
20	1.17	0.03	0.0460	0.0442
21	1.23	0.0	−0.0017	−0.0035

shown in Table 5.11 where the sum of all the f_i^+'s is 0.0096 rather than zero. This number is quite small and has a small effect but if desired the f_{-14}^+ and f_{15}^+ values can be each decreased by 0.0048 to make the sum equal to zero and thus ensure conservation of energy. For the test case of the triangular heat flux with exact data, the difference is only in the third decimal place. Some of the $\hat{q}_M$ values are given in Table 5.12 along with the exact q_M values. The uncorrected column is for the f_i^+ values of Table 5.10 and the corrected column is for the 0.0048 corrections to f_{-14}^+ and f_{15}^+. The corrected and uncorrected values are quite close to each other and also close to those given by the usual implementation of the whole domain regularization procedure which is illustrated by Figure 5.15.

Because the sum of the filter coefficients oscillates about zero with decreasing magnitude for large values of absolute i (as can be seen from Table 5.11), it is possible to terminate the summation for a smaller value of terms than 30. A sum of 17 has been noted to do quite well particularly when the extreme f_i^+ values are adjusted to give conservation of energy. This 17-sum algorithm uses much less computer time than the whole domain regularization or even a sequential method. For linear cases the digital filter concept is highly recommended for heat flux sensors that are repeatedly employed (if the computing costs are a serious consideration).

5.6 OPTIMAL CONSIDERATIONS

In the sequential function specification method there are parameters to choose such as the time step and number of future temperatures, r. In the sequential regularization method, the regularization parameter, α, must also be chosen. In the whole domain regularization method only the time step and α are needed. (The method in the literature for selecting α is described in Section 4.5.3.3.) One purpose of this section is to provide a basis for optimal choice of these parameters.

Another objective is to erect a framework for the comparison of estimation algorithms. If the concepts are further extended, insight is gained into the optimal design of IHCP estimation algorithms.

The organization of this section is first to consider optimal ideas for the function specification procedure and then the whole domain regularization procedure is studied.

It is not possible to consider all the geometries and all the variations of the algorithms. There are too many geometries such as finite and semi-infinite bodies and plane, cylindrical, and spherical coordinates. Furthermore, the bodies can be homogeneous or composite. New algorithms can always be proposed. Hence, it is reasonable to demonstrate only the basic concepts for the most common algorithms.

5.6.1 Optimal Function Specification Algorithm

The basic geometry considered is the plate heated at $x=0$ and insulated at $x=L$. The sensor is at the insulated surface. This case is considered here and in the next subsection.

In this subsection the function specification method with the constant heat flux approximation is investigated.

The optimality criterion is the minimization of the mean squared error, $\mathscr{S}^2$. This error is discussed in Section 4.8 and elsewhere in the book. From Eq. (4.8.9) the mean squared error $\mathscr{S}^2$ is

$$\mathscr{S}^2 = V(\hat{q}) + \mathscr{D}^2 \tag{5.6.1}$$

where $V(\hat{q})$ is the variance of the estimated q component and $\mathscr{D}$ is a measure of the deterministic error. This equation, unlike Eq. (4.8.9), does not have an M subscript because the asymptotic value of $\mathscr{S}^2$ is needed. It is assumed that the calculations have proceeded over a sufficiently large number of steps so that $\mathscr{S}^2$ has converged to its asymptotic value. (See the discussion in Example 4.6.)

The variance of the estimator $\hat{q}_M$ depends on the statistics of the measurement errors. For simplicity the first four standard assumptions are used. These assumptions are: the temperature errors are additive, have zero mean, have constant variance, and are uncorrelated. The errors are assumed to be only in the temperatures, not in the locations of the sensors, for example. For computer data acquisition, the measured values of temperature tend to be correlated. This is particularly true as the time steps between the measured values become small. Methods of modeling this case are discussed in Section 6.9 in Beck and Arnold[6] and in Beck.[10] Even though correlated errors and nonconstant variance

SEC. 5.6 OPTIMAL CONSIDERATIONS

might be better assumptions in some cases, the basic concepts of optimality remain the same. Hence, the use of the standard assumptions serves to illustrate the procedures as well as provide results that have intrinsic worth.

The variance of $\hat{q}_M$ for the standard statistical assumptions is given by [see Eq. (4.8.11)]

$$V(\hat{q}_M) = \sigma_Y^2 \sum_{j=1}^{M+r-1} \left(\frac{\delta \hat{q}_j}{\delta Y_r}\right)^2 \qquad (5.6.2)$$

The $\delta \hat{q}_j/\delta Y_r$ values are the heat fluxes calculated for all the Y_i components set equal to zero except the rth Y which is set equal to unity. As Eq. (5.6.2) is written, the variance has units of $(W/m^2)^2$, σ_Y has units of K, and $\delta \hat{q}_j/\delta Y_r$ has units of W/m^2-K. In order to present more general results, however, Eq. (5.6.2) is put into a dimensionless form. Equation (5.6.2) is divided by q_0^2 where q_0 is a nominal surface heat flux; q_0 can be chosen to be any convenient constant value such as the flux in the step rise in the q test case or the peak flux for the triangular heat flux test case. Then Eq. (5.6.2) can be written as

$$\frac{V(\hat{q}_M)}{q_0^2} = \left(\frac{\sigma_Y}{q_0 L/k}\right)^2 \sum_{j=1}^{M+r-1} \left[\frac{\delta \hat{q}_j/q_0}{\delta Y_r/(q_0 L/k)}\right]^2 \qquad (5.6.3)$$

where L is the plate thickness and k is the thermal conductivity. This dimensionless equation can be written in the form

$$\sigma_{\hat{q}}^+ = \sigma_Y^+ \left[\sum_{j=1}^{M+r-1} \left(\frac{\delta \hat{q}_j^+}{\delta Y_r^+}\right)^2\right]^{1/2} \qquad (5.6.4a)$$

which is called the dimensionless standard deviation of $\hat{q}_M$, and

$$\sigma_{\hat{q}}^+ \equiv \frac{[V(\hat{q}_M)]^{1/2}}{q_0} \qquad (5.6.4b)$$

Other dimensionless quantities in Eq. (5.6.4a) are

$$\sigma_Y^+ \equiv \frac{\sigma_Y}{(q_0 L/k)}, \quad \delta Y_r^+ \equiv \frac{\delta Y_r}{(q_0 L/k)}, \quad \delta \hat{q}_j^+ \equiv \frac{\delta \hat{q}_j}{q_0} \qquad (5.6.5a, b, c)$$

Figure 5.22 displays results for the dimensionless standard deviation of $\hat{q}_M$ as a function of the number of future time steps, r, and dimensionless calculational time steps,

$$\Delta t^+ = \frac{\alpha \Delta t}{L^2} \qquad (5.6.6)$$

for the function specification method. Notice that the plot is semi-logarithmic with the range of the standard deviation going from about 0.5 to 10^4 and the values of Δt^+ from 0.01 to 0.25. The effects of increasing r are to decrease the standard deviation of $\hat{q}$ and to make smaller time steps possible. Also shown in Figure 5.22 is the Burggraf result which is higher than all the curves except the $r=2$ curve. This indicates that the Burggraf result is relatively sensitive to

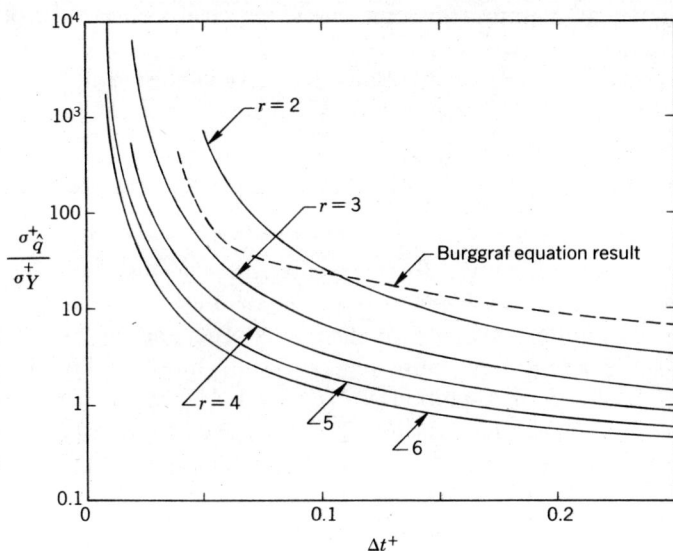

FIGURE 5.22 Standard deviation of $\hat{q}$ for sequential function specification method with $q = C$ assumption.

measurement errors. (The Burggraf results were obtained by using Eq. (2.5.32); the variance is the sum of the squares of the D's.) All the results shown in Figure 5.22 are for the asymptotic values which occur for large M values as mentioned below Eq. (5.6.1).

Values for the deterministic error, $\mathscr{D}$, are also needed. One measure of the deterministic error is Eq. (4.8.16) which is repeated as

$$\mathscr{D}^+ = \left[\sum_{\substack{i=1 \\ i \neq r}}^{\infty} \left(\frac{\delta \hat{q}_i}{\delta q_r}\right)^2 + \left(\frac{\delta \hat{q}_r}{\delta q_r} - 1\right)^2 \right]^{1/2} \tag{5.6.7}$$

where $\mathscr{D}^+$ is the dimensionless deterministic errors,

$$\mathscr{D}^+ \equiv \frac{\mathscr{D}}{\delta q_r} \tag{5.6.8}$$

For the linear IHCP algorithms (all those in Chapters 4 and 5 are linear), $\mathscr{D}^+$ is independent of the magnitude of δq_r. Then the $\delta \hat{q}_i / \delta q_r$ values can be calculated using the input temperatures given by

$$Y_1 = Y_2 = \cdots = Y_{r-1} = 0, \quad Y_r = \Delta \phi_0, \quad Y_{r+1} = \Delta \phi_1, \ldots, Y_{r+i} = \Delta \phi_i \tag{5.6.9}$$

since these values correspond to

$$q_i = \begin{cases} 1 & i = r \\ 0 & i \neq r \end{cases} \tag{5.6.10}$$

Plots of $\mathscr{D}^+$ versus Δt^+ for various r values are given in Fig. 5.23. There are several differences between these curves and those for $\sigma_{\hat{q}}$ given in Figure 5.22.

SEC. 5.6 OPTIMAL CONSIDERATIONS

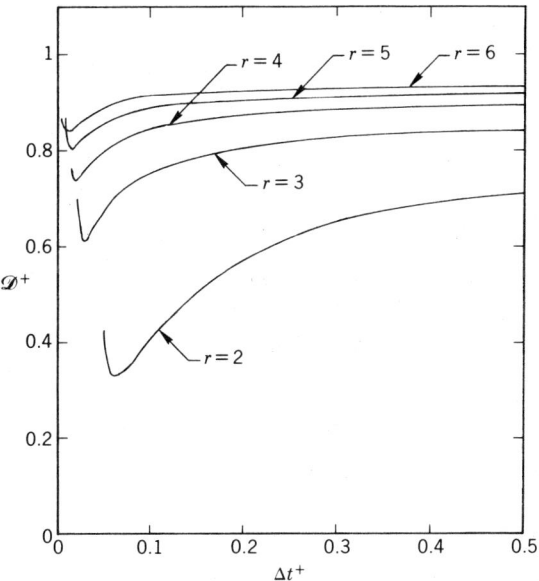

FIGURE 5.23 Deterministic error of $\hat{q}$ for sequential function specification method with $q = C$ assumption.

In Figure 5.23, the smallest r's have the smallest deterministic errors (for sufficiently large Δt^+), but Figure 5.22 shows that the small r's have the largest sensitivity to errors (i.e., largest $\sigma_{\hat{q}}$'s). Also note that in Figure 5.22 increasing the time step has the effect of reducing $\sigma_{\hat{q}}$, whereas in Figure 5.23 (except for small Δt's) the effect on $\mathscr{D}$ is to increase its value. A final difference is that Figure 5.23 reveals a minimum of $\mathscr{D}$ for a fixed r.

The minimum with respect to Δt^+ for each r curve needs further discussion. For a given number of future time steps, r, the minimum indicates an optimal time step when there are no random measurement errors. (For no measurement errors, $\mathscr{S}^2$ given by Eq. (5.6.1) reduces to just $\mathscr{D}^2$.) For time steps less than those at the several minima of the curves, the $\mathscr{D}^+$ values increase rapidly as Δt^+ becomes smaller. Hence, it is better to err on the large side in the choice of a Δt^+ value.

The two terms on the right of Eq. (5.6.1) can be made dimensionless as mentioned before to get

$$\mathscr{S}^+ = [(\sigma_{\hat{q}}^+)^2 + (\mathscr{D}^+)^2]^{1/2} \tag{5.6.11}$$

$$\mathscr{S}^+ \equiv \frac{\mathscr{S}}{\delta q_r} \tag{5.6.12}$$

For consistency the nominal q_0 value in Eqs. (5.6.3) and (5.6.4) can be replaced by δq_r which is used in Eq. (5.6.8). The δq_r value is a measure of the change in the surface heat flux from one time to another. This is illustrated below. The

$\sigma_{\hat{q}}^+$ symbol appearing in Eq. (5.6.11) is given by Eq. (5.6.4a) and is proportional to σ_Y^+. For shorter notation $(\sigma_Y^+)^2$ is denoted

$$R^2 = \left(\frac{\sigma_Y k}{\delta q_r L}\right)^2 = (\sigma_Y^+)^2 \qquad (5.6.13)$$

Plots of $\mathscr{S}^+$ versus Δt^+ are displayed in Figure 5.24 for $r=3$ and several values of R^2. Minimum values indicate optimal conditions. As R^2 increases, the optimal values are associated with both increased values of Δt^+ and $\mathscr{S}^+$.

The minimum values of several curves similar to Figure 5.24 have been employed to get the optimal curves of Figure 5.25. The curved line for $r=3$, for example, was obtained using the minimum points of Figure 5.24. To make the optimal results more convenient to use, Table 5.13 is given. This table is based on Figure 5.25 for the R values of 0, 0.01, and 0.031. The $R=0$ case corresponds to the limiting case of no measurement errors and is not recommended for actual data. A much more realistic case is $R=0.01$ which is recommended if an estimate of R is not available or the standard statistical assumptions are not valid.

For no information regarding the measurement errors, an alternative procedure of setting $R=0.01$ is suggested. In this procedure the algorithm is first tried with r equal to a small value such as 2 or 3. If there is significant oscillation, the r value is gradually increased until the oscillations in the heat flux are small. If the necessary value exceeds some "large" value, such as 10, then the investi-

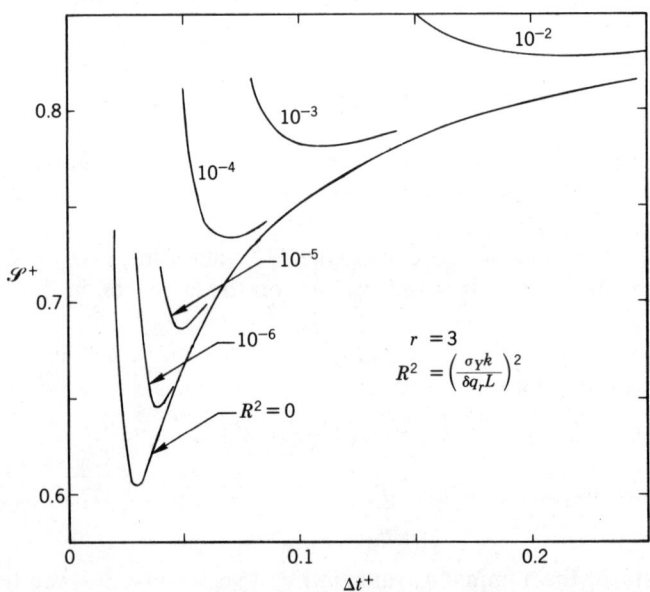

FIGURE 5.24 Plots showing optimal values for $\mathscr{S}^+$ and Δt^+ for $r=3$. Function specification method.

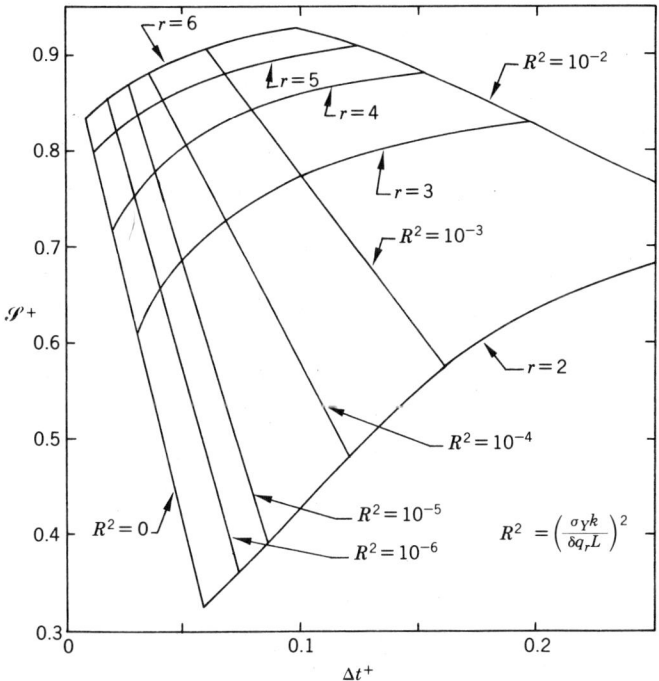

FIGURE 5.25 Optimal conditions for function specification method for plate with sensor at $x=L$. $q=C$ approximation.

TABLE 5.13 Optimal Conditions for Function Specification Method

Δt^+	$R=0$		$R=0.01$		$R=0.031$	
	r	$\mathscr{S}^+$	r	$\mathscr{S}^+$	r	$\mathscr{S}^+$
0.01	6	0.83				
0.012	5	0.8				
0.02	4	0.72				
0.03	3	0.61	6	0.83		
0.04			5	0.85		
0.05			4	0.81		
0.06	2	0.325			6	0.9
0.07			3	0.72	5	0.875
0.08					4	0.84
0.1					3	0.775
0.14			2	0.48		
0.16					2	0.575
1.5			1			

gator should double the time step and start the search for r again. This increasing of the time step is recommended because the requirement of an extremely larger r value implies that there is insufficient information in the data to resolve the surface heat flux variations with the chosen (small) time step.

EXAMPLE 5.4. Find the optimal value of r for the triangular heat flux test case assuming the standard statistical assumptions and $\sigma_Y^+ = \sigma_Y k/q_N L = 0.0017$ and $\Delta t^+ = \alpha \Delta t/L^2 = 0.06$. For this case the largest changes in q^+ are about $\delta q^+ = \delta q/q_N = 0.06$.

Solution. The value of R is needed, $(R = \sigma_Y k/\delta q_r L)$. The δq_r value is about $0.06 q_N$ and thus

$$R = \frac{\sigma_Y k}{0.06 q_N L} = \frac{1}{0.06} \sigma_Y^+ = \frac{0.0017}{0.06} = 0.0283$$

This R value is not given in Table 5.13 but is only 10% less than the $R = 0.031$ which is given and which gives the optimal value of 6; $r = 5$ is probably close to the optimal also. Figure 5.13 displays calculated heat fluxes for $r = 3$ and 4; $r = 5$ and 6 would further improve the results. □

EXAMPLE 5.5. In an actual experiment (Reference 6, p. 402) thermocouples are located at the insulated surfaces of two Armco iron specimens of 0.0254 m thickness. The thermal diffusivity is 0.000020 m^2/s and the thermal conductivity is 75 W/m-K. The measurement times steps are 0.3 s. The statistical assumption of uncorrelated measurement errors is known to be invalid. Select a calculational time step and an r value.

Solution. The dimensionless time step is calculated from the measured time step of 0.3 s to get

$$\Delta t^+ = \frac{\alpha \Delta t}{L^2} = \frac{(0.00002)(0.3)}{(0.0254)^2} = 0.0093$$

Since the standard statistical assumptions are not valid, the $R = 0.01$ column of Table 5.13 is used but it does not give dimensionless time steps less than 0.03. If time steps of 0.9 s are chosen, $\Delta t^+ = 0.028$ which is near 0.03. The $\Delta t^+ = 0.03$ and $R = 0.01$ value of r in Table 5.13 is 6. Hence, for this case, the time step for estimating q is 0.9 s and the recommended r value is 6. □

5.6.2 Optimal Whole Domain Regularization Method

The whole domain regularization method is now discussed for the same basic plate geometry considered in Section 5.6.1. Values are reported in this section only for the midrange; see Figures 5.16 and 5.17. These whole domain regularization results are typical of those obtained by the sequential regularization method.

For the regularization methods, the regularization parameter, α, must be specified. The first-order regularization method is used as in the previous examples of this chapter. It is given mathematically by Eq. (5.4.1).

In the same manner that Table 5.13 was constructed, Table 5.14 was developed for the first-order whole domain regularization procedure. Notice that

SEC. 5.6 OPTIMAL CONSIDERATIONS

TABLE 5.14 Optimal Conditions for Whole Domain Regularization Method

	$R=0.01$		$R=0.031$	
Δt^+	α^+	$\mathscr{S}^+$	α^+	$\mathscr{S}^+$
0.01	10^{-4}	0.86	0.001	0.98
0.03	10^{-4}	0.83	0.001	0.92
0.04	10^{-4}	0.8	0.001	0.88
0.05	10^{-4}	0.77	0.001	0.86
0.06	7×10^{-5}	0.69	0.001	0.84
0.08	3×10^{-5}	0.6	0.001	0.78
0.1	2×10^{-5}	0.49	0.001	0.73
0.12	10^{-5}	0.4	0.001	0.67
0.14	10^{-5}	0.3	0.001	0.64

the time steps are relatively independent of α^+ which is defined by

$$\alpha^+ \equiv \frac{\alpha k^2}{L^2} \qquad (5.6.14)$$

An approximate relation for the dimensionless regularization parameter is

$$\alpha^+ = R^2 = \left(\frac{\sigma_Y k}{\delta q_r L^2}\right)^2 \qquad (5.6.15)$$

or with units

$$\alpha = \left(\frac{\sigma_Y}{\delta q_r}\right)^2 \qquad (5.6.16)$$

As in the function specification method, if the value of R is not known or the standard statistical assumptions are not valid, the $R=0.01$ value is reasonable. (This is based on experience.)

A comparison of Tables 5.13 and 5.14 shows for the same Δt^+ and R values that the whole domain regularization method gives slightly lower values of $\mathscr{S}^+$ than the function specification method, the difference being less than 10% in most cases. Hence the whole domain regularization method can provide better heat flux estimates than the function specification method but the differences are small. The whole domain regularization is computationally inefficient (unless it is implemented as a digital filter). If one is interested in obtaining the best possible heat flux estimates, then the whole domain regularization method may be appropriate; the sequential regularization method, however, gives nearly the same results. Due to its simplicity and only slightly inferior performance, the function specification procedure may be preferred over the regularization methods.

EXAMPLE 5.6. Repeat Example 5.4 for the whole domain regularization method. Find an optimal value of α^+.

Solution. Using the R value of 0.0283 in Eq. (5.6.15) gives $\alpha^+ = 0.0008$ which is quite close to the $\alpha^+ = 0.001$ values used in the examples for the triangular heat flux test case. □

EXAMPLE 5.7. Repeat Example 5.5 for the whole domain regularization method.

Solution. The Δt^+ value is 0.028 and thus Eq. (5.6.15) gives

$$\alpha^+ = R^2$$

Since the standard statistical assumptions are not valid, a value of $R = 0.01$ is used and then Eqs. (5.6.14) and (5.6.15) give

$$\alpha = 0.0000 \frac{L^2}{k^2}$$

$$= 0.0001 \left(\frac{0.0254}{75}\right)^2 = 1.15 \times 10^{-12} \text{ m}^4\text{-K}^2/\text{W}^2$$

Note that the temperatures are in K and that the heat flux has units of W/m². □

The time step did enter the above selection of α. Suppose that the $R = 0.01$ value is appropriate for $\Delta t = 0.9$ s. It is interesting to see the effect of choosing the time step of 0.3 s. With the time step decreased by a factor of 3, the δq_r value is also decreased by this factor. Then Eq. (5.6.16) shows that α is increased by a factor of 9.

REFERENCES

1. Keltner, N. R. and Beck, J. V., Unsteady Surface Element Method, *J. Heat Transfer* **103**, 759–764 (1981).
2. Beck, J. V. and Keltner, N. R., Transient Thermal Contact of Two Semi-Infinite Bodies Over a Circular Area, *Spacecraft Radiative Transfer and Temperature Control*, T. E. Horton, Ed. Vol. 83 of Progress in Astronautics and Aeronautics, AIAA, New York, 1982, pp. 61–82.
3. Litkouhi, B. and Beck, J. V., Intrinsic Thermocouple Analysis Using Multinode Unsteady Surface Element Method, AIAA Paper No. 83-1437, 1983.
4. Stolz, G., Jr., Numerical Solutions to an Inverse Problem of Heat Conduction for Simple Shapes, *J. Heat Transfer* **82**, 20–26 (1960).
5. Beyer, W. H., Ed., *Handbook of Tables for Probability and Statistics*, 2nd ed., The Chemical Rubber Co., Cleveland, OH, 1968, p. 484.
6. Beck, J. V. and Arnold, K. J., *Parameter Estimation in Engineering and Science*, Wiley, New York, 1977.
7. Farnia, K., Computer-Assisted Experimental and Analytical Study of Time/Temperature-Dependent Thermal Properties of the Aluminum Alloy 2024-T351, Ph.D. Dissertation, Dept. of Mechanical Engineering, Michigan State University, 1976.
8. Twomey, S., On the Numerical Solution of Fredholm Integral Equations of the First Kind by the Inversion of the Linear System Produced by Quadrature, *J. Assoc. Comp. Mach.* **10**, 97–101 (1963).
9. Twomey, S., The Application of Numerical Filtering to the Solution of Integral Equations Encountered in Indirect Sensing Measurements, *J. Franklin Inst.* **279**, 95–109 (1965).

10. Beck, J. V., Criteria for Comparison of Methods of Solution of the Inverse Heat Conduction Problem, *Nucl. Eng. Des.* **53**, 11–22 (1979).

PROBLEMS

5.1. a. Calculate the $\phi(r, t)$ function for a solid cylinder for $r=0$. Give the values for a steel cylinder 0.664 cm in diameter. The dimensionless time step is $\Delta t_a^+ = \alpha \Delta t/a^2 = 0.05$ where $a =$ radius of the cylinder. Give ϕ values in SI units for 6 time steps and give the times in seconds. Use values given in Table 1.3 and the properties given in Table 5.4.
b. Repeat the problem for a cylinder 4 cm in diameter.
c. Repeat part (a) for brick.

5.2. a. A solid steel sphere is 4 cm in diameter. Calculate the center temperatures for ten time steps of $\Delta t = 1.818$ s for the heat flux given by
$$t<0, \quad q=0$$
$$0<t<4\Delta t, \quad q=100{,}000 \text{ W/m}^2$$
$$0<t<8\Delta t, \quad q=200{,}000 \text{ W/m}^2$$
$$t>8\Delta t, \quad q=0 \text{ W/m}^2$$
The initial temperature is 50°C.
b. Add to the temperatures some random errors that have zero mean, have a standard deviation of 2 C, are uncorrelated, and are normal.

5.3. Show that the expressions
$$T_M = T_0 + \sum_{n=1}^{M} \nabla q_n \phi_{M-n+1}, \quad \nabla q_n \equiv q_n - q_{n-1}$$
and
$$T_M = T_0 + \sum_{n=1}^{M} q_n \Delta \phi_{M-n}$$
are equivalent where M is an arbitrary positive integer.

5.4. A plate of thickness $2L$ initially at the constant temperature of T_0 is dropped into a fluid at T_∞. The boundary conditions for one-half the plate can be given by
$$\frac{\partial T}{\partial x}=0 \text{ at } x=0, \quad -L\leq x \leq L, \quad -k\frac{\partial T}{\partial x}\bigg|_L = h[T(L, t) - T_\infty]$$
where h is the heat transfer coefficient.
a. Using the solution of the problem for a step change in the surface heat flux at $x=L$ and insulated at $x=0$, denoted $\Phi(x, t)$, derive
$$T(L, t) = T_0 + \int_0^t h[T_\infty - T(L, \lambda)] \frac{\partial \Phi(L, t-\lambda)}{\partial t} d\lambda \quad \text{(a)}$$
where h is assumed constant. (How is $\phi(x, t)$ related to $\Phi(x, t)$?)

b. Equation (a) can be considered an integral equation for the unknown function $T(L, t)$. By using exact matching, derive the following:

$$T_M = \left(1 + \frac{h\Delta\Phi_0}{2}\right)^{-1} \left[T_0 + h\left(\frac{T_\infty - T_{M-1}}{2}\right)\Delta\Phi_0 \right.$$

$$\left. + h \sum_{n=1}^{M-1} \left(T_\infty - \frac{T_{n-1} + T_n}{2}\right) \Delta\Phi_{M-n}\right], \quad M = 1, 2, \ldots$$

where $T_M = T(L, t_M)$.

c. The foregoing analysis is not restricted to a flat plate. Describe some other geometries for which this analysis applies. This problem is related to the surface element method.[1-3]

5.5. Modify the analysis of Problem 5.4 to treat the case of $T'_\infty = T_\infty(t)$.

5.6. Starting with Eq. (a) of Problem 5.4, derive a sequential function specification algorithm for estimating $T_\infty(t)$ given the temperature history $T(L, t)$. Use $r = 2$ and assume that $T_\infty(t)$ is temporarily constant.

5.7. The Stolz method can be investigated by considering a single temperature error, δY_r, and a special sensitivity matrix. The Stolz method can be represented by the set of n algebraic equations

$$\begin{bmatrix} a & & & & \\ b & a & & & \\ b & b & a & & \\ b & b & b & a & \\ \vdots & \vdots & \vdots & \vdots & \\ b & b & b & b & \cdots & a \end{bmatrix} \begin{bmatrix} \delta q_1 \\ \delta q_2 \\ \vdots \\ \delta q_n \end{bmatrix} = \begin{bmatrix} 0 \\ 0 \\ \vdots \\ \delta Y_r \\ \vdots \\ 0 \end{bmatrix}$$

where δY_r is the rth entry in the right vector. Show that

$$\delta q_r = \frac{\delta Y_r}{a}, \quad \delta q_{r+1} = -\frac{b}{a^2}\delta Y_r$$

$$\delta q_j = \left(1 - \frac{b}{a}\right) \delta Y_r \delta q_{j-1}, \quad j = r+2, r+3, \ldots$$

Using these expressions, show that

$$\delta q_j = -\frac{b}{a^2} \delta Y_r \left(1 - \frac{b}{a}\right)^{j-1-r}, \quad j = r+1, r+2, \ldots$$

5.8. This problem uses the results of Problem 5.7. For the right side of the matrix equation equal to $\mathbf{Y}$ where

$$\mathbf{Y} = [Y_1 \ Y_2 \ \cdots \ Y_r + \delta Y_r \ Y_{r+1} \ \cdots \ Y_n]$$

PROBLEMS

show that

$$q = q_0 + \delta q$$

where q_0 is the solution with $\delta Y_r = 0$ and δq is the solution in Problem 5.7. Under what conditions does a bounded value of δY_r result in an unbounded response in q_j? Show that your answer is equivalent to

$$\phi_2 > 3\phi_1$$

when $\phi_{i+1} - \phi_i \approx$ constant for $i = 2, 3, \ldots$. Use this criterion to find the stability limit for the Stolz procedure for the finite plate of Table 1.1 and the semi-infinite body of Table 1.2.

5.9. a. Using the criterion for stability for the Stolz method, $\phi_2 > 3\phi_1$, given in Problem 5.8, find the stability limit for a solid cylinder with the sensor at the center. Use Table 1.3.
Answer: $\Delta t^+ \approx 0.23$.
b. Find the stability limit for the Stolz method for the same geometry as in part (a) by using the temperature impulse test case.
c. Give conclusions regarding your work for parts (a) and (b).

5.10. Repeat Problem 5.9 for the center point in a sphere. Use Table 1.4. Answer part (a), $\Delta t^+ \approx 0.19$.

5.11. Write a computer program for Eq. (5.3.4). Let r be variable from 1 to 5 and make the maximum value of M be 40. The Y_i and ϕ_i values are inputs.

5.12. a. Use the program developed in Problem 5.11 to verify the results of Figure 5.7.
b. Use the program developed in Problem 5.11 to verify the results of Figure 5.10.
c. Use the program developed in Problem 5.11 to verify the results of Figure 5.12.

5.13. a. Use the program developed in Problem 5.11 to investigate the stability of the function specification method for the flat plate with the sensor at the insulated surface. Let $r = 2$.
b. Repeat part (a) for $r = 3$.
c. Repeat part (a) for $r = 4$.

5.14. Solve Problem 5.13 for a solid cylinder with the sensor at the center.

5.15. Solve Problem 5.13 for a solid sphere with the sensor at the center.

5.16. Write a computer program for Eq. (5.3.4a). The K_i values are inputs. Write the program to treat $r = 3$ and solve for the flat plate geometry with the sensor at $x = L$ and heated at $x = 0$. The time steps are $\Delta t^+ = 0.05$. Investigate the effect of arbitrary K_i values by modifying those that are obtained for the function specification method which are $K_1 = 0.292$,

$K_2 = 8.56$, and $K_3 = 31.8$. For example, try
a. $K_1 = 0.292$, $K_2 = 8.56$, and $K_3 = 15.9$
b. $K_1 = 0.292$, $K_2 = 8.56$, and $K_3 = 47.4$
c. $K_1 = 0$, $K_2 = 0$, and $K_3 = 31.8$

Consider the heat impulse test to check for conservation of energy. What conclusions can you draw?

5.17. Repeat Problem 5.16 for the temperature impulse test case.

5.18. Write a computer program for the function specification method for the IHCP with q temporarily constant for $r = 3$ and two interior temperature histories.

5.19. a. Calculate the dimensionless components of $\mathbf{X}^T\mathbf{X}$ for $n = 10$ and $\Delta t^+ = 0.05$ for the flat plate case with the sensor at $x = L$ and the plate heated at $x = 0$. Also display the components for $\mathbf{X}^T\mathbf{X} + \alpha \mathbf{H}_1^T \mathbf{H}_1$ and for $\alpha = 0.001$. Plot several rows for $\mathbf{X}^T\mathbf{X}$ and $\mathbf{X}^T\mathbf{X} + \alpha \mathbf{H}_1^T \mathbf{H}_1$, including the last row, versus the column numbers. What conclusions can you draw from your results?
b. Repeat part (a) for the case of a solid sphere with the sensor at the center. Use $\Delta t_a^+ = 0.05$.

5.20. a. Write a computer program for the whole domain first-order regularization method for the IHCP.
b. Use the program to reproduce the results shown in Figure 5.15.
c. Use the program to reproduce the results shown in Figure 5.16.
d. Use the program to reproduce Figure 5.17.

5.21. Modify the program developed in Problem 5.20 to solve the zeroth-order regularization method and repeat Problem 5.20.

5.22. Write a computer program for the zeroth-order sequential regularization method for $r = 5$ and compare the gain coefficients with those obtained using the function specification method for $\Delta t^+ = 0.05$ for a flat plate with the sensor at $x = L$ and heated at $x = 0$. (Use Table 4.1.) Vary α over a large range of values.

5.23. An experiment for a flat plate heated at $x = 0$ and insulated at $x = 1$ in. is discussed in Reference 6, p. 402. The plate is Armco iron with $k = 43.3$ Btu/hr-ft-F and $\rho c = 55.6$ Btu/ft^3-F. The initial temperature is 81.36°F. Calculate the surface heat flux using the function specification method and the following actual data.

time	Temp., °F	time	Temp., °F
0	81.36	16.2	85.85
1.8	81.52	18.0	86.91
3.6	81.36	19.8	88.05

PROBLEMS

time	Temp., °F	time	Temp., °F
5.4	81.36	21.6	88.79
7.2	81.85	23.4	89.69
9.0	82.42	25.2	89.77
10.8	83.48	27.0	89.93
12.6	84.48	28.8	90.09
14.4	85.28		

    a. Use $r=3$.
    b. Use $r=4$.
    c. Use $r=5$.

5.24. Repeat Problem 5.23 using the whole domain first-order regularization method. Select several values of α.

5.25. Repeat Problem 5.23 using the sequential first-order regularization method with $r=5$ and several α values.

5.26.   a. Find the digital filter coefficients for Problem 5.23 for the function specification method for $r=4$.
      b. Find the digital filter coefficients for Problem 5.23 for the whole domain regularization method. Select an optimal α value. The midrange of times is to be used to obtain the filter coefficients.

5.27. Using a computer program for the sequential regularization method investigate the effect of the regularization order. Consider the zeroth, first and second orders. Let α vary over a wide range. Investigate both the deterministic and random errors.

5.28. For the whole domain regularization method generate a figure analogous to Figure 5.22. Consider values of $\alpha = 10^{-6}$, 0.001, and 1.

5.29. For the whole domain regularization method generate a figure analogous to Figure 5.23. Consider values of $\alpha = 10^{-6}$, 0.001, and 1.

CHAPTER 6

DIFFERENCE METHODS FOR THE SOLUTION OF THE ONE-DIMENSIONAL INVERSE HEAT CONDUCTION PROBLEM

6.1 INTRODUCTION

Chapter 3 indicated that difference methods offer considerable potential for solving a wide variety of nonlinear direct heat conduction problems. This statement is also true for the inverse heat conduction problem. In particular, the nonlinearity associated with temperature dependent thermal properties can be easily accommodated with difference methods. The recommended procedure is to locally linearize the problem by evaluating all thermal properties at temperatures corresponding to the previous time step. Thermal properties are seldom known to sufficient accuracy to justify iteration. This approach is called quasi-linearization.

Chapter 6 focuses entirely on difference methods for the solution of the IHCP. The techniques presented are equally applicable to lumped capacitance, finite control volume, and finite element techniques. Section 6.2 expands earlier discussions on sensitivity coefficients and how they can be efficiently calculated by difference methods. Section 6.3 considers exactly matching the calculated temperatures with measured values for a single sensor for each time step; Section 6.4 considers multiple sensors. Section 6.5 presents techniques for simultaneously estimating several heat flux components. Sections 6.6, 6.7 and 6.8 apply difference techniques to the function specification method. Section 6.9 considers the sequential regularization method. Section 6.10 introduces space marching techniques. Section 6.11 gives several examples and Section 6.12 lists some computer programs.

6.2 SENSITIVITY COEFFICIENTS AND THEIR CALCULATION BY DIFFERENCE METHODS

The concept of sensitivity coefficients was introduced in Chapter 1 and has been used extensively throughout this book. Since the key to understanding the material in this chapter is a thorough understanding of sensitivity coefficients, some concepts about them will be expanded. Two sensitivity coefficients occur repeatedly in this chapter: (1) that due to a step q_M in heat flux for infinite time duration and (2) that due to a pulse q_M in heat flux for time duration $t_M - t_{M-1}$. Both the heat flux time variation and the corresponding sensitivity coefficient variation with time are shown schematically in Figure 6.1. Actual step function sensitivity coefficients are shown in Figures 1.7 and 1.11 for planar slabs with an insulated surface and slabs that are semi-infinite, respectively. Since the duration of the step function is infinite, the step function sensitivity coefficient grows without bound. The curve labeled A in Figure 6.1 is for a step that begins at time t_{M-1}, that labeled C is for the same step but beginning at time t_M. The symbol Z is used to denote the step function sensitivity coefficient; this Z is the same as the ϕ in Chapters 3–5 if the thermal properties are constant. For a sensor at depth x_k, the temperature response due to the step in heat flux can be written as $T(x_k, t, t_{M-1}, \hat{\mathbf{q}}_{M-1}, q_M)$; t_{M-1} denotes the time at which the heat flux step or pulse begins. The components of the heat flux vector $\hat{\mathbf{q}}_{M-1}$, which are $\hat{q}_1, \hat{q}_2, \ldots, \hat{q}_{M-1}$, contain all the information about the time variation of the heat flux prior to t_{M-1}, and hence contain all the (initial) temperature information necessary to continue the temperature calculations to $t_M, t_{M+1}, \ldots$. The step function sensitivity coefficient is defined by

$$\frac{\partial T(x_k, t, t_{M-1}, \hat{\mathbf{q}}_{M-1}, q_M)}{\partial q_M} = Z(x_k, t, t_{M-1}, \hat{\mathbf{q}}_{M-1}, q_M) \qquad (6.2.1)$$

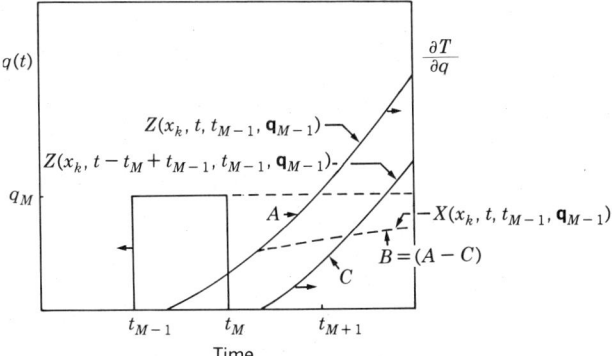

FIGURE 6.1 Schematic of sensitivity coefficient for step in heat flux and pulse in heat flux. A, sensitivity coefficient for step in heat flux at time t_{M-1}. B, sensitivity coefficient for pulse in heat flux, beginning at time t_{M-1}. C, sensitivity coefficient for step in heat flux at time t_M.

where:
- x_k is sensor location
- t is arbitrary time
- t_{M-1} is the time at the beginning of the heat flux step
- $\hat{\mathbf{q}}_{M-1}$ is the previous heat flux history
- q_M is the magnitude of the heat flux step

For a linear problem, Z is independent of q_M; that is, T is linear in q_M. There are applications in which the thermal properties are held constant over a few (typically 1–10) future time steps while allowing some property variation over the entire problem time. For this case, Z is independent of q_M but it does depend on t_{M-1} and $\hat{\mathbf{q}}_{M-1}$ through the temperature profile at t_{M-1}. If the thermal properties are independent of temperature over the entire problem time, then Z is independent of q_M and $\hat{\mathbf{q}}_{M-1}$ and the important time variable is $t - t_{M-1}$, the time from initiation of the step. These three important sensitivity coefficient cases are summarized in Table 6.1. By definition, a sensitivity coefficient is identically zero prior to the initiation of the heat flux step or pulse.

The pulse sensitivity coefficient will be denoted by X with the same arguments as those for the step function sensitivity coefficient. For quasi-linear and constant property cases, superposition can be used to calculate X from Z.

$$X(x_k, t, t_{M-1}, \hat{\mathbf{q}}_{M-1}) = Z(x_k, t, t_{M-1}, \hat{\mathbf{q}}_{M-1})$$
$$- Z[x_k, t - (t_M - t_{M-1}), t_{M-1}, \hat{\mathbf{q}}_{M-1}] \quad (6.2.2)$$

The two terms on the right of Eq. (6.2.2) are shown in Figure 6.1 as curves A and C; for constant thermal properties, curve C is identical to curve A except it is displaced in time by an amount $t_M - t_{M-1}$. Curve B is (A−C) and is the pulse sensitivity coefficient X which is analogous to $\Delta\phi$ of Chapters 3 to 5.

When using difference methods, discrete values of X and Z are important. At time $t = t_M$,

$$X(x_k, t_M, t_{M-1}, \hat{\mathbf{q}}_{M-1}) = Z(x_k, t_M, t_{M-1}, \hat{\mathbf{q}}_{M-1}) \quad (6.2.3)$$

because $Z(x_k, t_{M-1}, t_{M-1}, \hat{\mathbf{q}}_{M-1}) = 0$. For the constant property case, a convenient subscript notation can be used.

$$Z_{k, j-M+1} = Z(x_k, t_j - t_{M-1}) \quad (6.2.4)$$

This same notation will also be used for the quasi-linear property case where it is implicitly understood that all properties are evaluated at the temperature

TABLE 6.1 Notation for Step Function Sensitivity Coefficients

Case	Restrictions	Notation
I	None, general case	$Z(x_k, t, t_{M-1}, \hat{\mathbf{q}}_{M-1}, q_M)$
II	Quasi-linear thermal properties	$Z(x_k, t, t_{M-1}, \hat{\mathbf{q}}_{M-1})$
III	Constant thermal properties	$Z(x_k, t - t_{M-1}) = \phi(x_k, t - t_{M-1})$

SEC. 6.2 SENSITIVITY COEFFICIENTS

profile at t_{M-1}. Using Eq. (6.2.2) and the notation of Eq. (6.2.4), one can show that for constant properties,

$$Z_{k,r} = \sum_{j=1}^{r} X_{k,j}$$

which simply states that for the linear problem the response due to a step is the sum of the response due to a series of pulses of constant magnitude and distributed over the same time interval.

For nonlinear problems, the sensitivity coefficients can be calculated from differences of the temperatures at the same time and location for two values of the heat flux. For example, the temperature distribution is calculated for two different values of heat flux, q^* and $(1+\varepsilon)q^*$, where ε is a small parameter of order 0.001. Then the sensitivity coefficient is given by

$$X(x_k, t, t_{M-1}, \hat{\mathbf{q}}_{M-1}) = \frac{T[x_k, t, t_{M-1}, \hat{\mathbf{q}}_{M-1}, q^*(1+\varepsilon)] - T(x_k, t, t_{M-1}, \hat{\mathbf{q}}_{M-1}, q^*)}{\varepsilon q^*}$$

(6.2.5)

The only difference in the X and Z sensitivity coefficient calculations is the boundary condition:

Sensitivity Coefficient	Boundary Condition
X(pulse)	$q(t) = \begin{cases} q^*, & t_{M-1} \leq t \leq t_M \\ 0, & \text{all other } t \end{cases}$
Z(step)	$q(t) = q^*, \quad t \geq t_{M-1}$

(6.2.6)

The evaluation of Eq. (6.2.5) requires two different temperature calculations (or direct solutions). It is computationally more efficient to calculate X directly from its own partial differential equation, which can be derived from the transient heat conduction equation. For example, consider the problem discussed in Section 3.3 for the quasi-linear case* for $t > t_{M-1}$;

Temperature, T

$$-k \frac{\partial T}{\partial x}\bigg|_{x=0} = q(t)$$

$$\rho c \frac{\partial T}{\partial t} = \frac{\partial}{\partial x}\left(k \frac{\partial T}{\partial x}\right)$$

$$-k \frac{\partial T}{\partial x}\bigg|_{x=L} = q_L(t)$$

$$T(x, t_{M-1}) = F(x)$$

Sensitivity Coefficient, Z

$$-k \frac{\partial Z}{\partial x}\bigg|_{x=0} = 1$$

$$\rho c \frac{\partial Z}{\partial t} = \frac{\partial}{\partial x}\left(k \frac{\partial Z}{\partial x}\right) \quad (6.2.7)$$

$$-k \frac{\partial Z}{\partial x}\bigg|_{x=L} = 0$$

$$Z(x, t_{M-1}) = 0$$

*The quasi-linear case implies that k, ρ, and c are evaluated for the temperature profile at t_{M-1}.

A comparison of the above two sets of equations for T and Z indicates that they both satisfy the same partial differential equation. The boundary and initial conditions are different and those for Z are considerably simpler. If X is to be calculated instead of Z, then the foregoing $x=0$ boundary condition must be modified as follows

$$-k\left.\frac{\partial X}{\partial x}\right|_{x=0} = \begin{cases} 1 & \text{for pulse duration} \\ 0 & \text{for all other times} \end{cases} \tag{6.2.8}$$

The differential equation and the remaining boundary and initial conditions for X are identical to those for Z. Due to the similarity of the T and Z differential equations, considerable computational savings are possible. This point is explored in detail in a subsequent section.

6.3 SINGLE TEMPERATURE SENSOR, FUNCTION SPECIFICATION ($q=C$), SINGLE FUTURE TIME STEP (EXACT MATCHING OF DATA)

A single temperature sensor is considered to be located at a depth x_k below the active surface. If the heat flux q_M is constant over the time interval $t_{M-1} \leq t \leq t_M$, the value of q_M that forces a matching of the computed temperature at x_k with the measured temperature can be calculated. For realistic temperature data containing errors and small dimensionless times, this approach produces significant oscillations in the computed heat flux and would probably not be used in practice. The method is studied here because it aids in the understanding of inverse problems and also because an exact matching of temperature data is a building block that can be used in the development of other techniques.

6.3.1 Modification of Difference Equations of the Direct Heat Conduction Problem for the Solution of the IHCP

Difference methods for solving direct heat conduction problems were introduced in Section 3.3. By analogy with Eq. (3.3.45), the difference equations for a one-dimensional direct heat conduction problem with a nonuniform grid can be written in symbolic form as

$$\begin{bmatrix} b_1 & -a_1 & & & & \\ -c_2 & b_2 & -a_2 & & & \\ & & \ddots & & & \\ & & -c_{N-1} & b_{N-1} & -a_{N-1} & \\ & & & -c_N & b_N \end{bmatrix} \begin{bmatrix} T_1^M \\ T_2^M \\ \vdots \\ T_{N-1}^M \\ T_N^M \end{bmatrix} =$$

$$\begin{bmatrix} b_1' & a_1' & & & & \\ c_2' & b_2' & a_2' & & & \\ & & \ddots & & & \\ & & c_{N-1}' & b_{N-1}' & a_{N-1}' & \\ & & & c_N' & b_N' \end{bmatrix} \begin{bmatrix} T_1^{M-1} \\ T_2^{M-1} \\ \vdots \\ T_{N-1}^{M-1} \\ T_N^{M-1} \end{bmatrix} + \begin{bmatrix} \frac{\Delta t}{(\rho c \Delta x)_1} q_M \\ 0 \\ \vdots \\ \frac{\Delta t}{(\rho c \Delta x)_N} q_{N,M} \end{bmatrix} \tag{6.3.1}$$

SEC. 6.3 SINGLE TEMPERATURE SENSOR

The coefficients a_j, b_j, c_j, a'_j, b'_j, and c'_j depend on grid spacing and thermal properties and are independent of the constant value of heat flux q_M for the linear and quasi-linear cases. For the purpose of discussion, only five nodes are used in what follows. Equation (6.3.1) can be written more compactly as

$$\begin{bmatrix} b_1 & -a_1 & 0 & 0 & 0 \\ -c_2 & b_2 & -a_2 & 0 & 0 \\ 0 & -c_3 & b_3 & -a_3 & 0 \\ 0 & 0 & -c_4 & b_4 & -a_4 \\ 0 & 0 & 0 & -c_5 & b_5 \end{bmatrix} \begin{bmatrix} T_1^M \\ T_2^M \\ T_3^M \\ T_4^M \\ T_5^M \end{bmatrix} = \begin{bmatrix} d'_1 + g_1 q_M \\ d_2 \\ d_3 \\ d_4 \\ d_5 \end{bmatrix} \quad (6.3.2)$$

The d's represent all of the right-hand side of Eq. (6.3.1) except the q_M term. In the direct problem the T_j^M, $j=1,\ldots,5$ values are found from a simultaneous solution of Eq. (6.3.2). Note the explicit dependence of the temperature on the heat flux q_M.

Instead of solving the direct problem given by Eq. (6.3.2), suppose q_M is an unknown and a temperature sensor is located at node 3 ($x_k/L=0.5$). The direct problem unknown T_3^M is now a known measured temperature Y_M. For the IHCP, Eq. (6.3.2) becomes

$$\begin{bmatrix} b_1 & -a_1 & g_1 & 0 & 0 \\ -c_2 & b_2 & 0 & 0 & 0 \\ 0 & -c_3 & 0 & -a_3 & 0 \\ 0 & 0 & 0 & b_4 & -a_4 \\ 0 & 0 & 0 & -c_5 & b_5 \end{bmatrix} \begin{bmatrix} T_1^M \\ T_2^M \\ \hat{q}_M \\ T_4^M \\ T_5^M \end{bmatrix} = \begin{bmatrix} d'_1 \\ d_2 + a_2 Y_M \\ d_3 - b_3 Y_M \\ d_4 + c_4 Y_M \\ d_5 \end{bmatrix} \quad (6.3.3)$$

Equation (6.3.3) represents a system of five linear algebraic equations for the five unknowns (T_1^M, T_2^M, $\hat{q}_M$, T_4^M, T_5^M) and can be solved by any appropriate technique for linear algebraic equations. As indicated in previous chapters, this formulation is linear in the unknown heat flux and should not require iteration. It has been demonstrated by Beck and Wolf[1] that the foregoing formulation is unstable for $\theta=1/2$ and lumped thermal capacitance. Hence, when the data from a single temperature sensor are matched exactly, $\theta=1/2$ should be avoided. Alifanov[16] used the same technique with $\theta=1$. (See Sect. 3.3.3 for definition of θ.)

6.3.2 Sensitivity Coefficient Approach for Exactly Matching Data From a Single Sensor

In Section 6.3.1 it was demonstrated that matching data exactly from a single temperature sensor produces a system of linear algebraic equations. However, the system of equations represented by Eq. (6.3.3) is no longer of the tridiagonal form. (See Section 3.3.4.) Hence, a direct solution of Eq. (6.3.3) is not the most computationally efficient approach. Because of the linearity of the IHCP, it is possible to utilize the tridiagonal structure of the algebraic equations (for one-dimensional problems) and develop a very efficient algorithm.

The temperature field $T(x, t)$ depends in a continuous manner on the unknown heat flux q_M (constant over the interval $t_{M-1} \leq t \leq t_M$). This dependence is written as $T(x, t, t_{M-1}, \mathbf{q}_{M-1}, q_M)$ where $\mathbf{q}_{M-1}$ is the vector of all previous heat flux values and t_{M-1} indicates the time that the heat flux step begins. Because the temperature field is continuous in q_M, it can be expanded in a Taylor series about an arbitrary but known value of heat flux q^*,

$$T(x, t, t_{M-1}, \mathbf{q}_{M-1}, q_M) = T(x, t, t_{M-1}, \mathbf{q}_{M-1}, q^*)$$
$$+ (q_M - q^*) \left. \frac{\partial T(x, t, t_{M-1}, \mathbf{q}_{M-1}, q_M)}{\partial q_M} \right|_{q_M = q^*}$$
$$+ \frac{(q_M - q^*)^2}{2!} \left. \frac{\partial^2 T(x, t, t_{M-1}, \mathbf{q}_{M-1}, q_M)}{\partial q_M^2} \right|_{q_M = q^*} + \cdots$$
(6.3.4)

For linear problems, only the first derivative in Eq. (6.3.4) is nonzero, thus the following is an exact result for location x at time t_M

$$T(x, t_M, t_{M-1}, \mathbf{q}_{M-1}, q_M) = T(x, t_M, t_{M-1}, \mathbf{q}_{M-1}, q^*)$$
$$+ (q_M - q^*) X(x, t_M, t_{M-1}, \mathbf{q}_{M-1}) \quad (6.3.5)$$

where the sensitivity coefficient is defined by

$$X(x, t_M, t_{M-1}, \mathbf{q}_{M-1}) = \left. \frac{\partial T(x, t_M, t_{M-1}, \mathbf{q}_{M-1}, q_M)}{\partial q_M} \right|_{q_M = q^*} \quad (6.3.6)$$

Note that X does not depend on q_M for a linear problem.

Equation (6.3.5) is a prescription for calculating the temperature field corresponding to q_M provided the temperature field is known for a given q^* and a means exists for calculating the sensitivity coefficient X. If the problem is amenable to an analytical solution, X can be calculated as indicated in Section 1.6. Difference methods for obtaining X are discussed in Section 6.2 in connection with Eq. (6.2.5).

For the IHCP in which the sensor data are matched exactly, the left-hand side of Eq. (6.3.5) is replaced by the experimental temperature; hence, Eq. (6.3.5) becomes

$$Y_M = \overset{*}{T}{}_k^M + (\hat{q}_M - q^*) X_{k,1} \quad (6.3.7)$$

Solving Eq. (6.3.7) for the estimate of the unknown heat flux, $\hat{q}_M$, gives:

$$\hat{q}_M = q^* + \frac{Y_M - \overset{*}{T}{}_k^M}{X_{k,1}} \quad (6.3.8)$$

where $\overset{*}{T}{}_k^M$ is the temperature at t_M for the sensor node k with $q = q^*$ over $t_{M-1} \leq t \leq t_M$. The calculational procedure is to assume an arbitrary value of heat flux q^*, calculate the temperature field $T(x, t_M, t_{M-1}, \mathbf{q}_{M-1}, q^*)$, and knowing the sensitivity coefficients $X(x, t_M, t_{M-1}, \mathbf{q}_{M-1})$, calculate the heat flux q_M from Eq. (6.3.8) that exactly matches the temperature data Y_M. Once q_M is

SEC. 6.3 SINGLE TEMPERATURE SENSOR

known, the complete temperature field can be calculated from Eq. (6.3.5). Equation (6.3.8) is equivalent to the Stolz equation given by Eq. (5.3.1) if $q^* = 0$.

An efficient way of calculating sensitivity coefficients is to solve the differential equation for X that is analogous to Eq. (6.2.7). Since the T and X calculations are similar, considerable computational savings are possible. The development of an efficient algorithm requires a thorough understanding of the following *tridiagonal matrix algorithm* (TDMA):

Linear equation form:

$$b_1 u_1 - a_1 u_2 = d_1$$
$$-c_j u_{j-1} + b_j u_j - a_j u_{j+1} = d_j, \quad j=2, 3, \ldots, N-1$$
$$-c_N u_{N-1} + b_N u_N = d_N \tag{6.3.9}$$

Forward elimination (starting at node N):

$$\omega_N = \frac{1}{b_N}, \quad e_N = \omega_N c_N, \quad f_N = \omega_N d_N$$

$$\omega_j = \frac{1}{b_j - a_j e_{j+1}}, \quad j = N-1, N-2, \ldots, 1$$

$$f_j = \omega_j(d_j + a_j f_{j+1}), \quad e_j = \omega_j c_j, \quad j = N-1, N-2, \ldots, 2$$

$$f_1 = \omega_1(d_1 + a_1 f_2) \tag{6.3.10}$$

Back substitution:

$$u_1 = f_1$$
$$u_j = e_j u_{j-1} + f_j, \quad j = 2, 3, \ldots, N \tag{6.3.11}$$

In comparing the two problems for T and X, the following conclusions can be drawn:

T		X
a_j, b_j, c_j	←same for both→	a_j, b_j, c_j
ω_j, e_j	←same for both because they→ depend only on a_j, b_j, c_j	ω_j, e_j
$d_1 = d_1' + g_1 q_M$		$d_1 = g_1$
$d_j \neq 0, j = 2, 3, \ldots, N$		$d_j = 0, j = 2, 3, \ldots, N$

Note that the difference equations for X can be derived by differentiating the difference equations for T with respect to q_M. Going through the forward elimination portion of the algorithm for X, it can be shown that all the f_n's are identically zero except for f_1 and

$$f_1 = g_1 \omega_1 \tag{6.3.12}$$

Consequently, the sensitivity coefficients are calculated as follows:

$$X_{1,1} = f_1 = g_1 \omega_1 \tag{6.3.13}$$

$$X_{j,1} = e_j X_{j-1,1}, \quad j=2, 3, \ldots, N \tag{6.3.14}$$

The algorithm for exactly matching the temperature data from a single sensor can be summarized as follows:

1. Assume an arbitrary value of q^* and calculate the entire temperature field $\overset{*}{T} = T(x, t_M, t_{M-1}, \mathbf{q}_{M-1}, q^*)$. $q^* = q_{M-1}$ is a common choice; however, $q^* \equiv 0$ is also perfectly acceptable for the linear problem and will reduce the total number of calculations.

2. Using the same matrix coefficients a_n, b_n, c_n, ω_n, and e_n as were used for the $\overset{*}{T}$ calculation in step 1, calculate the sensitivity coefficients from Eqs. (6.3.13) and (6.3.14)

$$X_{1,1} = f_1 = g_1 \omega_1 \tag{6.3.13}$$

$$X_{j,1} = e_j X_{j-1,1}, \quad j=2, 3, \ldots, N \tag{6.3.14}$$

3. Calculate the heat flux that exactly matches the experimental temperature data by using the Taylor series expansion

$$\hat{q}_M = q^* + \frac{Y_M - \overset{*}{T}{}_k^M}{X_{k,1}} \tag{6.3.8}$$

4. Calculate the complete temperature field from the Taylor series expansion

$$\hat{T}_j^M = \overset{*}{T}{}_j^M + (\hat{q}_M - q^*) X_{j,1} \tag{6.3.15}$$

Table 6.2 presents an arithmetic operation count to determine the relative increase in computations of the inverse problem over the direct problem. The operational counts can be used to evaluate the computational efficiency for a particular algorithm. Table 6.3 presents the number of cycles on the CDC 7600 computer required for the four arithmetic operations; one cycle is 27 ns. Let

TABLE 6.2 Operational Counts for One Time Step of Inverse Algorithm that Exactly Matches Single Temperature Sensor Data (N=total number of nodes)

Step	Equation	+ or −	×	÷	Comments
1	(6.3.10) (6.3.11)	$3(N-1)$	$5(N-1)$	N	Direct calculation
2	(6.3.13) (6.3.14)		N		
3	(6.3.8)	2		1	Inverse calculations only
4	(6.3.15)	$2N$	N		
Sum(2+3+4)		$2(N+1)$	$2N$	1	

SEC. 6.3 SINGLE TEMPERATURE SENSOR

N_a, N_m, and N_d represent the cycle times required for a single addition, multiplication, and division, respectively. Using the information in Tables 6.2 and 6.3, the approximate fractional increase in computation time for the inverse problem relative to the direct problem is given by:

$$\frac{\Delta \tau}{\tau} = \frac{2(N+1) + 2N(\tau_m/\tau_a) + (\tau_d/\tau_a)}{3(N-1) + 5(N-1)(\tau_m/\tau_a) + N(\tau_d/\tau_a)} \quad (6.3.16)$$

"Computation index" is probably a more appropriate term for $\Delta\tau/\tau$ because Eq. (6.3.16) does not account for the fact that some cycle time is lost between the end of one operation and the beginning of the next operation. For a particular computer, $\Delta\tau/\tau$ depends only on the total number of nodes N. Using the computation times given in Table 6.3, $\Delta\tau/\tau$ for various values of N are presented in Table 6.4. The IHCP causes less than a 40% increase in computations in comparison to the direct problem.

Note that Eq. (6.3.16) does not depend on the node number of the temperature sensor; a temperature sensor at node 1 requires the same number of computations as a backface sensor at node N. Some computational improvement is

TABLE 6.3 Relative Speeds of Various Arithmetic Operations on the CDC 7600 Computer[2]

Operation	No. of Machine Cycles	Relative Time (τ/τ_a)
+ or −	4	1.0
×	5	1.25
÷	20	5.0

TABLE 6.4 Computation Index for Exactly Matching Data from Single Temperature Sensor (for CDC 7600 computer, Table 6.3). No. of elements = $N-1$

N	$\Delta\tau/\tau = \dfrac{4.5N+7}{14.25N-9.25}$
11	0.383
21	0.350
31	0.339
41	0.333
81	0.324

possible provided the algorithm is restructured. This restructuring is possible because it is not necessary to know the sensitivity coefficients $X_{j,1}$ and $\overset{*}{T}{}_{j}^{M}$ for $x > x_K$ (the sensor depth). If the sensor is near node 1, the active surface, a restructuring can offer considerable savings. The computational algorithm is reordered in the following manner:

1. Assume a value for q^* and perform the forward elimination portion of the tridiagonal matrix algorithm. (Note that the only calculation in this step that depends on the heat flux is f_1.)

2. Calculate $\overset{*}{T}{}_{j}^{M}$ from the back substitution portion of the tridiagonal matrix algorithm.

$$\overset{*}{T}{}^{M} = f_1$$
$$\overset{*}{T}{}_{j}^{M} = e_j \overset{*}{T}{}_{j-1}^{M} + f_j, \quad j=2, 3, \ldots, K \qquad (6.3.17)$$
$$\overset{*}{T}(x_K, t_M) = \overset{*}{T}{}_{K}^{M}$$

3. Calculate the sensitivity coefficient $X_{j,1}$ in exactly the same way as for the previous algorithm

$$X_{1,1} = g_1 \omega_1 \qquad (6.3.13)$$

$$X_{j,1} = e_j X_{j-1,1}, \quad j=2, 3, \ldots, K \qquad (6.3.14)$$

4. Calculate the heat flux that exactly matches the experimental data by using the Taylor series expansion,

$$q_M = q^* + \frac{Y_M - \overset{*}{T}{}_{K}^{M}}{X_{K,1}} \qquad (6.3.8)$$

5. With the heat flux q_M calculated from step 4, recalculate $f_1 = \omega_1(d_1 + a_1 f_2)$ and repeat the back substitution portion of the tridiagonal matrix algorithm.

$$T_1^M = f_1$$
$$T_j^M = e_j T_{j-1}^M + f_j, \quad j=2, 3, \ldots, N \qquad (6.3.18)$$

Note that it is not possible to calculate the temperature field from the Taylor series result, Eq. (6.3.15), because the sensitivity coefficients for depths greater than the temperature sensor depth were not calculated in step 2.

An operational count demonstrates that the restructured algorithm is more efficient than the one first proposed. Table 6.5 summarizes the operational count for the new algorithm. Note that when steps 1 and 5 are combined, they are equivalent to one direct problem solution plus one additional evaluation of $f_1 = \omega_1(d_1 + a_1 f_2)$.

The computation index for the restructured algorithm is

$$\frac{\Delta \tau}{\tau} = \frac{K+2+(2K+1)(\tau_m/\tau_a)+(\tau_d/\tau_a)}{3(N-1)+5(N-1)(\tau_m/\tau_a)+N(\tau_d/\tau_a)} \qquad (6.3.19)$$

SEC. 6.3 SINGLE TEMPERATURE SENSOR

TABLE 6.5 Operational Count for Restructured Algorithm that Exactly Matches Single Temperature Sensor Data (K=sensor node number, N=number of nodes, $K \leqslant N$)

Step	Equation	+ or −	×	÷	Comments
1	(6.3.11)				One equivalent direct problem
5	(6.3.10)				solution
1+5		$3(N-1)$	$5(N-1)$	N	
2	(6.3.17)	$K-1$	$K-1$		$\dot{T}_n^M$ calculation
3	(6.3.13)		K		$X_{n,1}$ calculation
	(6.3.14)				
4	(6.3.8)	2		1	
5	—	1	2		$f_1 = \omega_1(d_1 + a_1 f_2)$
Sum (inverse)		$K+2$	$2K+1$	1	Extra inverse calculations

Figure 6.2 presents the results of Eq. (6.3.19) for the CDC 7600 computer. If the temperature sensor is located near the active surface (small numerical value of K), then the restructured algorithm is very efficient and is faster than the originally proposed algorithm.

The foregoing type of minimization of calculations for the IHCP was first developed by Blackwell.[3] However, the original approach did not make explicit use of the sensitivity coefficient concept.

In real-world problems, there will be errors in the temperature measurements. If the temperature data are matched exactly, the temperature errors produce heat flux errors. Consequently, exact matching of the temperature data

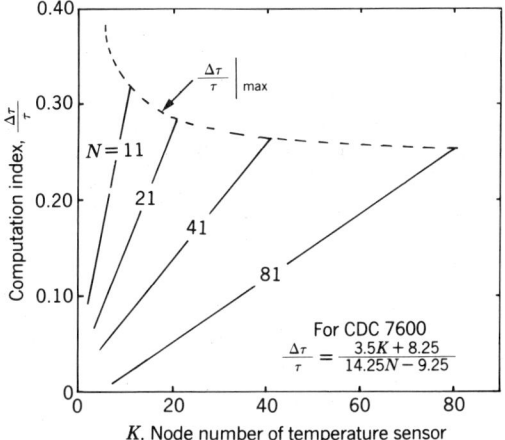

FIGURE 6.2 Computation index for algorithm that exactly matches single temperature sensor data (K=sensor node number, N=number of nodes, $K \leqslant N$).

is not recommended unless the temperature data are smoothed prior to calculating the heat flux and/or relatively large time steps are taken ($\alpha \Delta t/x_K^2 > 1$). The function specification method using future temperatures (Chapter 4) can be used with difference methods and allows smaller computational time steps to be used without encountering stability problems; this method is discussed in Sections 6.6 and 6.7.

6.4 MULTIPLE TEMPERATURE SENSORS, FUNCTION SPECIFICATION ($q=C$), SINGLE FUTURE TIME STEP

Multiple temperature sensors are recommended in order to obtain as much experimental information as possible. These include sensors at different depths as well as several sensors at the same depth. For multiple sensors, only one sensor can be matched exactly. An alternative is to determine the value of q_M, constant over the interval $t_{M-1} \leqslant t \leqslant t_M$, that minimizes the least squares error for all sensors considered. If there is a total of J sensors, then the least squares error is defined as

$$S = \sum_{k=1}^{J} [Y_{k,M} - T(x_k, t_M, t_{M-1}, \mathbf{q}_{M-1}, q_M)]^2 \qquad (6.4.1)$$

where k is the sensor index and M is the time index in $Y_{k,M}$.* It is important to remember that k is the sensor index, not the finite difference node index. Differentiating Eq. (6.4.1) with respect to q_M, gives

$$\frac{\partial S}{\partial q_M} = 0 = 2 \sum_{k=1}^{J} [Y_{k,M} - T(x_k, t_M, t_{M-1}, \mathbf{q}_{M-1}, q_M)][-X(x_k, t_M, t_{M-1}, \mathbf{q}_{M-1})]$$
$$(6.4.2)$$

where X is defined by

$$X(x_k, t_M, t_{M-1}, \mathbf{q}_{M-1}) = \frac{\partial T(x_k, t_M, t_{M-1}, \mathbf{q}_{M-1}, q_M)}{\partial q_M} = X_{k,1} \qquad (6.4.3)$$

As in Section 6.3, an arbitrary value of heat flux q^* is assumed and the Taylor series expansion given by Eq. (6.3.5) is applied:

$$T(x_k, t_M, t_{M-1}, \mathbf{q}_{M-1}, q_M) = \overset{*}{T}_k^M + (q_M - q^*) X_{k,1} \qquad (6.4.4)$$

Substituting Eq. (6.4.4) into Eq. (6.4.2), replacing q_M by $\hat{q}_M$ and solving for $\hat{q}_M$, gives

$$\hat{q}_M = q^* + \frac{\sum_{k=1}^{J} X_{k,1}(Y_{k,M} - \overset{*}{T}_k^M)}{\sum_{k=1}^{J} X_{k,1}^2} \qquad (6.4.5)$$

---

*The double subscript notation $Y(x_k, t_M) = Y_{k,M}$ corresponds to the classical (x, t) notation.

SEC. 6.4 MULTIPLE TEMPERATURE SENSORS

If $J=1$, then Eq. (6.4.5) reduces to Eq. (6.3.8). Again, note that $q^* \equiv 0$ is a valid assumption.

The computation index in Figure 6.2 is very nearly valid for the multiple sensor case provided K is the node number of the sensor located the greatest distance below the active surface. The only additional calculation of the multiple-sensor problem compared to the single-sensor problem (with the sensor located at node K) is the evaluation of Eq. (6.4.5) instead of Eq. (6.3.8). This additional calculation is generally trivial in comparison to the total number of calculations. Hence, Figure 6.2 is an approximate indicator for the multiple temperature sensor case.

An alternative to the use of Eq. (6.4.5) for the multiple-sensor problem for a single future time step is to calculate the heat flux $\tilde{q}_{k,M}$ that exactly matches the individual (kth) measured temperature, $Y_{k,M}$, and then take a weighted average of all $\tilde{q}_{k,M}$'s. This average can be derived by using a Taylor series slightly different from Eq. (6.4.4). Expanding the temperature field for q_M in terms of $\tilde{q}_{k,M}$

$$T(x_k, t_M, t_{M-1}, \mathbf{q}_{M-1}, q_M) = T(x_k, t_M, t_{M-1}, \mathbf{q}_{M-1}, \tilde{q}_{k,M})$$
$$+ (q_M - \tilde{q}_{k,M}) \left. \frac{\partial T(x_k, t_M, t_{M-1}, \mathbf{q}_{M-1}, q_M)}{\partial q_M} \right|_{q_M = \tilde{q}_{k,M}}$$
(6.4.6)

But exactly matching the kth sensor at t_M gives

$$T(x_k, t_M, t_{M-1}, \mathbf{q}_{M-1}, \tilde{q}_{k,M}) = Y_{k,M} \tag{6.4.7}$$

and

$$\left. \frac{\partial T(x_k, t_M, t_{M-1}, \mathbf{q}_{M-1}, q_M)}{\partial q_M} \right|_{q_M = \tilde{q}_{k,M}} = X(x_k, t_M, t_{M-1}, \mathbf{q}_{M-1}) = X_{k,1} \tag{6.4.8}$$

The minimum least squares condition, Eq. (6.4.2), becomes

$$\sum_{k=1}^{J} X_{k,1}(\hat{q}_M - \tilde{q}_{k,M}) X_{k,1} = 0 \tag{6.4.9}$$

Solving for $\hat{q}_M$,

$$\hat{q}_M = \sum_{k=1}^{J} w_k \tilde{q}_{k,M} \tag{6.4.10}$$

where the heat flux weighting factors are related to the sensitivity coefficients through

$$w_k = \frac{X_{k,1}^2}{\sum_{k=1}^{J} X_{k,1}^2}, \quad \sum_{k=1}^{J} w_k = 1 \tag{6.4.11}$$

Equation (6.4.10) provides an alternative interpretation to this solution of the multiple sensor, single future time step IHCP. The procedure is to treat each sensor independently (over a single time step) and determine the heat flux $\tilde{q}_{k,M}$ that exactly matches the single data point $Y_{k,M}$. This heat flux is related to the assumed arbitrary heat flux q^* through the Taylor series expansion

$$\tilde{q}_{k,M} = q^* + \frac{Y_{k,M} - \overset{*}{T}{}_k^M}{X_{k,1}}, \quad k = 1, 2, \ldots, J \tag{6.4.12}$$

After evaluating Eq. (6.4.12), these $\tilde{q}_{k,M}$ heat flux values are weighted (averaged) according to Eq. (6.4.10). From the discussion of sensitivity coefficients in Section 1.6, the sensors closest to the active surface will have the largest sensitivity coefficients and hence will automatically be weighted more heavily. For a problem with constant thermal properties, the sensitivity coefficients $X_{k,1}$ and hence heat flux weighting coefficients w_k need be calculated only once. If the thermal properties are treated as being quasi-linear in temperature, both $X_{k,1}$ and w_k must be recalculated at each problem time step.

EXAMPLE 6.1. Using the information in Table 1.1 for the response of a planar slab with an insulated inactive surface to a step in heat flux, calculate the heat flux weighting coefficients of Eq. (6.4.11) for two sensors at dimensionless depths of $x_1/L = 0$ and $x_2/L = 0.5$ and for two dimensionless time steps of $\Delta t^+ = \alpha \Delta t/L^2 = 0.05$ and 0.5. Note that Table 1.1 is also shown in graphical form in Fig. 1.7.

Solution. Although Table 1.1 contains the dimensionless temperature response, the dimensionless sensitivity coefficients $\partial T/\partial (qL/k)$ are identical to T^+.

$$\Delta t^+ = \frac{\alpha \Delta t}{L^2} = 0.05 \qquad \qquad \frac{\alpha \Delta t}{L^2} = 0.5$$

$$X_{1,1}^+ = 0.252313 \qquad \qquad X_{1,1}^+ = 0.831876$$
$$X_{2,1}^+ = 0.0153659 \qquad \qquad X_{2,1}^+ = 0.458333$$
$$w_1 = 0.9963 \ (x/L = 0) \qquad w_1 = 0.7671$$
$$w_2 = 0.0037 \ (x/L = 0.5) \qquad w_2 = 0.2329 \qquad \square$$

In comparing the w_k's for the foregoing example, several important conclusions can be drawn. For $\Delta t^+ = 0.05$ the sensor at $x^+ = 0.5$ yields very little information because its sensitivity coefficient is small relative to that for the surface-mounted sensor. For $\Delta t^+ = 0.5$, the time step is sufficiently large that sensor 2 has time to respond; however, sensor 1 is still weighted more heavily because of the square of the sensitivity coefficients.

Because temperature measurements invariably contain errors, the heat flux components determined by exactly matching a single sensor contain amplified errors. Even for the multiple sensors considered in Example 6.1, sensors located far apart tend to offer very little smoothing of the heat flux; consequently, other approaches should be considered. Smoothing of the temperature data prior to the solution of the IHCP certainly reduces the oscillations in the heat flux; however, this approach does not make use of the fact that a pulse of magnitude q_M over the interval $t_{M-1} \leq t \leq t_M$ has an influence on the sensor

SEC. 6.5 WHOLE DOMAIN ESTIMATION WITH DIFFERENCE METHODS

response for $t > t_M$. One approach that makes use of all temperature information simultaneously is the whole domain estimation procedure which is the subject of the next section.

6.5 WHOLE DOMAIN ESTIMATION WITH DIFFERENCE METHODS

The whole domain procedure has been discussed in Sections 4.4.2. This procedure is most applicable when the physics of the problem suggests that the heat flux variation with time is of a particular analytical form, for example, forms such as given by Eqs. (4.4.1)–(4.4.5). This list is by no means exhaustive. A definite advantage of the whole domain estimation procedure is that stability problems are minimized. However, more computations are required.

The whole domain procedure will be introduced first by means of a specific example and then will be generalized. Suppose a parabola adequately represents the time variation of the heat flux,

$$q(t) = \beta_1 + \beta_2 t + \beta_3 t^2 \qquad (6.5.1)$$

The parameters $\beta_1, \beta_2, \beta_3$ are to be estimated so that the computed temperatures agree (within certain limits) with the experimentally measured temperatures. The extent of the fit or agreement is determined by the ordinary least squares criteria. If Y_i is the experimentally measured temperature and T_i is the predicted temperature, both at time t_i and at the same location, then the least squares error is given by

$$S = \sum_{i=1}^{r} (Y_i - T_i)^2 \qquad (6.5.2)$$

The predicted temperatures depend implicitly on the parameters $(\beta_1, \beta_2, \beta_3)$

$$T_i = T_i(\beta_1, \beta_2, \beta_3) \qquad (6.5.3)$$

The object is to choose $(\beta_1, \beta_2, \beta_3)$ such that the ordinary least squares error S is minimized. This condition is determined by

$$\frac{\partial S}{\partial \beta_j} = -2 \sum_{i=1}^{r} (Y_i - \hat{T}_i) \frac{\partial T_i}{\partial \beta_j} = 0, \quad j = 1, 2, 3 \qquad (6.5.4)$$

Assuming that the computed temperature is continuous in the parameters $(\beta_1, \beta_2, \beta_3)$, the temperature can be expanded in a Taylor series about trial values of the parameters. If $(\beta_1^v, \beta_2^v, \beta_3^v)$ represent the trial values of the parameters and $(\beta_1^{v+1}, \beta_2^{v+1}, \beta_3^{v+1})$ represent new or improved estimates of the parameters, then

$$\hat{T}_i^{v+1} = \hat{T}_i^v + (\beta_1^{v+1} - \beta_1^v) \left.\frac{\partial T_i}{\partial \beta_1}\right|_{\beta = \beta^v} + (\beta_2^{v+1} - \beta_2^v) \left.\frac{\partial T_i}{\partial \beta_2}\right|_{\beta = \beta^v} + (\beta_3^{v+1} - \beta_3^v) \left.\frac{\partial T_i}{\partial \beta_3}\right|_{\beta = \beta^v}$$

$$(6.5.5)$$

234　CHAP. 6　ONE-DIMENSIONAL INVERSE HEAT CONDUCTION PROBLEM

Note that higher-order terms are ignored in the Taylor series expansion; however, the expansion is exact for linear problems. Substituting Eq. (6.5.5) into each of the three equations represented by (6.5.4) and simplifying,

$$\begin{bmatrix} \sum_{i=1}^{r} \frac{\partial T_i}{\partial \beta_1} \frac{\partial T_i}{\partial \beta_1} & \sum_{i=1}^{r} \frac{\partial T_i}{\partial \beta_2} \frac{\partial T_i}{\partial \beta_1} & \sum_{i=1}^{r} \frac{\partial T_i}{\partial \beta_3} \frac{\partial T_i}{\partial \beta_1} \\ \sum_{i=1}^{r} \frac{\partial T_i}{\partial \beta_1} \frac{\partial T_i}{\partial \beta_2} & \sum_{i=1}^{r} \frac{\partial T_i}{\partial \beta_2} \frac{\partial T_i}{\partial \beta_2} & \sum_{i=1}^{r} \frac{\partial T_i}{\partial \beta_3} \frac{\partial T_i}{\partial \beta_2} \\ \sum_{i=1}^{r} \frac{\partial T_i}{\partial \beta_1} \frac{\partial T_i}{\partial \beta_3} & \sum_{i=1}^{r} \frac{\partial T_i}{\partial \beta_2} \frac{\partial T_i}{\partial \beta_3} & \sum_{i=1}^{r} \frac{\partial T_i}{\partial \beta_3} \frac{\partial T_i}{\partial \beta_3} \end{bmatrix}^v \begin{bmatrix} \beta_1^{v+1} - \beta_1^v \\ \beta_2^{v+1} - \beta_2^v \\ \beta_3^{v+1} - \beta_3^v \end{bmatrix} =$$

$$\begin{bmatrix} \sum_{i=1}^{r} (Y_i - \hat{T}_i) \frac{\partial T_i}{\partial \beta_1} \\ \sum_{i=1}^{r} (Y_i - \hat{T}_i) \frac{\partial T_i}{\partial \beta_2} \\ \sum_{i=1}^{r} (Y_i - \hat{T}_i) \frac{\partial T_i}{\partial \beta_3} \end{bmatrix}^v \quad (6.5.6)$$

The superscripts v on the coefficient matrix and the right-hand-side vector indicate that all terms should be evaluated at $\beta = \beta^v$. Equation (6.5.6) is a linear system of three algebraic equations with three unknowns; $\beta_j^{v+1} - \beta_j^v$ is treated as an unknown correction factor. For a linear problem, the sensitivity coefficients are independent of the parameters, the coefficient matrix is in turn independent of parameters, and only one step (iteration) is required. Even though the heat flux in this example is linear in the parameters $(\beta_1, \beta_2, \beta_3)$, temperature-dependent thermal properties make the problem nonlinear and require iteration. For the nonlinear problem, one approach to the sensitivity coefficient calculation is to use differences. For example,

$$\frac{\partial T_i}{\partial \beta_1} \approx \frac{T_i[(1+\varepsilon)\beta_1, \beta_2, \beta_3] - T_i(\beta_1, \beta_2, \beta_3)}{\varepsilon \beta_1} \quad (6.5.7)$$

where ε is a small number, say 0.001. The evaluation of each sensitivity coefficient requires the solution of two direct problems but one is the standard condition; for the foregoing example with three parameters, a total of four direct problems must be solved for each iteration in Eq. (6.5.6). Each iteration will also require the simultaneous solution of three linear algebraic equations, Eq. (6.5.6). (See the comments below (4.5.19).)

At first glance, it might appear that some computational savings are possible if the sensitivity coefficients are calculated directly from a differential equation similar to Eq. (6.2.7). However, this is not the case for the nonlinear temperature-dependent thermal property case. For the whole domain problem, it may not be possible to locally linearize the problem by evaluating the thermal properties at the previous time step as it was for the sequential procedure of Section 6.3. Considering Eq. (6.2.7) as the basic model for the temperature field, the

SEC. 6.5 WHOLE DOMAIN ESTIMATION WITH DIFFERENCE METHODS

sensitivity-coefficient (T_{β_j}) differential equation is

$$-T_{\beta_j} \frac{\partial k}{\partial T} \frac{\partial T}{\partial x}\bigg|_{x=0} - k \frac{\partial T_{\beta_j}}{\partial x}\bigg|_{x=0} = t^{j-1}, \quad j=1, 2, 3 \quad (6.5.8)$$

$$\rho c \frac{\partial T_{\beta_j}}{\partial t} - \frac{\partial}{\partial x}\left(k \frac{\partial T_{\beta_j}}{\partial x}\right) = -T_{\beta_j} \frac{\partial}{\partial T}(\rho c) \frac{\partial T}{\partial t} + \frac{\partial}{\partial x}\left(T_{\beta_j} \frac{\partial k}{\partial T} \frac{\partial T}{\partial x}\right), \quad j=1, 2, 3 \quad (6.5.9)$$

$$-T_{\beta_j} \frac{\partial k}{\partial T} \frac{\partial T}{\partial x}\bigg|_{x=L} - k \frac{\partial T_{\beta_j}}{\partial x}\bigg|_{x=L} = 0 \quad j=1, 2, 3 \quad (6.5.10)$$

$$T_{\beta_j}(x, 0) = 0 \quad (6.5.11)$$

This sensitivity-coefficient differential equation is linear in T_{β_j} if the temperature field is known because terms like k, ρc, $\partial k/\partial T$, and $\partial(\rho c)/\partial T$ can be treated as known functions of position. However, the differential equation for T_{β_j} is no longer the same as that for T. Although the parameters ($\beta_1, \beta_2, \beta_3$) do not explicitly appear in Eqs. (6.5.8)–(6.5.11), they appear implicitly in the predicted temperature field $T(\beta_1, \beta_2, \beta_3)$. If the heat flux model is nonlinear in the parameters, then the parameters appear explicitly in the sensitivity coefficient boundary condition. For example, using Eq. (4.4.4) as the heat flux model,

$$q(t) = \beta_1(1 - e^{-\beta_2 t}) \quad (4.4.4)$$

then the boundary condition for T_{β_2} is

$$-\left(T_{\beta_2} \frac{\partial k}{\partial T} \frac{\partial T}{\partial x}\right)\bigg|_{x=0} - \left(k \frac{\partial T_{\beta_2}}{\partial x}\right)\bigg|_{x=0} = \beta_1 t e^{-\beta_2 t} \quad (6.5.12)$$

which is nonlinear in the parameters.

The foregoing procedure can be readily extended to heat flux models with a large number of parameters. In such a case, it is beneficial to use matrix notation; additional details on matrix calculus can be found in Chapter 6 of Beck and Arnold.[4] Let **Y** be the vector of the discrete temperature measurements and **T** be the vector of the corresponding computed temperatures at the given sensor location,

$$\mathbf{T} = \begin{bmatrix} T_1 \\ T_2 \\ \vdots \\ T_r \end{bmatrix} \quad \mathbf{Y} = \begin{bmatrix} Y_1 \\ Y_2 \\ \vdots \\ Y_r \end{bmatrix} \quad (6.5.13)$$

The subscripts on T and Y denote time and there are r discrete times. The matrix form of the least squares function, Eq. (6.5.2), is

$$S = (\mathbf{Y} - \mathbf{T})^T (\mathbf{Y} - \mathbf{T}) \quad (6.5.14)$$

The vector of p parameters, $\boldsymbol{\beta}$, is

$$\boldsymbol{\beta} = \begin{bmatrix} \beta_1 \\ \beta_2 \\ \vdots \\ \beta_p \end{bmatrix} \tag{6.5.15}$$

(with $p < r$) and it is to be determined so that S is a minimum. This procedure will involve the matrix derivative of Eq. (6.5.14); this derivative is defined as

$$\nabla_\beta \equiv \begin{bmatrix} \dfrac{\partial}{\partial \beta_1} \\ \dfrac{\partial}{\partial \beta_2} \\ \vdots \\ \dfrac{\partial}{\partial \beta_p} \end{bmatrix} \tag{6.5.16}$$

Differentiating Eq. (6.5.14) with respect to $\boldsymbol{\beta}$,

$$\nabla_\beta S = -2(\nabla_\beta \mathbf{T}^T)(\mathbf{Y} - \mathbf{T}) = 0 \tag{6.5.17}$$

The sensitivity coefficient matrix is defined as

$$\mathbf{X}(\boldsymbol{\beta}) = (\nabla_\beta \mathbf{T}^T)^T \tag{6.5.18}$$

or,

$$\mathbf{X} = \begin{bmatrix} \dfrac{\partial T_1}{\partial \beta_1} & \dfrac{\partial T_1}{\partial \beta_2} & \cdots & \dfrac{\partial T_1}{\partial \beta_p} \\ \vdots & \vdots & & \vdots \\ \dfrac{\partial T_r}{\partial \beta_1} & \dfrac{\partial T_r}{\partial \beta_2} & \cdots & \dfrac{\partial T_r}{\partial \beta_p} \end{bmatrix} \tag{6.5.19}$$

Substituting Eq. (6.5.18) into Eq. (6.5.17),

$$\mathbf{X}^T(\mathbf{Y} - \mathbf{T}) = 0 \tag{6.5.20}$$

The matrix form of the Taylor series expansion of the temperatures in terms of parameter space is

$$\mathbf{T}(\boldsymbol{\beta}^{\nu+1}) = \mathbf{T}^{\nu+1} = \mathbf{T}^\nu + (\nabla_\beta \mathbf{T}^T)^T|_{\boldsymbol{\beta} = \boldsymbol{\beta}^\nu}(\boldsymbol{\beta}^{\nu+1} - \boldsymbol{\beta}^\nu) + \cdots \tag{6.5.21}$$

Equation (6.5.21) is a matrix form of Eq. (6.5.5). Retaining only the linear terms in Eq. (6.5.21) and substituting into Eq. (6.5.20) gives the normal equations

$$\mathbf{X}^T(\boldsymbol{\beta}^\nu)\mathbf{X}(\boldsymbol{\beta}^\nu)(\boldsymbol{\beta}^{\nu+1} - \boldsymbol{\beta}^\nu) = \mathbf{X}^T(\boldsymbol{\beta}^\nu)(\mathbf{Y} - \mathbf{T}^\nu) \tag{6.5.22}$$

which is a matrix form of Eq. (6.5.6). The solution can be symbolically written as

$$\boldsymbol{\beta}^{\nu+1} - \boldsymbol{\beta}^\nu = \mathbf{P}(\boldsymbol{\beta}^\nu)\mathbf{X}^T(\boldsymbol{\beta}^\nu)(\mathbf{Y} - \mathbf{T}^\nu) \tag{6.5.23}$$

SEC. 6.6 SINGLE TEMPERATURE SENSOR

where

$$\mathbf{P}^{-1}(\boldsymbol{\beta}) = \mathbf{X}^T(\boldsymbol{\beta})\mathbf{X}(\boldsymbol{\beta}) \qquad (6.5.24)$$

For a linear problem, $\mathbf{X}(\boldsymbol{\beta})$ is independent of $\boldsymbol{\beta}$ and only one iteration of Eq. (6.5.22) is required. If the sensitivity coefficients are calculated using differences as in Eq. (6.5.7), a p-parameter problem will require $p+1$ direct problem solutions for each iteration. For p large and approaching r, as was the case for the sequential method of Section 6.3, the number of calculations required for the whole domain estimation becomes very large, even for the linear problem. Consequently, the whole domain estimation procedure is not recommended except when the functional form of the heat flux model is strongly suggested by the physics of the problem and the number of parameters is reasonably small.

6.6 SINGLE TEMPERATURE SENSOR, FUNCTION SPECIFICATION ($q=C$), r FUTURE TIME STEPS

The discussion in Section 1.6 pointed out that the sensitivity coefficient associated with a heat flux pulse of finite duration can be nonzero after the heat flux returns to zero. Consequently, $q(t)=q_M$, over the interval $t_{M-1} \leqslant t \leqslant t_M$ and zero otherwise, will influence sensor measurements $Y_M, Y_{M+1}, \ldots$. Therefore, powerful procedures for estimating q_M should use future temperature measurements. Beck[5] was the first to recognize the importance of future temperature information and apply it to the Duhamel's theorem solution of the IHCP. Beck and Wolf[1] were the first to combine difference methods and future temperatures for the solution of the IHCP. Beck et al.[6] applied sensitivity coefficient concepts and substantially reduced the number of computations required for difference methods with future temperatures. This section considers a single temperature sensor for the function specification method that assumes a constant heat flux and an arbitrary number of future time steps.

Suppose the IHCP has been solved up to time t_{M-1}; the estimated heat flux $\hat{q}_{M-1}$ and the entire temperature field $\hat{T}_i^{M-1}$ are known. Next, the time is advanced one step to t_M and an estimate of q_M is calculated. A single temperature sensor is located at a depth x_k below the active surface. A *temporary* assumption is made that the heat flux is constant over r future times. The IHCP is shown schematically in Figure 6.3. An estimate is sought of the value of q_M, constant over r future time steps, that minimizes the least squares error between the computed and measured sensor temperatures. The least squares function is

$$S = \sum_{i=1}^{r} [Y(x_k, t_{M+i-1}) - T(x_k, t_{M+i-1}, t_{M-1}, \mathbf{q}_{M-1}, q_M)]^2 \qquad (6.6.1)$$

Differentiating S with respect to q_M, replacing q_M by $\hat{q}_M$, and setting $\partial S/\partial q_M$ equal to zero yields,

$$0 = \frac{\partial S}{\partial q_M} = -2 \sum_{i=1}^{r} [Y(x_k, t_{M+i-1})$$
$$- T(x_k, t_{M+i-1}, t_{M-1}, \mathbf{q}_{M-1}, \hat{q}_M)] Z(x_k, t_{M+i-1}, t_{M-1}, \mathbf{q}_{M-1}) \quad (6.6.2)$$

238   CHAP. 6   ONE-DIMENSIONAL INVERSE HEAT CONDUCTION PROBLEM

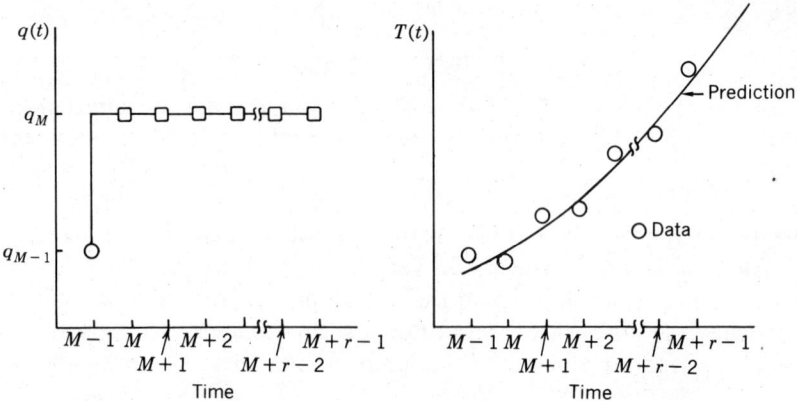

FIGURE 6.3 Pictorial representation of IHCP using r future temperatures.

where the sensitivity coefficient is defined by

$$Z_{k,i} = Z(x_k, t_{M+i-1}, t_{M-1}, \mathbf{q}_{M-1}) = \frac{\partial T(x_k, t_{M+i-1}, t_{M-1}, \mathbf{q}_{M-1}, q_M)}{\partial q_M} \quad (6.6.3)$$

Since q_M is assumed constant over r future time steps, the Z sensitivity coefficient is used instead of X; see Figure 6.1 for a comparison of the two sensitivity coefficients. Applying a Taylor series expansion similar to Eq. (6.3.4) gives

$$T(x_k, t_{M+i-1}, t_{M-1}, \mathbf{q}_{M-1}, q_M) = T(x_k, t_{M+i-1}, t_{M-1}, \mathbf{q}_{M-1}, q^*)$$

$$+ (q_M - q^*) Z(x_k, t_{M+i-1}, t_{M-1}, \mathbf{q}_{M-1}) \quad (6.6.4)$$

Figure 6.4 presents a sketch of the Taylor series expansion. In terms of index notation, Eq. (6.6.4) can be written as

$$T_k^{M+i-1} = \overset{*}{T}_k^{M+i-1} + (\hat{q}_M - q^*) Z_{k,i}, \quad i = 1, 2, \ldots, r \quad (6.6.5)$$

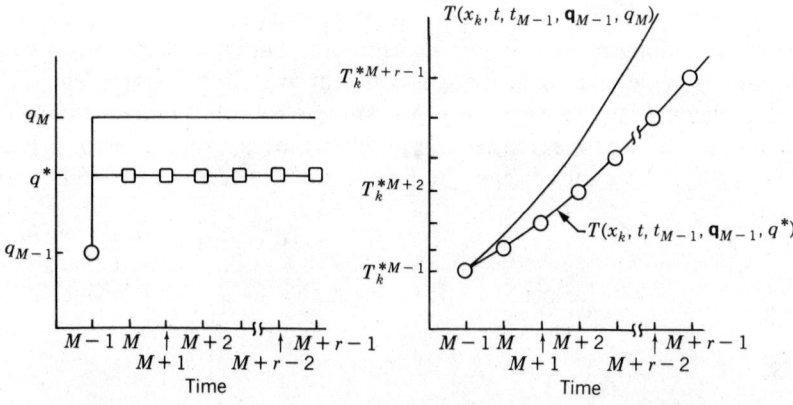

FIGURE 6.4 Reference condition in the calculation of q_M by differences.

SEC. 6.6 SINGLE TEMPERATURE SENSOR

Substituting Eq. (6.6.5) into Eq. (6.6.2) and solving for $\hat{q}_M$ gives,

$$\hat{q}_M = q_M^* + \frac{\sum_{i=1}^{r}(Y_{k,M+i-1} - \overset{*}{T}_k^{M+i-1})Z_{k,i}}{\sum_{j=1}^{r}(Z_{k,j})^2} \qquad (6.6.6)$$

Equation (6.6.6) can be written in terms of a gain coefficient as was Eq. (5.3.4a)

$$\hat{q}_M = q_M^* + \sum_{i=1}^{r}(Y_{k,M+i-1} - \overset{*}{T}_k^{M+i-1})K_{k,i} \qquad (6.6.7a)$$

where

$$K_{k,i} = \frac{Z_{k,i}}{\sum_{j=1}^{r} Z_{k,j}^2} \qquad (6.6.7b)$$

If thermal properties are independent of temperature, then the unit step function solution ϕ used in Chapter 5 is identical to the sensitivity coefficient Z solution of this chapter and the Z values need to be calculated only once for a given problem. If the thermal property variation with temperature is treated in a quasi-linear manner, then $Z \neq \phi$ and Z must be recalculated for each value of q_M. The subscript k in Eq. (6.6.4) is kept as a reminder that the sensitivity coefficients depend on the depth x_k of the sensor below the active surface; in the next section where the analysis is extended to include multiple sensors, the subscript k is even more important. It is necessary to remember that the constant heat flux assumption indicated in Figures 6.3 and 6.4 is a temporary assumption. The computed heat flux q_M is retained only over the time interval $t_{M-1} \leq t \leq t_M$. For each subsequent time interval, a new heat flux is calculated.

The detailed computational aspects of the algorithm given by Eq. (6.6.7) for the constant heat flux assumption, single temperature sensor, and r future times are summarized below:

1. Assume $q(t) = q^*$ over $t_{M-1} \leq t \leq t_{M+r-1}$ and perform the forward elimination portion of the TDMA. Starting with the last node N,

$$\omega_N = \frac{1}{b_N}, \quad e_N = \omega_N c_N, \quad f_N = \omega_N d_N$$

$$\omega_j = \frac{1}{b_j - a_j e_{j+1}}, \quad j = N-1, N-2, \ldots, 1$$

$$f_j = \omega_j(d_j + a_j f_{j+1}), \quad e_j = \omega_j c_j, \quad j = N-1, N-2, \ldots, 2$$

$$f_1 = \omega_1(d_1 + a_1 f_2), \quad d_1 = d_1(q^*)$$

Note that all $a_j, b_j, c_j, e_j,$ and ω_j are fixed during the calculation of both $\overset{*}{T}_j^{M+i-1}$ and $Z_{j,M+i-1}, i = 0, 1, \ldots, r-1$ because all thermal properties are held fixed at $\overset{*}{T}_j^M$.

240 CHAP. 6 ONE-DIMENSIONAL INVERSE HEAT CONDUCTION PROBLEM

2. Calculate $\overset{*}{T}{}^M_j, \overset{*}{T}{}^{M+1}_j, \ldots, \overset{*}{T}{}^{M+r-1}_j$ using the back substitution portion of the TDMA,

$$\left. \begin{array}{l} \overset{*}{T}{}^{M+i-1}_1 = f_{1,i} \\ \overset{*}{T}{}^{M+i-1}_j = e_j \overset{*}{T}{}^{M+i-1}_{j-1} + f_{j,i}, \quad j=2, 3, \ldots, N \end{array} \right\} \begin{array}{l} \text{Repeat for} \\ i = 0, 1, \ldots, r-1 \end{array}$$

The i subscript on $f_{j,i}$ is a reminder that this term must be recalculated for each value of future time (i). However, the e_j's are identical to those of step 1 and do not change for $i = 1, 2, \ldots, r$.

3. Calculate the step function sensitivity coefficients (Z) using the difference equations developed from Eq. (6.2.7). The TDMA coefficients a_j, b_j, c_j, e_j, and ω_j have already been calculated in step 1. Therefore only d_j and f_j need be recalculated and the back substitution must be performed.

$$\left. \begin{array}{l} Z_{1,M+i-1} = f_{1,i} \\ Z_{j,M+i-1} = e_j Z_{j-1,M+i-1} + f_{j,i}, \quad j=2, 3, \ldots, N \end{array} \right\} \begin{array}{l} \text{Repeat for} \\ i = 0, 1, \ldots, r-1 \end{array}$$

Note that the d_j and $f_{j,i}$ terms for the calculation of $Z_{j,M+i-1}$ are different from those for $\overset{*}{T}{}^{M+i-1}_j$. Also, the boundary conditions on Z are different from those on $\overset{*}{T}$.

4. Calculate $\hat{q}_M$ from (6.6.7a)

$$\hat{q}_M = q_M^* + \sum_{j=1}^{r} (Y_{k,M+i-1} - \overset{*}{T}{}^{M+i-1}_k) K_{k,i}$$

5. Knowing $\hat{q}_M$, reevaluate d_1 and f_1 and repeat the back substitution portion of TDMA for $\hat{T}(x, t_M, t_{M-1}, q_{M-1}, \hat{q}_M)$

$$\hat{T}^M_1 = f_1$$
$$\hat{T}^M_j = e_j \hat{T}^M_{j-1} + f_j, \quad j=2, 3, \ldots, N$$

EXAMPLE 6.2. Using the information in Table 1.1, calculate the K factors of Eq. (6.6.7) for $r=2$, $\Delta t^+ = \alpha \Delta t/L^2 = 0.05$, and a temperature sensor located at $x^+ = x/L = 0.5$. Repeat the calculation for $x^+ = x/L = 1.0$.

Solution.

$\Delta t^+ = 0.05, \quad x^+ = 0.5$
$Z_1 = 0.0153659$
$Z_2 = 0.0593109$

$$K_1 = \frac{0.0153659}{(0.0153659)^2 + (0.0593109)^2}$$
$$= 4.0933$$

$$K_2 = \frac{0.0593109}{(0.0153659)^2 + (0.0593109)^2}$$
$$= 15.80$$

$\Delta t^+ = 0.05, \quad x^+ = 1.0$
$Z_1 = 0.0002693$
$Z_2 = 0.0078853$

$$K_1 = \frac{0.0002693}{(0.0002693)^2 + (0.0078853)^2}$$
$$= 4.3261$$

$$K_2 = \frac{0.0078853}{(0.0002693)^2 + (0.007853)^2}$$
$$= 126.7$$

The $x^+ = 1.0$ and $t^+ = 0.05$ case is also included in Table 4.1. □

6.7 MULTIPLE TEMPERATURE SENSORS, FUNCTION SPECIFICATION ($q=C$), r FUTURE TIME STEPS

Multiple temperature sensors were considered in Section 6.4 for a single future time step. In this section, the analysis is extended to include an arbitrary number of future time steps with the temporary assumption of constant heat flux. The least squares error function must be modified to include a summation over the number of sensors. For J sensors and r future times, define a sum of squares function by

$$S = \sum_{k=1}^{J} \sum_{j=1}^{r} [Y_{k,M+j-1} - T(x_k, t_{M+j-1}, t_{M-1}, \mathbf{q}_{M-1}, q_M)]^2 \qquad (6.7.1)$$

The value of q_M, constant over r future time steps, that minimizes S is sought,

$$\frac{\partial S}{\partial q_M} = 0 = 2 \sum_{k=1}^{J} \sum_{j=1}^{r} [T(x_k, t_{M+j-1}, t_{M-1}, \mathbf{q}_{M-1}, q_M) - Y_{k,M+j-1}] Z_{k,j} \qquad (6.7.2)$$

where the step function sensitivity coefficient is defined by

$$Z_{k,j} = Z(x_k, t_{M+j-1}, t_{M-1}, \mathbf{q}_{M-1}) = \frac{\partial T(x_k, t_{M+j-1}, t_{M-1}, \mathbf{q}_{M-1}, q_M)}{\partial q_M} \qquad (6.7.3)$$

Expanding the temperature field in a Taylor series about an assumed heat flux q^*,

$$T_k^{M+j-1} = \mathring{T}_k^{M+j-1} + (q_M - q^*) Z_{k,j}, \quad j = 1, 2, \ldots, r-1 \qquad (6.7.4)$$

substituting Eq. (6.7.4) into Eq. (6.7.2) and solving for the desired heat flux,

$$\hat{q}_M = q^* + \frac{\sum_{k=1}^{J} \sum_{j=1}^{r} (Y_{k,M+j-1} - \mathring{T}_k^{M+j-1}) Z_{k,j}}{\sum_{k=1}^{J} \sum_{j=1}^{r} Z_{k,j}^2} \qquad (6.7.5)$$

Equation (6.7.5) can be written more compactly as

$$\hat{q}_M = q^* + \sum_{k=1}^{J} \sum_{j=1}^{r} K_{k,j}(Y_{k,M+j-1} - \mathring{T}_k^{M+j-1}) \qquad (6.7.6\text{a})$$

where

$$K_{k,j} = \frac{Z_{k,j}}{\sum_{k=1}^{J} \sum_{j=1}^{r} Z_{k,j}^2} \qquad (6.7.6\text{b})$$

EXAMPLE 6.3. Repeat Example 6.2 with the modification that two sensors ($J=2$) are located at $x^+ = 0.5$ and $x^+ = 1.0$. From Eq. (6.7.6b),

$$K_{k,j} = \frac{Z_{k,j}}{Z_{1,1}^2 + Z_{2,1}^2 + Z_{1,2}^2 + Z_{2,2}^2} = b Z_{k,j}$$

Solution. From Example 6.2,

$$Z_{1,1} = 0.01537 \qquad Z_{2,1} = 0.0002693$$
$$Z_{1,2} = 0.05931 \qquad Z_{2,2} = 0.007885$$
$$b = 262.0$$
$$K_{1,1} = 4.027 \qquad K_{2,1} = 0.07057$$
$$K_{1,2} = 15.54 \qquad K_{2,2} = 2.066 \qquad \square$$

The presence of a second sensor at $x^+ = 1$ does not add significant information as evidenced by the fact that $K_{1,1}$ and $K_{1,2}$ for this problem are very close to K_1 and K_2 for Example 6.2 (for $x^+ = 0.5$). This is because of the relative size of the sensitivity coefficients for $x^+ = 0.5$ and 1.0.

EXAMPLE 6.4. Repeat Example 6.3 for two identical sensors, both at $x^+ = 0.5$.

Solution.
$$b = 133.2$$
$$K_{1,1} = K_{2,1} = 2.047$$
$$K_{1,2} = K_{2,2} = 7.900 \qquad \square$$

6.8 SINGLE TEMPERATURE SENSOR, FUNCTION SPECIFICATION, LINEAR HEAT FLUX (CONNECTED SEGMENTS)

Up to this point, all the function specification methods in Chapter 6 have utilized the (temporary) assumption of a constant heat flux. Other assumed functional forms were considered in Chapters 4 and 5. All of the heat flux variations utilized with the convolution integral methods in Chapter 5 can also be utilized with the difference methods of this chapter. An example is the connected linear heat flux functional form shown in Fig. 4.8. This approach is based on a straight line extrapolation in time through the two points $\hat{q}_{M-1}$ and q_M. For equally spaced time increments, the heat flux variation can be expressed as

$$q_{M+i-1} = (1-i)\hat{q}_{M-1} + i q_M, \quad i = 0, 1, \ldots, r \qquad (6.8.1)$$

Note that Eq. (6.8.1) contains only a single unknown, q_M. Additional details can be found in Section 4.4.3.2. The following least squares error is minimized

$$S = \sum_{i=1}^{r} [Y(x_k, t_{M+i-1}) - T(x_k, t_{M+i-1}, t_{M-1}, \mathbf{q}_{M-1}, q_M, \ldots, q_{M+r-1})]^2 \qquad (6.8.2)$$

A comparison of Eqs. (6.8.2) and (6.6.1) indicates that the temperature field for the linear heat flux case depends on the r heat flux values $q_M, \ldots, q_{M+r-1}$; however, there is only one independent heat flux (q_M) since they are related through Eq. (6.8.1). Using the standard Taylor series expansion, the temperature field can be written as

$$T_k^{M+i-1} = \mathring{T}_k^{M+i-1} + \sum_{j=1}^{i} \Delta Z_{k, i-j+1}(q_{M+j-1} - q^*) \qquad (6.8.3)$$

SEC. 6.9 SECOND-ORDER SEQUENTIAL REGULARIZATION METHODS

Note that it is necessary to use the pulse sensitivity coefficient $X_{k,i-j+1} = \Delta Z_{k,i-j+1}$ because the heat flux is allowed to vary in steps over the r future time steps. Substituting Eq. (6.8.1) into Eq. (6.8.3) gives

$$T_k^{M+j-1} = \overset{+}{T}_k^{M+j-1} + q_M \psi_{k,j} - q_{M-1} \psi_{k,j-1} - q^* Z_{k,j} \qquad (6.8.4)$$

where the following relationships are used:

$$\sum_{j=1}^{i} \Delta Z_{k,i-j+1} = Z_{k,i}, \quad \sum_{j=1}^{i} j\Delta Z_{k,i-j+1} = \sum_{j=1}^{i} Z_{k,j} = \psi_{k,j} \qquad (6.8.5)$$

Substituting Eq. (6.8.4) into Eq. (6.8.2) yields

$$S = \sum_{i=1}^{r} (\overset{+}{T}_k^{M+i-1} - Y_{k,M+i-1} + q_M \psi_{k,i} - q_{M-1} \psi_{k,i-1} - q^* Z_{k,i})^2 \qquad (6.8.6)$$

The heat flux component q_{M-1} is treated as a quantity known from the previous time step. The value of q_M that minimizes S is found to be

$$\hat{q}_M = \frac{\sum_{i=1}^{r} \psi_{k,i}(Y_{k,M+i-1} - \overset{+}{T}_k^{M+i-1})}{\sum_{j=1}^{r} \psi_{k,j}^2} + \hat{q}_{M-1} \frac{\sum_{i=1}^{r} \psi_{k,i-1}\psi_{k,i}}{\sum_{j=1}^{r} \psi_{k,j}^2} + q^* \frac{\sum_{i=1}^{r} Z_{k,i}\psi_{k,i}}{\sum_{j=1}^{r} \psi_{k,j}^2}$$

$$(6.8.7)$$

A comparison of the constant heat flux result, Eq. (6.6.7), should be made with the linear heat flux result Eq. (6.8.7). Both are of the same general form but the coefficients are different and the latter depends explicitly on $\hat{q}_{M-1}$. The linear heat flux variation yields a more complicated equation than does the constant heat flux result. Experience has shown that the linear heat flux algorithm requires larger time steps for stability than does the constant heat flux algorithm for the same value of r.

6.9 SECOND-ORDER SEQUENTIAL REGULARIZATION METHODS

The general aspects of regularization methods were discussed in Section 4.5 and the convolution integral method was applied to first-order sequential regularization in Section 5.4.3. In this section, the equations for the second-order sequential regularization will be developed for the case of $r=3$ and then generalized by using matrix notation for an arbitrary r. The matrix derivation also applies to other orders of regularization.

The second-order regularization simulates the continuous term

$$\int_{t_{M-1}}^{t_{M+r-1}} \left(\frac{d^2 q}{dt^2}\right)^2 dt \qquad (6.9.1)$$

This integral is used in the cubic interpolatory spline[7] and the minimization of it yields the so-called minimum curvature property.

Following the material in Section 4.5, the standard least squares function is augmented as follows:

$$S = \sum_{i=1}^{r} (Y_{M+i-1} - T_k^{M+i-1})^2 + \alpha \sum_{i=1}^{r-2} (q_{M+i-1} - 2q_{M+i} + q_{M+i+1})^2 \quad (6.9.2)$$

with $r \geq 3$. Note that α is dimensional and has units of T/q^2. In order to keep the development simple, a value of $r=3$ is assumed; therefore, the second summation in Eq. (6.9.2) contains only one term. Differentiating Eq. (6.9.2) with respect to the three unknown heat flux components q_M, q_{M+1}, and q_{M+2} and setting the results equal to zero gives,

$$\sum_{i=1}^{3} (T_k^{M+i-1} - Y_{M+i-1}) \frac{\partial T_k^{M+i-1}}{\partial q_M} + \alpha(q_M - 2q_{M+1} + q_{M+2}) = 0 \quad (6.9.3a)$$

$$\sum_{i=2}^{3} (T_k^{M+i-1} - Y_{M+i-1}) \frac{\partial T_k^{M+i-1}}{\partial q_{M+1}} - 2\alpha(q_M - 2q_{M+1} + q_{M+2}) = 0 \quad (6.9.3b)$$

$$\sum_{i=3}^{3} (T_k^{M+i-1} - Y_{M+i-1}) \frac{\partial T_k^{M+i-1}}{\partial q_{M+2}} + \alpha(q_M - 2q_{M+1} + q_{M+2}) = 0 \quad (6.9.3c)$$

The lower limit on the summation does not always start with unity because a sensitivity coefficient is identically zero for all times prior to the initiation of the particular heat flux component. Sensitivity coefficients are defined through

$$\frac{\partial T_k^{M+i-1}}{\partial q_{M+j-1}} = X_{k, i-j+1} \quad (6.9.4)$$

The usual Taylor series expansion is written as

$$T_k^{M+i-1} = \overset{*}{T}_k^{M+i-1} + \sum_{j=1}^{i} X_{k, i-j+1}(q_{M+j-1} - q^*) \quad (6.9.5)$$

Substituting Eqs. (6.9.4) and (6.9.5) into Eq. (6.9.3), one obtains

$$\sum_{i=1}^{3} [\overset{*}{T}_k^{M+i-1} - Y_{M+i-1}$$

$$+ \sum_{j=1}^{i} X_{k,i-j+1}(q_{M+j-1} - q^*)] X_{k,i} + \alpha(q_M - 2q_{M+1} + q_{M+2}) = 0 \quad (6.9.6a)$$

$$\sum_{i=2}^{3} [\overset{*}{T}_k^{M+i-1} - Y_{M+i-1}$$

$$+ \sum_{j=1}^{i} X_{k,i-j+1}(q_{M+j-1} - q^*)] X_{k,i-1} - 2\alpha(q_M - 2q_{M+1} + q_{M+2}) = 0 \quad (6.9.6b)$$

SEC. 6.9 SECOND-ORDER SEQUENTIAL REGULARIZATION METHODS

$$\sum_{i=3}^{3} [\overset{*}{T}{}_k^{M+i-1} - Y_{M+i-1}$$

$$+ \sum_{j=1}^{i} X_{k,i-j+1}(q_{M+j-1} - q^*)] X_{k,i-2} + \alpha(q_M - 2q_{M+1} + q_{M+2}) = 0 \quad (6.9.6c)$$

After straightforward but lengthy algebra, Eq. (6.9.6) can be written as

$$\begin{bmatrix} \left(\sum_{i=1}^{3} X_{k,i} X_{k,i} + \alpha\right) & \left(\sum_{i=1}^{2} X_{k,i+1} X_{k,i} - 2\alpha\right) & \left(\sum_{i=1}^{1} X_{k,i+2} X_{k,i} + \alpha\right) \\ \left(\sum_{i=1}^{2} X_{k,i} X_{k,i+1} - 2\alpha\right) & \left(\sum_{i=1}^{2} X_{k,i} X_{k,i} + 4\alpha\right) & \left(\sum_{i=1}^{1} X_{k,i+1} X_{k,i} - 2\alpha\right) \\ \left(\sum_{i=1}^{1} X_{k,i} X_{k,i+2} + \alpha\right) & \left(\sum_{i=1}^{1} X_{k,i} X_{k,i+1} - 2\alpha\right) & \left(\sum_{i=1}^{1} X_{k,i} X_{k,i} + \alpha\right) \end{bmatrix}$$

$$\times \begin{bmatrix} \hat{q}_M - q^* \\ \hat{q}_{M+1} - q^* \\ \hat{q}_{M+2} - q^* \end{bmatrix} = \begin{bmatrix} \sum_{i=1}^{3} X_{k,i}(Y_{M+i-1} - \overset{*}{T}{}_k^{M+i-1}) \\ \sum_{i=1}^{2} X_{k,i}(Y_{M+i} - \overset{*}{T}{}_k^{M+i}) \\ \sum_{i=1}^{1} X_{k,i}(Y_{M+i+1} - \overset{*}{T}{}_k^{M+i+1}) \end{bmatrix} \quad (6.9.7)$$

The symmetrical coefficient matrix affords some computational savings. The regularization parameter α improves the diagonal dominance of the coefficient matrix and reduces the sensitivity to measurement errors. Since the equations are linear in $\hat{q}_j - q^*$, the value of q^* is completely arbitrary. If $q^* = 0$ is used, then $\overset{*}{T}$ is replaced by $T|_{q=0}$. In the sequential procedure, generally only $\hat{q}_M$ found from Eq. (6.9.7) is retained. The procedure is then repeated for the next time step.

For the second-order regularization, the smallest allowable value of future times is $r = 3$. Even for this simplest case, the algebra is rather lengthy. In order to extend this procedure to arbitrary r, matrix methods are used. The matrix analog of Eq. (6.9.2) is

$$S = (\mathbf{Y} - \mathbf{T})^T (\mathbf{Y} - \mathbf{T}) + \alpha (\mathbf{H}_2 \mathbf{q})^T (\mathbf{H}_2 \mathbf{q}) \quad (6.9.8)$$

The $\mathbf{Y}$ and $\mathbf{T}$ matrices have already been defined in Eq. (6.5.13). Since the procedure being developed is sequential, the heat flux vector contains only r elements,

$$\mathbf{q} = \begin{bmatrix} q_M \\ q_{M+1} \\ \vdots \\ q_{M+r-2} \\ q_{M+r-1} \end{bmatrix} \quad (6.9.9)$$

By analogy with Eq. (4.5.16c), the $\mathbf{H}_2$ matrix is

$$\mathbf{H}_2 = \begin{bmatrix} 1 & -2 & 1 & 0 & 0 & & & & & 0 \\ 0 & 1 & -2 & 1 & 0 & & & & & 0 \\ 0 & 0 & 1 & -2 & 1 & 0 & & & & 0 \\ 0 & 0 & 0 & & & & & & & \vdots \\ \vdots & & & \ddots & & & & & & 0 \\ 0 & \cdot & \cdot & & & & 0 & 1 & -2 & 1 \\ 0 & \cdot & \cdot & & & & 0 & 0 & 0 & 0 \\ 0 & \cdot & \cdot & & & & 0 & 0 & 0 & 0 \end{bmatrix} \updownarrow r-2 \quad (6.9.10)$$

For $r=3$, only the first row of $\mathbf{H}_2$ is nonzero. The matrix equivalent of the Taylor series expansion, Eq. (6.8.3), is

$$\mathbf{T} = \dot{\mathbf{T}} + \mathbf{X}(\mathbf{q} - \mathbf{q}^*) \quad (6.9.11)$$

where the elements of $\dot{\mathbf{T}}$ are defined in a manner completely analogous to that for $\mathbf{T}$, and $\mathbf{q}^*$ is a vector of r elements all equal to q^*, and $\mathbf{X}$ is the sensitivity coefficient matrix,

$$\mathbf{X} = \begin{bmatrix} X_{k,1} & 0 & 0 & 0 & \cdot & \cdot & 0 \\ X_{k,2} & X_{k,1} & 0 & 0 & & & 0 \\ X_{k,3} & X_{k,2} & X_{k,1} & 0 & & & 0 \\ \cdot & \cdot & \cdot & \cdot & \cdot & & 0 \\ X_{k,r} & X_{k,r-1} & \cdot & & \cdot & X_{k,2} & X_{k,1} \end{bmatrix} \quad (6.9.12)$$

Note that $\mathbf{X}$ is lower triangular and each element is defined by Eq. (6.9.4). The matrix derivative of S with respect to the heat flux vector $\mathbf{q}$ is given by

$$\nabla_q S = 2[\nabla_q(\mathbf{Y} - \mathbf{T})^T](\mathbf{Y} - \mathbf{T}) + 2\alpha \nabla_q(\mathbf{H}_2\mathbf{q})^T(\mathbf{H}_2\mathbf{q}) = 0 \quad (6.9.13)$$

From Eq. (6.9.11),

$$\nabla_q(\mathbf{Y} - \mathbf{T})^T = -\nabla_q \mathbf{T}^T = -\mathbf{X}^T \quad (6.9.14)$$

Also,

$$\nabla_q(\mathbf{H}_2\mathbf{q})^T = \nabla_q(\mathbf{q}^T \mathbf{H}_2^T) = \mathbf{H}_2^T \quad (6.9.15)$$

Substituting Eqs. (6.9.11), (6.9.14), and (6.9.15) into Eq. (6.9.13) and simplifying:

$$\mathbf{X}^T \mathbf{X}(\hat{\mathbf{q}} - \mathbf{q}^*) + \alpha \mathbf{H}_2^T \mathbf{H}_2 \hat{\mathbf{q}} = \mathbf{X}^T(\mathbf{Y} - \dot{\mathbf{T}}) \quad (6.9.16)$$

Equation (6.9.16) can be put in a different form by noting that

$$\alpha \mathbf{H}_2^T \mathbf{H}_2 \mathbf{q}^* = \mathbf{0} \quad (6.9.17)$$

If $\mathbf{1}$ is a vector with all elements equal unity, then for $\mathbf{q}^* = q^*\mathbf{1}$,

$$\mathbf{H}_2 \mathbf{q}^* = q^* \mathbf{H}_2 \mathbf{1} = \mathbf{0} \quad (6.9.18)$$

The matrix product $\mathbf{H}_2\mathbf{1}$ is the vector, the elements of which are the sum of the elements of each corresponding row of $\mathbf{H}_2$. The elements of $\mathbf{H}_2\mathbf{1}$ individually sum to zero. Equation (6.9.16) can be written as the matrix normal equation

$$(\mathbf{X}^T\mathbf{X} + \alpha\mathbf{H}_2^T\mathbf{H}_2)(\hat{\mathbf{q}} - \mathbf{q}^*) = \mathbf{X}^T(\mathbf{Y} - \dot{\mathbf{T}}) \qquad (6.9.19)$$

Equation (6.9.19) is a matrix form of Eq. (6.9.7) and similarities can be observed. If $q^* = 0$, then $\dot{\mathbf{T}}$ is replaced by $\hat{\mathbf{T}}|_{\mathbf{q}=0}$. With matrix manipulation subroutines available, Eq. (6.9.19) is much preferred over a generalization of Eq. (6.9.7) to arbitrary r. The solution to Eq. (6.9.19) can be written symbolically as

$$\hat{\mathbf{q}} = \mathbf{q}^* + (\mathbf{X}^T\mathbf{X} + \alpha\mathbf{H}_2^T\mathbf{H}_2)^{-1}\mathbf{X}^T(\mathbf{Y} - \dot{\mathbf{T}}) \qquad (6.9.20)$$

It is common in the sequential procedure to retain only the $\hat{q}_M$ component of $\hat{\mathbf{q}}$ and repeat the process for each successive time step. In the linear version of the sequential regularization, the thermal properties are maintained constant at values corresponding to the temperature field at time t_{M-1}. If the thermal properties are independent of temperature over the range of interest, the whole domain regularization equations are identical to those of Eq. (6.9.20) provided r is the total number of heat flux components being estimated. Since the sequential regularization method requires the solution of a system of r algebraic equations for each value of q_M, it is computationally more costly than the other sequential methods presented in this chapter but for small values of r such as $r \leqslant 5$ the increase in computation may not be large.

The individual elements of $\mathbf{X}$ are calculated by means similar to those discussed in Sections 6.2 and 6.3. Namely, $X_{k,j-i+1}$ is determined from the difference solution of a direct problem in which the initial profile is zero, the heat flux is unity during the time interval $t_{M+i-2} \leqslant t \leqslant t_{M+i-1}$ and zero thereafter, and the inactive boundary is perfectly insulated.

For a discussion of the choice of the regularization parameter α, see Chapter 5.

Although Eq. (6.9.19) is derived for a *second*-order regularization, it also applies to other orders. For a zeroth-order regularization, for example, $\mathbf{H}_2$ is replaced by $\mathbf{H}_0$ which is developed in Section 4.5.

6.10 SPACE MARCHING TECHNIQUES FOR ONE-DIMENSIONAL PROBLEMS

The difference methods presented in the previous sections can be termed "time marching" because starting with the initial temperature profile, the entire temperature field is calculated at the next time step. One marches ahead in time. For the inverse problem, two boundary conditions are known at the sensor location; namely, temperature, and heat flux. The knowledge of two conditions at the same spatial location suggests that the IHCP is related to an initial value problem instead of a boundary value problem. This section explores several space marching techniques which are applicable to "initial" value problems.

In the analysis that follows, a planar geometry with constant thermal properties is assumed; these restrictions are for simplicity only.

For the IHCP, the geometry can be divided into inverse and direct regions, as shown in Figure 1.4. The direct region is the classical boundary value problem; the experimental temperature measurements are the boundary conditions for the interface between the direct and inverse regions. The boundary conditions are known at the remaining boundary. Within the direct region, it is possible to solve this well-posed problem. Once the temperature field is known for the direct region, the heat flux at the sensor location can be determined by differentiating the temperature profile. In the discussions that follow, all temperatures and heat fluxes are assumed known within the direct region; hence, both temperature and heat flux are known at the inverse boundary.

6.10.1 Analytical Solution

Consider the constant property heat conduction equation for a planar geometry

$$\frac{\partial^2 T(x,t)}{\partial x^2} = \frac{1}{\alpha}\frac{\partial T(x,t)}{\partial t} \qquad (6.10.1)$$

If the time derivative is replaced by a backwards difference, Eq. (6.10.1) can be converted from a partial differential equation to an ordinary differential equation

$$\frac{d^2 T^i(x)}{dx^2} - \frac{1}{\alpha \Delta t} T^i(x) = -\frac{1}{\alpha \Delta t} T^{i-1}(x) \qquad (6.10.2)$$

where the superscript i denotes time index. The boundary conditions for Eq. (6.10.2) are

$$T^i(x=E) = Y_i$$

$$q_{E,i} = -k \left. \frac{dT^i(x)}{dx} \right|_{x=E}$$

Equation (6.10.2) has the appearance of an initial value problem because the function and its first derivative are both specified at $x=E$. If the initial temperature profile (T_0) is uniform throughout the body, a very simple solution of Eq. (6.10.2) exists for $i=1 (t=\Delta t)$

$$T^1(z) - T_0 = (Y_1 - Y_0)\cosh\left(\frac{z}{\sqrt{\alpha \Delta t}}\right) + q_{E,1}\sqrt{\alpha \Delta t}\,\sinh\left(\frac{z}{\sqrt{\alpha \Delta t}}\right)$$

where $z = E - x$. The unknown flux at $x=0$, $(z=E)$, is found by differentiating this equation with respect to x to obtain

$$Q_1 = \frac{q_{0,1}E}{k} = \frac{q_{E,1}E}{k}\cosh\left(\frac{E}{\sqrt{\alpha \Delta t}}\right) + (Y_1 - Y_0)\frac{E}{\sqrt{\alpha \Delta t}}\sinh\left(\frac{E}{\sqrt{\alpha \Delta t}}\right) \qquad (6.10.3)$$

SEC. 6.10 SPACE MARCHING TECHNIQUES

Equation (6.10.3) demonstrates the significance of the dimensionless time step $\Delta\tau_E \equiv \alpha\Delta t/E^2$ for inverse problems.

If an error is made in the measurement of Y_1, there will be errors in both $q_{E,1}$ and $q_{0,1}$. A *lower bound* of this error can be estimated by replacing Y_1 by $Y_1 + \delta Y_1$ and ignoring the error in $q_{E,1}$. The resulting error in $q_{0,1}$ is found to be

$$\frac{\delta Q_1}{\delta Y_1} = \frac{1}{\sqrt{\Delta\tau_E}} \sinh\left(\frac{1}{\sqrt{\Delta\tau_E}}\right)$$

which is tabulated in Table 6.6; these results are consistent with earlier observations that small values of dimensionless time $\Delta\tau_E$ can produce large errors in surface flux.

The foregoing process can be repeated for $t = 2\Delta t$ by solving

$$\frac{d^2 T^2(z)}{dz^2} - \frac{1}{\alpha\Delta t} T^2(z) = -\frac{1}{\alpha\Delta t}\left[T_0 + (Y_1 - Y_0)\cosh\left(\frac{z}{\sqrt{\alpha\Delta t}}\right)\right.$$
$$\left. + \frac{q_{E,1}}{K}\sqrt{\alpha\Delta t}\sinh\left(\frac{z}{\sqrt{\alpha\Delta t}}\right)\right]$$

An analytical solution of this equation is possible; however, due to the complexity of the inhomogeneous terms, the algebra is more complicated and solutions at subsequent times even more so. Consequently, this approach is not pursued further. The important points to remember from this analysis are that for a given value of time, it is possible to march in space, and that the important length scale in the dimensionless time step $\Delta\tau_E$ is the sensor distance below the heated surface, E.

6.10.2 Method of D'Souza

D'Souza[8,9] used the pure implicit difference approximation to Eq. (6.10.1); the time derivative was replaced by a backward difference and the spatial

TABLE 6.6 Dimensionless Heat Flux Error for Single Temperature Error δY_1

$\Delta\tau_E$	$\delta Q_1/\delta Y_1$
10	0.1017
1	1.175
0.5	2.737
0.25	7.254
0.125	23.843
0.0625	109.160
0.03125	809.618

derivative by a central difference

$$\frac{T^i_{j-1} - 2T^i_j + T^i_{j+1}}{\Delta x^2} = \frac{1}{\alpha}\left(\frac{T^i_j - T^{i-1}_j}{\Delta t}\right) \tag{6.10.4}$$

Solving for T^i_{j-1} in Eq. (6.10.4) yields

$$T^i_{j-1} = \left(2 + \frac{1}{p}\right)T^i_j - T^i_{j+1} - \frac{1}{p}T^{i-1}_j \quad \begin{cases} j = K, K-1, \ldots, 2 \\ i = 1, 2, \ldots, n \end{cases} \tag{6.10.5}$$

where $p = \alpha \Delta t/\Delta x^2$. Using the space-time grid shown in Figure 6.5, the calculation procedure is started at node $j = K$ and sequentially steps through time $i = 1, 2, \ldots, n$. Because the temperature field is known along both the spatial grid lines K and $K+1$, Eq. (6.10.5) is an explicit relationship even though it was derived from the fully implicit equations for direct problems. After the temperature is calculated all along space line $K-1$, the procedure is repeated for $K-2, \ldots, 1$. A finite difference equation for node $j = 1$ yields an equation for calculating the surface heat flux.

The D'Souza procedure is intended to be implemented on a computer as just outlined. Insight can be gained, however, by obtaining some algebraic expressions for a small number of nodes, K, from the heated surface to the sensor. Using the pure implicit procedure, equations are given for each node. For node 1, the heated surface, the equation at time t_i is

$$q^i_0 + k\frac{T^i_2 - T^i_1}{\Delta x} = \rho c \frac{\Delta x}{2} \frac{T^i_1 - T^{i-1}_1}{\Delta t} \tag{6.10.6a}$$

The equation for an interior node j is given by Eq. (6.10.4) and the equation for node K is

$$k\frac{T^i_{K-1} - Y_i}{\Delta x} - q^i_E = \rho c \frac{\Delta x}{2}\frac{Y_i - Y_{i-1}}{\Delta t} \tag{6.10.6b}$$

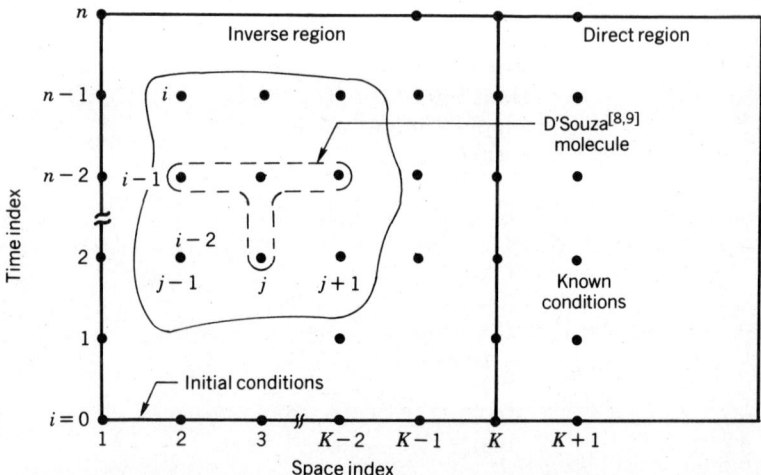

FIGURE 6.5 Space-time grid for space marching methods.

SEC. 6.10 SPACE MARCHING TECHNIQUES

The heat flux q_0^i is unknown and is to be estimated while q_E^i is the heat flux leaving the inverse region shown in Figure 1.4; this latter heat flux is entering the direct region and is found from calculations for the direct region. In Eq. (6.10.6b) the temperatures T_K^{i-1} and T_K^i are replaced by Y_{i-1} and Y_i, respectively. Equation (6.10.6b) is solved for T_{K-1}^i for $i = 1, 2, \ldots, n$ to get

$$T_{K-1}^i = \frac{q_E^i \Delta x}{k} + Y_i + \frac{(\Delta x)^2}{2\alpha \Delta t}(Y_i - Y_{i-1}) \tag{6.10.7}$$

For $K > 2$, Eq. (6.10.7) is introduced into Eq. (6.10.4) written for $j = K - 1$ to obtain an expression for T_{K-2}^i, $i = 1, 2, \ldots$. For $K = 2$, Eq. (6.10.7) gives expressions for T_1^i and (by replacing i by $i-1$) T_1^{i-1} which are introduced into Eq. (6.10.6a) to obtain

$$Q_0^i \equiv \frac{q_0^i E}{k} = \frac{\nabla Y_i}{\Delta \tau_E} + \frac{1}{4}\frac{\nabla^2 Y_i}{(\Delta \tau_E)^2} + Q_E^i + \frac{\nabla Q_K^i}{2\Delta \tau_E} \tag{6.10.8a}$$

$$Q_E^i \equiv \frac{q_E^i E}{k} \tag{6.10.8b}$$

since $\Delta x = E$ for this case.

The symbol ∇ is the backward difference operator,

$$\nabla Y_i = Y_i - Y_{i-1}, \quad \nabla^2 Y_i = Y_i - 2Y_{i-1} + Y_{i-2} \tag{6.10.8c,d}$$

$$\nabla Q_E^i = Q_E^i - Q_E^{i-1} \tag{6.10.8e}$$

If the foregoing procedure is used for $K = 3$, and thus $\Delta x = E/2$, the heat flux expression is

$$Q_0^i = \frac{\nabla Y_i}{\Delta \tau_E} + \frac{3}{16}\frac{\nabla^2 Y_i}{(\Delta \tau_E)^2} + \frac{1}{128}\frac{\nabla^3 Y_i}{(\Delta \tau_E)^3} + Q_E^i + \frac{1}{2}\frac{\nabla Q_E^i}{\Delta \tau_E} + \frac{1}{32}\frac{\nabla^2 Q_E^i}{(\nabla \tau_E)^2} \tag{6.10.9}$$

Several comments regarding Eqs. (6.10.8a) and (6.10.9) are now made. First, this is a procedure that utilizes exact matching of the calculated temperature with the measured temperature and thus is sensitive to measurement errors. Second, these equations are numerical analogs of the exact solution which is given by the Burggraf equation, Eq. (2.5.20). The coefficients of Eqs. (6.10.8a) and (6.10.9) approach those of Eq. (2.5.20) as the number of nodes increases; the first difference coefficients in both of the former equations are the same as each other and those in Eq. (2.5.20) whereas the $\nabla^2 Y_i$ starts at 0.25 [given by Eq. (6.10.8a)], decreases to 0.1875 in Eq. (6.10.9), and additional nodes bring the coefficient closer to the exact value of 0.1667. Third, only past temperatures are used to obtain the present heat flux estimates. It has repeatedly been noted that future temperatures are needed to obtain accurate estimates with small time steps. Fourth, if an analogous procedure were utilized starting with the explicit equation [i.e., left side of Eq. (6.10.4) evaluated at $i-1$], forward difference operators would replace the backward difference operators of Eqs. (6.10.8a) and (6.10.9), ensuring that only future temperatures are used. Fifth, as more nodes are used, higher-order differences are needed, increasing sensitivity to temperature measurement errors.

6.10.3 Method of Weber

Weber[10] worked with the hyperbolic form of the constant property heat conduction equation

$$\gamma^{-2}\frac{\partial^2 T}{\partial t^2}+\frac{\partial T}{\partial t}=\alpha\frac{\partial^2 T}{\partial x^2} \qquad (6.10.10)$$

where γ is a normalized thermal wave speed. Central differences in both time and space were used,

$$\frac{\gamma^{-2}}{\Delta t^2}(T_j^{i-1}-2T_j^{i+1}+T_j^{i+1})+\frac{1}{2\Delta t}(T_j^{i+1}-T_j^{i-1})=\frac{\alpha}{\Delta x^2}(T_{j+1}^i-2T_j^i+T_{j-1}^i) \qquad (6.10.11a)$$

Solving for T_{j-1}^i, Eq. (6.10.11a) yields,

$$T_{j-1}^i=-\left(\frac{1}{2p}-\sigma\right)T_j^{i-1}+2(1-\sigma)T_j^i+\left(\frac{1}{2p}+\sigma\right)T_j^{i+1}-T_{j+1}^i \qquad (6.10.11b)$$

$$p\equiv\frac{\alpha\Delta t}{\Delta x^2},\quad \sigma\equiv\frac{\gamma^{-2}}{\alpha}\left(\frac{\Delta x}{\Delta t}\right)^2 \qquad (6.10.11c,d)$$

This algorithm is explicit; it can be started at node $j=K$ and time $i=2, 3, \ldots, n-1$. The value $i=n$ is not permissible because T_K^{n+1} is not defined. Consequently the maximum time index for the grid location j is one less than that for the grid location $j+1$. This grid pattern is shown in Figure 6.6. If the surface temperature is desired at time M, then the temperature must be measured out to time $i=M+K$. The algorithm given by Eq. (6.10.11b) has the appearance of using K future times where K is the number of spatial grid points. Refining the grid has the effect of increasing the number of future times. For the methods presented in Sections 6.3 and 6.4, the number of future times is dependent on the physics of the problem and should be relatively independent of the number of spatial grid points.

Weber indicated that the parameter σ should be chosen small, without indicating how small. In fact, the algorithm works with $\sigma=0$.

Algebraic expressions similar to those for the D'Souza procedure can also be found for Weber's. Following the same procedure as in Section 6.10.2 one can derive for two nodes; that is, $K=2$, and with $\sigma=0$

$$Q_0^i=\frac{Y_{i+1}-Y_{i-1}}{2\Delta\tau_E}+\frac{1}{4}\frac{Y_{i+2}-2Y_i+Y_{i-2}}{(2\Delta\tau_E)^2} \qquad (6.10.12a)$$

SEC. 6.10 SPACE MARCHING TECHNIQUES

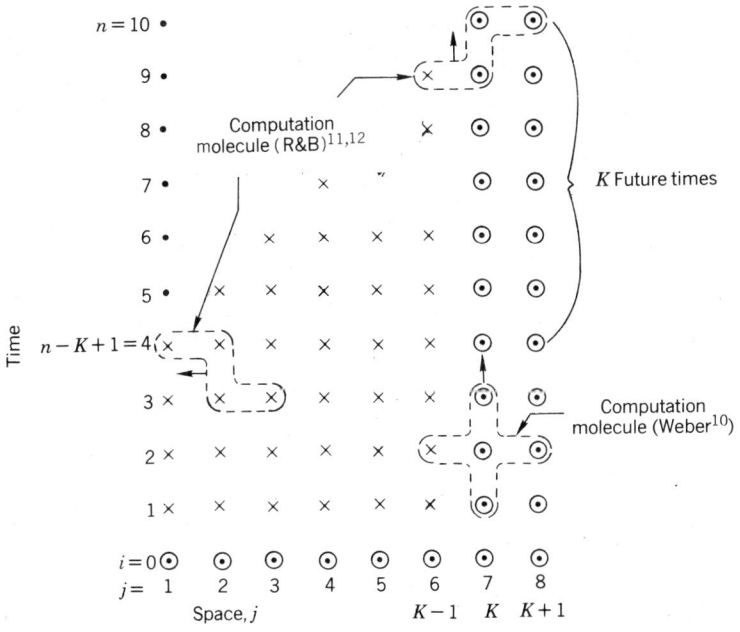

FIGURE 6.6 Space-time grid for Weber[10] and Raynaud and Bransier[11,12] methods; $K=7$ (node points), $n=10$ (temperature measurements), ⊙ known, × calculated.

and for three nodes ($K=3$),

$$Q_0^i = \frac{Y_{i+1} - Y_{i-1}}{2\Delta\tau_E} + \frac{3}{16}\frac{Y_{i+2} - 2Y_i + Y_{i-2}}{(2\Delta\tau_E)^2} + \frac{1}{128}\frac{Y_{i+3} - 3Y_{i+1} + 3Y_{i-1} - Y_{i-2}}{(2\Delta\tau_E)^3}$$

(6.10.12b)

where Q_E^i is a set equal to zero. These equations are quite similar to Eqs. (6.10.8a) and (6.10.9), found for D'Souza's method, and comments one, two, and five made in connection with the latter equations also apply for Eq. (6.10.12). However, Eq. (6.10.12) has two advantages over Eqs. (6.10.8a) and (6.10.9): (1) both past and future temperatures are used and (2) central differences are used that are much less sensitive to measurement errors.

6.10.4 Method of Raynaud and Bransier

The method of Raynaud and Bransier[11,12] consists of a two-step marching procedure. The first consists of a space marching step followed by a time marching step; the final result is an average of the two procedures. For the space march, the heat conduction equation is differenced as follows:

$$\rho c \frac{\Delta x}{\Delta t}(\tilde{T}_j^{i+1} - \tilde{T}_j^i) = -\underbrace{\frac{k}{\Delta x}(\tilde{T}_j^i - \tilde{T}_{j-1}^i)}_{\text{(explicit)}} + \underbrace{\frac{k}{\Delta x}(\tilde{T}_{j+1}^{i+1} - \tilde{T}_j^{i+1})}_{\text{(implicit)}}$$

(6.10.13)

Note that the conduction terms are mixed explicit/implicit. Solving for $\tilde{T}^i_{j-1}$ from Eq. (6.10.13),

$$\tilde{T}^i_{j-1} = \left(1 - \frac{1}{p}\right)\tilde{T}^i_j + \left(1 + \frac{1}{p}\right)\tilde{T}^{i+1}_j - \tilde{T}^{i+1}_{j+1} \qquad (6.10.14)$$

The explicit algorithm for $\tilde{T}^i_{j-1}$ starts with node $j = K$ and steps through the time lines $i = 2, 3, \ldots, n-1$. In general, the maximum value of i is $n+j-K$ which is the same as the Weber algorithm; see Figure 6.6 for details.

The second part of the algorithm approximates the heat conduction equation as follows:

$$\rho c \frac{\Delta x}{\Delta t}(\tilde{\tilde{T}}^{i+1}_j - \tilde{\tilde{T}}^i_j) = \underbrace{\frac{-k}{\Delta x}(\tilde{\tilde{T}}^{i+1}_j - \tilde{\tilde{T}}^{i+1}_{j-1})}_{\text{(implicit)}} + \underbrace{\frac{k}{\Delta x}(\tilde{\tilde{T}}^i_{j+1} - \tilde{\tilde{T}}^i_j)}_{\text{(explicit)}} \qquad (6.10.15)$$

Again, the conduction terms are mixed explicit/implicit. Solving for $\tilde{\tilde{T}}^{i+1}_{j-1}$ from Eq. (6.10.15),

$$\tilde{\tilde{T}}^{i+1}_{j-1} = \left(1 + \frac{1}{p}\right)\tilde{\tilde{T}}^{i+1}_j + \left(1 - \frac{1}{p}\right)\tilde{\tilde{T}}^i_j - \tilde{\tilde{T}}^i_{j+1} \qquad (6.10.16)$$

The $\tilde{\tilde{T}}^{i+1}_{j-1}$ algorithm is explicit and time marching. Starting with $i = 1$, the space index j steps through $K-1, K-2, \ldots, 1$. The final step of the algorithm is an average of the foregoing two procedures.

$$T^i_j = \tfrac{1}{2}(\tilde{T}^i_j + \tilde{\tilde{T}}^i_j) \qquad (6.10.17)$$

Although there is a time marching component to the Raynaud and Bransier algorithm, the combined effect of the two steps has the same overall appearance as the Weber algorithm given in Figure 6.6 but it is even less sensitive to the measurement errors than Weber's method. The apparent number of future times is the same as the number of spatial grid points (K).

Raynaud and Bransier also considered temperature-dependent properties.

6.10.5 Method of Hills and Hensel

Hensel and Hills[13] wrote the heat conduction equation as a pair of first order equations

$$\rho c \frac{\partial T}{\partial t} = -\frac{\partial q}{\partial x} \qquad (6.10.18a)$$

and

$$q = -k \frac{\partial T}{\partial x} \qquad (6.10.18b)$$

and then converted them to finite difference equations individually; the two unknowns were $T(x, t)$ and $q(x, t)$. For a discussion of the purported advantages

SEC. 6.10 SPACE MARCHING TECHNIQUES

of this approach, see References 13 and 14. Some of the basic features of this method can be understood by working with the combined form of Eq. (6.10.18). The so-called leapfrog difference scheme was used,

$$\frac{\Delta x}{2\Delta t}(T_j^{i+1} - T_j^{i-1}) = \frac{\alpha}{\Delta x}(T_{j+1}^i - 2T_j^i + T_{j-1}^i) \quad (6.10.19)$$

Although Richtmyer and Morton[15] indicate that the foregoing difference technique is unconditionally unstable for direct problems, it can still be useful for inverse problems. If Eq. (6.10.19) is solved for T_{j-1}^i,

$$T_{j-1}^i = -\frac{1}{2p}T_j^{i-1} + 2T_j^i + \frac{1}{2p}T_j^{i+1} - T_{j+1}^i \quad (6.10.20)$$

See Figure 6.7 for the computational molecule for Eq. (6.10.20). (A "computational molecule" is a schematic representation of a difference equation such as Eq. (6.10.20).) The preceding algorithm is explicit and is identical to the Weber algorithm, Eq. (6.10.11b), for $\sigma = 0$. Although Eq. (6.10.20) is valid for $i = 1$, Hensel and Hills[13] chose to replace the centered time difference in Eq. (6.10.19) with a forward time difference

$$\frac{\Delta x}{\Delta t}(T_j^2 - T_j^1) = \frac{\alpha}{\Delta x}(T_{j+1}^1 - 2T_j^1 + T_{j-1}^1) \quad (6.10.21)$$

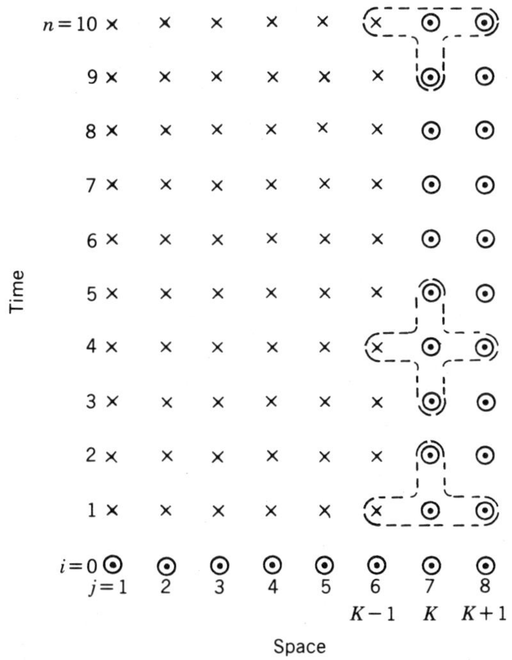

FIGURE 6.7 Space-time grid for Hills and Hensel[13,14] method; $K = 7$ (node points), $n = 10$ (temperature measurement times), ⊙'s known, ×'s to be calculated.

Solving for T_{j-1}^1,

$$T_{j-1}^1 = \left(2 - \frac{1}{p}\right) T_j^1 - T_{j+1}^1 + \frac{1}{p} T_j^2 \qquad (6.10.22)$$

The computational molecule for Eq. (6.10.22) is shown in Figure 6.7.

Equation (6.10.20) is not applicable for $i = n$. Instead of using the approach of Weber and generating the solution up to the diagonal line of Figure 6.6, Hensel and Hills[13] replaced the time derivative with a backward difference

$$\rho c \frac{\Delta x}{\Delta t} (T_j^n - T_j^{n-1}) = \frac{k}{\Delta x} (T_{j-1}^n - 2T_j^n + T_{j+1}^n) \qquad (6.10.23)$$

Solving for T_{j-1}^n,

$$T_{j-1}^n = \left(2 + \frac{1}{p}\right) T_j^n - \frac{1}{p} T_j^{n-1} - T_{j+1}^n \qquad (6.10.24)$$

which is the same equation used by D'Souza. The computational molecule for Eq. (6.10.24) is shown in Figure 6.7.

The depth of the discussion presented here does not do justice to the analysis presented by Hensel and Hills[13] and Hills and Hensel.[14] They extended the analysis to consider temperature-dependent thermal properties, nonuniform space-and-time grid, planar, cylindrical, and spherical geometries, and an estimate of the variance of the calculated surface heat flux provided the variances of the temperature measurements and heat flux at the inverse boundary are known. Also, their algorithm has the capability of considering a digital filter of the measured temperatures.

6.10.6 Comparison with Prior Methods

A major weakness of most of the methods in Section 6.10 is the lack of a statistical basis. Another is that the methods do not readily extend to multiple interior sensors and to two- and three-dimensional cases. The function specification and regularization methods can have a statistical basis and the same formalism applied to one-dimensional problems can be directly extended to two- and three-dimensional cases. Moreover, quite different problems such as those involving integral equations or ablation can be treated by the latter methods whereas the space marching techniques may or may not be appropriate.

6.11 NUMERICAL CALCULATIONS

This section presents numerical calculations that bring out the salient features of several difference methods presented in this chapter.

Most methods in this chapter require the calculation of various heat flux sensitivity coefficients ($\partial T/\partial q$). The most powerful of these use differences.

SEC. 6.11 NUMERICAL CALCULATIONS

It is appropriate to ask what difference method is the most accurate in the calculation of sensitivity coefficients. Although an exhaustive study has not been performed, Figure 6.8 provides some insight into this question; this figure considers a sensor at the midplane of a planar slab. The sensitivity coefficient is evaluated at a dimensionless time equal to the dimensionless time step of the difference method; that is, $t^+ = \Delta t^+$. The backward difference ($\theta = 1$) method is superior for Δt^+ near unity and larger whereas the central difference method is superior for $\Delta t^+ < 0.1$.

As pointed out earlier, the sensitivity coefficient calculation time step Δt_X can be smaller than the problem time step Δt. This should produce more accurate sensitivity coefficients.

Chapter 5 presented several example calculations related to a triangular heat flux pulse. This same problem will be utilized here for a number of examples. The triangular heat flux pulse is maximum at $t^+ = 0.6$ and then falls linearly to $q^+ = 0$ at $t^+ = 1.2$. The analytical solution of this problem was presented in Chapter 5 [see Eqs. (5.2.4)–(5.2.6). Table 5.1 tabulates the exact dimensionless temperature response (to six decimal places) for a sensor located

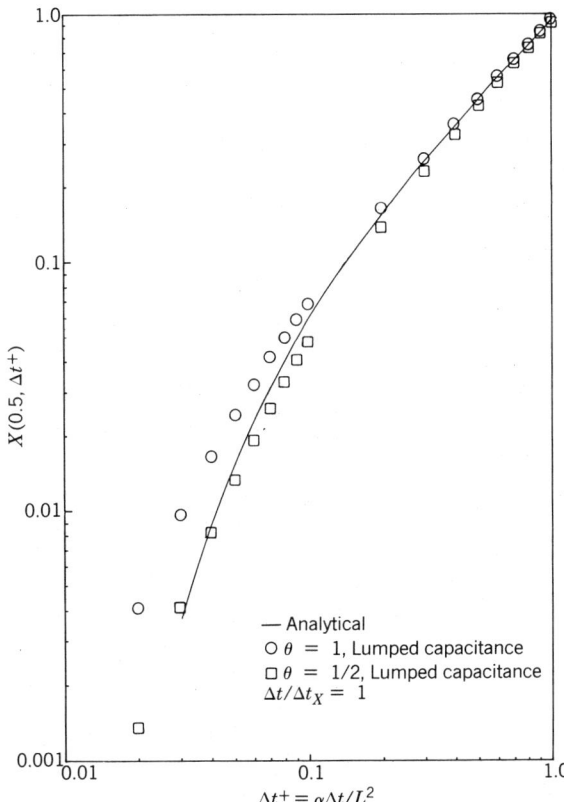

FIGURE 6.8 Calculation of planar slab sensitivity coefficient by difference methods for $x_k/L = 0.5$.

TABLE 6.7 Temperature Response to a Triangular Heat Flux Pulse. Additive Normal Uncorrelated Random Errors with Standard Deviation of 0.0017

t^+	$T^+ = \dfrac{T - T_0}{q_N L/K}$
−0.24	0.00034
−0.18	0.00281
−0.12	0.00135
−0.06	−0.00090
0.0	−0.00020
0.0600	−0.004241
0.1200	−0.000639
0.1800	0.001307
0.2400	0.007161
0.3000	0.012874
0.3600	0.021764
0.4200	0.038954
0.4800	0.054821
0.5400	0.072952
0.6000	0.098381
0.6600	0.127722
0.7200	0.155529
0.7800	0.193275
0.8400	0.223066
0.9000	0.248541
0.9600	0.275499
1.0200	0.293646
1.0800	0.311805
1.1400	0.330811
1.2000	0.342215
1.2600	0.348997
1.3200	0.350227
1.3800	0.355722
1.4400	0.359318
1.5000	0.361609

at the insulated inactive surface. Temperature measurement errors simulated by using a random number generator with normal distribution and standard deviation of $\sigma_Y^+ = 0.0017$ are given in Table 6.7. This standard deviation is approximately 5% of the maximum dimensionless temperature rise for the triangular pulse. All difference calculations that follow are performed using

SEC. 6.11 NUMERICAL CALCULATIONS

$\theta = 1$ and lumped capacitance. Similar results are expected for $\theta = 1/2$ and/or other capacitance distribution schemes.

Figure 6.9 compares results for exact and inexact sensor data, both with $r = 3$ for the function specification method, $q = C$. The results are quite good for the exact ($\sigma_Y^+ < 10^{-6}$) temperature data. However, the inclusion of temperature errors causes scatter in the computed heat flux. The results for $\sigma_Y^+ = 0.0017$ are similar to those using Duhamel's theorem that are given in Figure 5.13; the $\sigma_{\hat{q}}^+ / \sigma_Y^+$ ratio is 23.2 for Figure 6.9, which is close to the 25.4 value of Figure 5.13.

Figure 6.10 compares the function specification method ($q = C$) for $r = 4$ and 5 for data containing errors. Both values of r yield satisfactory values of heat flux considering that the data contain errors. As r increases, the ability to follow sharp changes in q is impaired, as evidenced by the calculations near $t^+ = 0, 0.6,$ and 1.2. The optimal choices of r and Δt are discussed in Section 5.6.1.

The results shown in Figures 6.9 and 6.10 are nearly identical to those obtained using the convolution approach, Figures 5.12 and 5.13. This is expected because the IHCP algorithms are the same; the method of solving the heat conduction equation and the evaluation of the sensitivity coefficients are different. The random errors are also different.

The same example problem was also solved using both first- and second-order sequential regularization with $\alpha_1 = \alpha_2 = 0.001$; these results are shown in

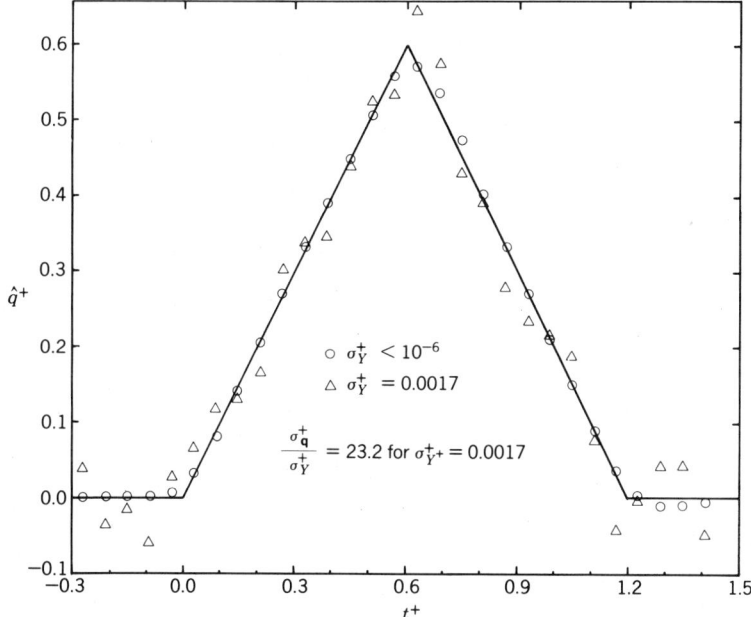

FIGURE 6.9 Comparison of exact ($\sigma_Y^+ < 10^{-6}$) and inexact ($\sigma_Y^+ = 0.0017$) data for function specification method, $q = C$, 20 elements, $r = 3$, $\Delta t / \Delta t_X = 2$, $\Delta t^+ = 0.06$, $\theta = 1$, lumped capacitance, $x/L = 1.0$.

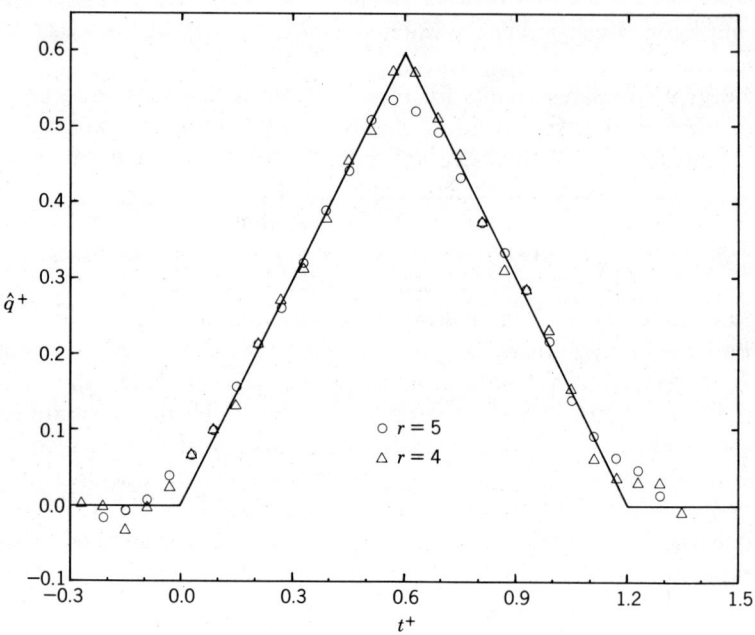

FIGURE 6.10 Function specification method, $q=C$, 20 elements, $\sigma_Y^+ =0.0017$, $\Delta t/\Delta t_X=2$, $\Delta t^+ =0.06$, $\theta=1$, lumped capacitance, $x/L=1.0$.

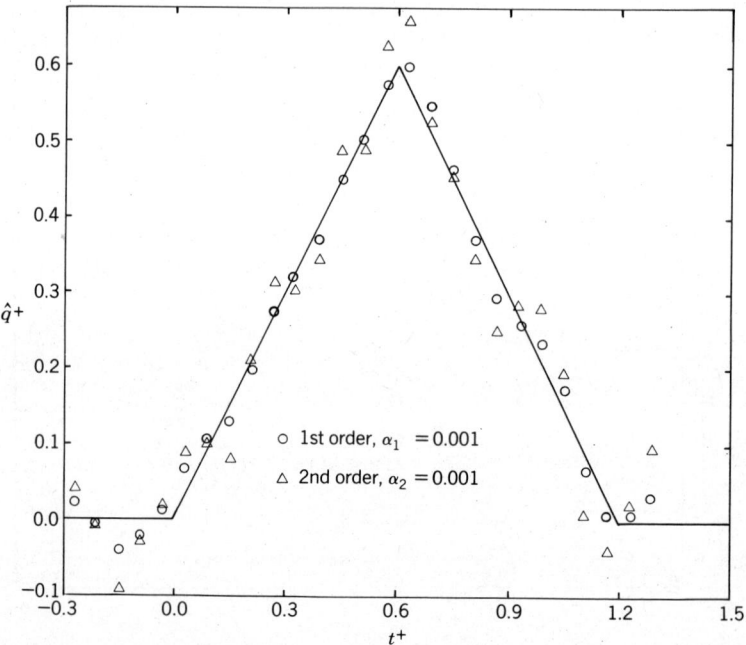

FIGURE 6.11 Comparison of first- and second-order sequential regularization with $\alpha=0.001$, 20 elements, $\sigma_Y^+ =0.0017$, $\Delta t/\Delta t_X=2$, $\Delta t^+ =0.06$, $r=5$, $\theta=1$, lumped capacitance, $x/L=1.0$.

SEC. 6.11 NUMERICAL CALCULATIONS

Figure 6.11. The first-order regularization appears superior to the second-order by a considerable margin. The choice of α was based on the results in Chapter 5 for the convolution integral approach. The analysis for the optimum choice of α_2 has not been performed. However, Figure 6.12 compares results for $\alpha_2 = 5 \times 10^{-3}$ with $\alpha_2 = 5 \times 10^{-4}$ and shows very little difference.

Numerous examples were presented in Chapter 5 illustrating how a single temperature error affects the heat flux calculation at times both before and after the temperature error. Similar results are shown in Figure 6.13 and were determined by setting $Y_1 = 1.0$ and all other temperature measurements identically zero. The second-order sequential regularization exhibits much greater swings than either the function specification ($q = C$) or first-order sequential regularization. This partially explains why the second-order regularization performed poorly in Figure 6.11. The function specification method has the smallest swings.

It has been demonstrated that comparable results can be obtained for both function specification ($q = C$) and first-order sequential regularization methods. The presence of the regularization parameter α requires that trial calculations be performed for each new class of problems solved by the regularization method. Both the function specification and first-order regularization methods have given reasonable results.

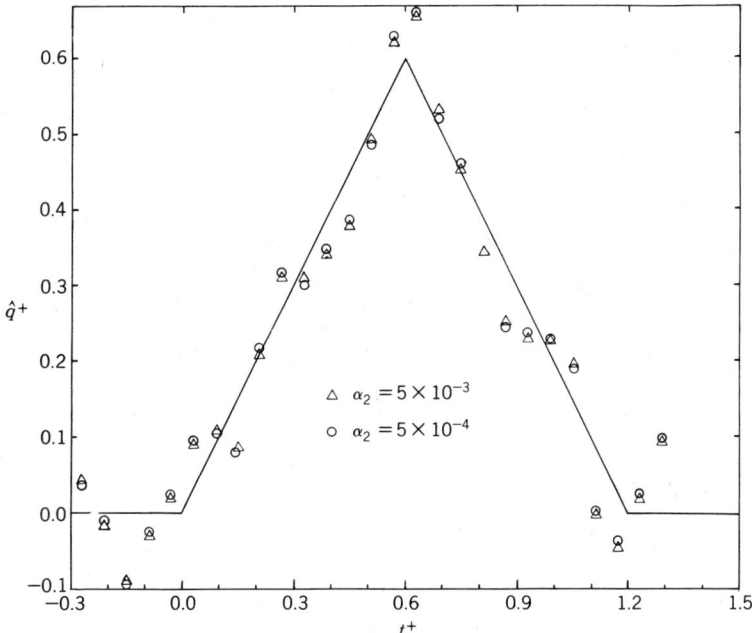

FIGURE 6.12 Influence of α_2 on second-order sequential regularization, 20 elements, $\sigma_Y^+ = 0.0017$, $\Delta t/\Delta t_X = 2$, $\Delta t^+ = 0.06$, $r = 5$, $\theta = 1$, lumped capacitance, $x/L = 1.0$.

262 CHAP. 6 ONE-DIMENSIONAL INVERSE HEAT CONDUCTION PROBLEM

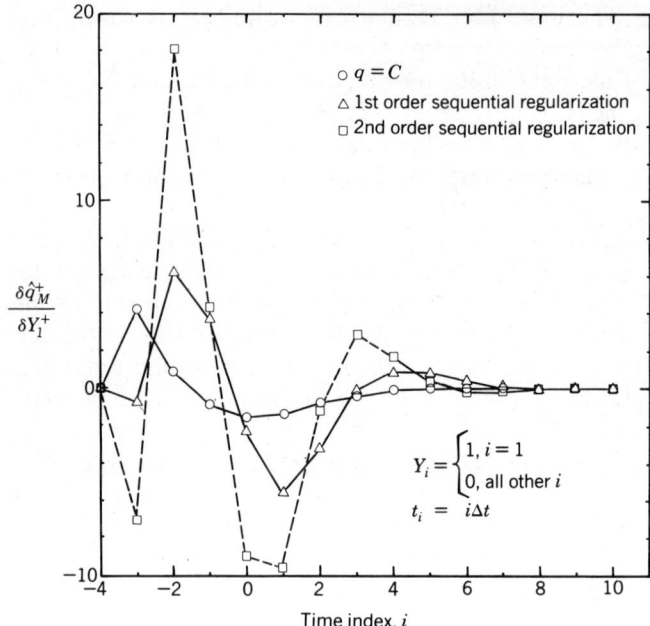

FIGURE 6.13 Dimensionless heat flux error due to a single measurement error at $t^+ = \Delta t^+ = 0.06$, $r = 5$, 20 elements, $\Delta t/\Delta t_X = 2$, $\theta = 1$, lumped capacitance, $x/L = 1.0$.

6.12 COMPUTER PROGRAMS

A great many computer programs have been written for the inverse heat conduction problem. The purpose of this section is to mention some of these. A review of some computer programs is given by Beck.[17]

A list of some computer programs is given in Table 6.8. All of the computer programs, except the last one listed, utilize some difference procedure for solving the transient heat conduction equation and thus can treat temperature-variable properties. Each program has unique features and capabilities. Brief descriptions of the programs listed in Table 6.8 are given next.

CONTA was developed by Beck[18] and was supported by both Michigan State University and Sandia National Laboratories. It uses the function specification procedure with the temporary assumption of a constant q. The sensitivity coefficients are calculated in a very efficient manner.[6] The program can treat multiple internal sensors in composite plates.

HCODE was developed by General Electric[19] in connection with nuclear reactor safety analyses. It can treat composite cylindrical rods. A modification of the function specification method is used.

HEATINV[20] was developed at the Naval Surface Weapons Center and is the only difference-based program listed that uses whole domain regularization.

TABLE 6.8 List of Computer Programs for the Inverse Heat Conduction Problem

Program Name	Developer/ Organization	IHCP Algorithm	Difference Method	Integral Method
CONTA	Beck; Michigan State Univ.– Sandia, Albuq.	Sequential function specification	Crank-Nicolson	No
HCODE	Muzzy, Avila, Root; GE San Jose	Modified sequential function specification	Crank-Nicolson	No
HEATINV	Bell, Wardlaw; Naval Surf. Weap. Center	0th-	1st order whole domain regularization	Crank-Nicolson	No
INVERT 1.0	Snider; EGG Idaho	Sequential function specification	Generalized Crank-Nicolson	No
ORMDIM	Bass, Drake, Ott; ORNL	Sequential function specification	Finite element	No
ORINC	Ott, Hedrick; ORNL	Uses temperatures only at time t_M	Backward difference	No
SODDIT	Blackwell; Sandia, Albuq.	Sequential function specification	Finite control volume, backward difference	No
SMICC	Hills, Hensel; New Mexico State	Space marching	Yes	No
ARIES	Drake; ORNL	0th-, 1st-, and 2nd-order whole domain regularization		Yes

A combination of zeroth- and first-order regularization is employed. A sequential regularization procedure, instead of the whole domain method, would be more computationally efficient.

INVERT 1.0 was written by Snider[21] at EG&G, Idaho, and uses the function specification method. It can treat composite, cylindrical rods.

ORMDIM was developed at Oak Ridge National Laboratory by Bass et al.[22] It is a two-dimensional, finite element-based program.

ORINC[23] was also developed at Oak Ridge National Laboratory. It is for composite cylinders. The fully implicit method of solution of the difference equations is employed. The temperatures only up to time t are used to calculate the flux at time t.

SODDIT was developed at Sandia National Laboratories, Albuquerque, by Blackwell.[24] The difference equations are derived using the finite control volume method. The sequential function method is used and arbitrary one-dimensional geometries are possible.

SMICC is a program developed by Hills and Hensel[14] that uses space marching and is described in Section 6.10.5. It also provides estimates for the resolving power and accuracy of the calculated heat flux components. Hills and Hensel are at New Mexico State University and the research was funded in part by Sandia National Laboratories.

ARIES[25] is the only program listed in Table 6.8 that is based on an integral model and thus is limited to linear problems. The integral model does give the flexibility to treat a variety of inverse problems, however. The user has the option of choosing zeroth, first or second (or some combination thereof) whole domain regularization.

Copies of some of the reports may be obtained from:

National Technical Information Service
U.S. Department of Commerce
5285 Port Royal Road
Springfield, VA 22161

and possibly from the authors. In some cases the reports contain listings of the programs.

REFERENCES

1. Beck, J. V. and Wolf, H., The Nonlinear Inverse Heat Conduction Problem, ASME Paper 65-HT-40, presented at the ASME/AIChE Heat Transfer Conference and Exhibit, Los Angeles, CA, August 8–11, 1965.
2. CDC 7600/CYBER 70 Model 76, Computer Systems Hardward Reference Manual, Control Data Corp., Publication no. 60367200, Minneapolis, 1972.
3. Blackwell, B. F., Efficient Technique for the Numerical Solution of the One-Dimensional Inverse Problem of Heat Conduction, *Numer. Heat Transfer* **4**, 229–238 (1981).
4. Beck, J. V. and Arnold, K. J., *Parameter Estimation in Engineering and Science*, Wiley, New York, 1977.
5. Beck, J. V., Surface Heat Flux Determination Using an Integral Method, *Nucl. Eng. Des.* **7**, 170–178 (1968).
6. Beck, J. V., Litkouhi, B., and St. Clair, C. R., Effective Sequential Solution of the Nonlinear Inverse Heat Conduction Problem, *Numer. Heat Transfer* **5**, 275–286 (1982).
7. deBoor, C., *A Practical Guide to Splines*, Springer-Verlag, New York, 1978.
8. D'Souza, N., Inverse Heat Conduction Problem for Prediction of Surface Temperatures and Heat Transfer from Interior Temperature Measurements, Report No. SRC-R-74, Space Research Corporation, Montreal, December 1973.
9. D'Souza, N., Numerical Solution of One-Dimensional Inverse Transient Heat Conduction by Finite Difference Method, ASME Paper No. 75-WA/HT-81, presented at Winter Annual Meeting, Houston, TX, Nov. 30–Dec. 4, 1975.
10. Weber, C. F., Analysis and Solution of the Ill-Posed Inverse Heat Conduction Problem, *Int. J. Heat Mass Transfer*, **24**(11), 1783–1792 (1981).

11. Raynaud, M., Determination du Flux Surfacique Traversant Une Paroi Soumise a Un Incendie au Moyen D'Une Methods D'Inversion, Laboratoire D'Aerothermique Groupe 'Echanges Thermiques' Universite Pierre et Marie Curie, Paris, France, August 1983.
12. Raynaud, M., and Bransier, J., A New Finite Difference Method for Non Linear Inverse Heat Conduction Problem, to be published in *Numerical Heat Transfer*.
13. Hensel, E. C. and Hills, R. G., A Space Marching Finite Difference Algorithm for the One Dimensional Inverse Conduction Heat Transfer Problem, ASME Paper No. 84-HT-48, presented at 22nd National Heat Transfer Conference, Niagara Falls, NY, August 6–8, 1984.
14. Hills, R. G. and Hensel, E. C., SMICC, the Space Marching Inverse Conduction Code, SAND 84-1563, Sandia National Laboratory, Albuquerque, NM, 1985.
15. Richtmyer, R. D. and Morton, K.W., *Difference Methods for Initial Value Problems*, 2nd ed., Interscience Publishers, New York, 1967.
16. Alifanov, O. M., Inverse Boundary Value Problems of Heat Conduction, *J. Eng. Phys.* **29**(1), 821–830 (1975).
17. Beck, J. V., Review of Six Inverse Heat Conduction Computer Codes, ANL/RAS/LWR 81-1, Argonne National Laboratory, Argonne, Illinios, February 1981.
18. Beck, J. V., User's Manual for CONTA—Program for Calculating Surface Heat Fluxes From Transient Temperatures Inside Solids, Sandia National Laboratory, Contractor Report, SAND83-7134, December, 1983.
19. Muzzy, R. J., Avila, J. H., and Root, R. E., Topical Report: Determination of Transient Heat Transfer Coefficients and the Resultant Surface Heat Flux from Internal Temperature Measurements, General Electric, GEAP-20731, 1975.
20. Bell, J. B. and Wardlaw, A. B., Numerical Solution of an Ill-Posed Problem Arising in Wind Tunnel Heat Transfer Data Reduction, NSWC TR 82-32, Naval Surface Weapons Center, Dahlgren, Virginia 22448, December 1981.
21. Snider, D. M., Invert 1.0—A Program for Solving the Nonlinear Inverse Heat Conduction Problem for One-Dimensional Solids, EGG-2068, EG&G Idaho, Inc., Idaho Falls, Idaho 83415, February 1981.
22. Bass, B. R., Drake, J. B., Ott, L. J., ORMDIN: A Finite Element Program for Two-Dimensional Nonlinear Inverse Heat Conduction Analysis, NUREG/CR-1709, ORNL/NUREG/CSD/TM-17, U.S. Nuclear Regulatory Commission, Washington, D.C., December 1980.
23. Ott, L. J. and Hedrick, R. A., ORINC: A One Dimensional Implicit Approach to the Inverse Heat Conduction Problem, NTIS Report ORNL/NUREG-23, Oak Ridge National Laboratories, Oak Ridge, TN., 1977.
24. Blackwell, B. F., User's Manual for the Sandia One-Dimensional Direct and Inverse Thermal (SODDIT) Code, Sandia National Laboratories internal report of Div. 7537, 1983.
25. Drake, J. B., ARIES: A Computer Program for the Solution of First Kind Integral Equations with Noisy Data, K/CSD/TM-43, Computer Sciences at Oak Ridge Gaseous Diffusion Plant, Post Office Box P, Oak Ridge, Tennessee 37830, October 1983.

PROBLEMS

6.1. Prove that Eq. (6.2.5) is valid for constant properties.

6.2. Write the difference equations for the solution of Eq. (6.2.7) for the step sensitivity coefficient Z for constant properties and one-dimensional planar geometry. Use the time integration of your choice. Demonstrate that the same difference equations can be obtained by differentiating Eq. (3.3.45) with respect to q_M.

6.3. In the algorithm for a single temperature sensor that exactly matches the temperature data, can the X sensitivity coefficient be replaced by Z in Eq. (6.3.8)? Why?

6.4. Write an efficient computer subroutine to solve systems of linear algebraic equations that are tridiagonal in form; see Eqs. (6.3.10) and (6.3.11).

6.5. Verify the algebra for Eqs. (6.3.12)–(6.3.14), the algorithm for calculating the sensitivity coefficients.

6.6. Calculate the sensitivity coefficients $Z_{k,1}$ and $Z_{k,2}$ for $x_k/L = 0.5$ and 1 for $\Delta t^+ = 0.05$ and 0.1 respectively and for planar geometry with a perfectly insulated surface. Use the pure implicit method. Compare your results with the tabular values given in Table 1.1.

6.7. Prove that $\sum_{k=1}^{N} w_k = 1$ in Eq. (6.4.11).

6.8. Repeat Example 6.1 for $x_k/L = 1$ and for $\Delta t^+ = 0.05$ and 0.5.

6.9. In the analysis of Section 6.6, the Z (step) sensitivity coefficient was used in the Taylor series expansion Eq. (6.6.4). Why Z and not X?

6.10. Repeat Example 6.2 for $\Delta t^+ = 0.1$.

6.11. Extend the analysis of Section 6.9 to include multiple temperature sensors.

6.12. Compare the computational molecules for all the space marching methods of Section 6.10.

6.13. As a heat transfer consultant, you have been asked to comment on the data quality from the following inverse problem: A thermocouple is located 0.001 m below the surface of a 0.005 m thick copper slab. The known back face boundary condition is a specified heat flux of 50 kW/m². The person describing the experiment to you "thinks" the unknown heat flux you are trying to estimate is of the order of magnitude of 1 kW/m². Will this experiment yield meaningful information?

6.14. a. For the D'Souza procedure, derive a two-node expression for the surface heat flux for a solid cylinder with the sensor at the center. Derive the difference equations from an energy balance.
b. Repeat part (a) for three nodes.
c. Compare the expressions with that obtained using the exact expression given by Burggraf.

6.15. Solve Problems 6.14 for a solid sphere.

6.16. For the Weber procedure, derive a two-node expression for the surface heat flux for a solid cylinder with the sensor at the center. The wave speed is infinite. Derive the difference equations from an energy balance. Compare the equation with the exact expression given by Burggraf.

6.17. Solve Problem 6.16 for a solid sphere.

CHAPTER 7

MULTIPLE HEAT FLUX ESTIMATION

7.1 INTRODUCTION

In the previous chapters the case of the single unknown surface heat flux history was considered. In this chapter the case of the multiple heat flux IHCP is treated.

A multiple heat flux case arises when both surfaces of a one-dimensional body are exposed to unknown heat flux histories. See Figure 7.1a for a plate heated on both sides. The unknown heat flux histories are $q_1(t)$ and $q_2(t)$. Figure 7.1b depicts either a hollow cylinder or sphere. Again there are two unknown heat flux histories. In both Figures 7.1a and 7.1b there must be at

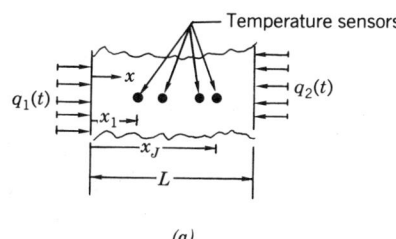

(a)

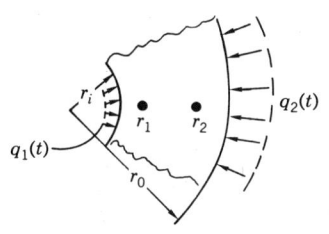

(b)

FIGURE 7.1 Some one-dimensional cases of bodies exposed to two unknown heat flux histories, $q_1(t)$ and $q_2(t)$. (a), plate; (b), hollow cylinder or sphere.

267

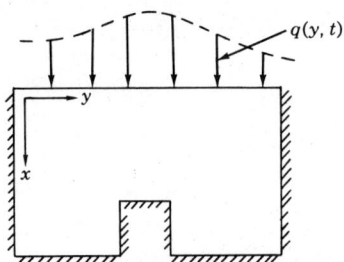

FIGURE 7.2 Two-dimensional body heated by space- and time-variable surface heat flux.

least two temperature sensors that are not at the same location. The heat flows are one-dimensional, and the heat flux histories $q_1(t)$ and $q_2(t)$ are independent. Hence any time variation in $q_1(t)$ has no effect on $q_2(t)$. Thus this problem is little different from those covered in previous chapters.

It is not necessary that the two heat fluxes are restricted to one-dimensional heat flow. For example, the plate of Figure 7.1a might not have parallel faces; also the radii in Figure 7.1b might not be concentric. Another case would be a square with two adjacent sides each with a different heat flux history.

A more typical multiple heat flux IHCP is for the case of a body exposed to a heat flux that is both space- and time-variable (see Figure 7.2). Thus the temperature distribution is two-dimensional. Any coordinate system can be used and the analysis is not restricted to two-dimensional cases. The body can be composed of several materials and can be irregular. The only requirement is that a method of solving the direct problem (*known* surface heat flux time and space variation) is available.

In this chapter the method of solving the direct problem is not of interest. Rather a method of solution is assumed such as finite differences, finite elements, or Duhamel's theorem. For simplicity the problems are considered to be linear but nonlinear cases can be treated by using the modifications discussed in Chapter 6.

Few papers on the two-dimensional inverse heat conduction problem have been written. One by Bass et al.[1] gives a description of a computer program for a solid cylinder with radial and angular dependence; finite elements were used. An analytical solution was presented by Imber.[2] Some two-dimensional inverse problems for which the geometry is found (and thus are not the IHCP) have been examined by Hsieh and co-workers.[3,4]

The scope of this chapter is to discuss the two heat flux case illustrated in Figure 7.1 and to cover the multiple heat flux case. Both the sequential function specification and sequential regularization methods are employed. There is less emphasis on displaying numerical results than in previous chapters.

7.2 TWO INDEPENDENT HEAT FLUXES CASE

The temperatures in a one-, two-, or three-dimensional body with temperature-independent thermal properties can be given in the standard form of

SEC. 7.2 TWO INDEPENDENT HEAT FLUXES CASE

$$\mathbf{T} = \mathbf{T}|_{\mathbf{q}=0} + \mathbf{X}\mathbf{q} \qquad (7.2.1)$$

The symbol $\mathbf{T}|_{\mathbf{q}=0}$ means the calculated temperature vector $\mathbf{T}$ with $\mathbf{q}$, given by Eq. (7.2.3a), set equal to $\mathbf{0}$. [See Eq. (3.2.22).] For the case of two q components at each time, J temperature sensors, and r future times, the various components of $\mathbf{T}$ are

$$\mathbf{T} = \begin{bmatrix} \mathbf{T}(M) \\ \mathbf{T}(M+1) \\ \vdots \\ \mathbf{T}(M+r-1) \end{bmatrix}, \quad \mathbf{T}(i) = \begin{bmatrix} T_1(i) \\ T_2(i) \\ \vdots \\ T_J(i) \end{bmatrix} \qquad (7.2.2\mathrm{a},\mathrm{b})$$

where $\mathbf{T}$ is a $Jr \times 1$ matrix; that is, Jr vector,

$$\mathbf{q} = \begin{bmatrix} \mathbf{q}(M) \\ \mathbf{q}(M+1) \\ \vdots \\ \mathbf{q}(M+r-1) \end{bmatrix}, \quad \mathbf{q}(i) = \begin{bmatrix} q_1(i) \\ q_2(i) \end{bmatrix} \qquad (7.2.3\mathrm{a},\mathrm{b})$$

where $\mathbf{q}$ is a $rp \times 1$ matrix with $p=2$ in this case,

$$\mathbf{X} = \begin{bmatrix} \mathbf{a}(1) & & & \\ \mathbf{a}(2) & \mathbf{a}(1) & & \\ \mathbf{a}(3) & \mathbf{a}(2) & \mathbf{a}(1) & \\ \vdots & \vdots & & \ddots \\ \mathbf{a}(r) & \mathbf{a}(r-1) & \cdots & \mathbf{a}(1) \end{bmatrix} \qquad (7.2.4)$$

where $\mathbf{X}$ is a $Jr \times pr$ matrix and

$$\mathbf{a}(i) = \begin{bmatrix} a_{11}(i) & a_{12}(i) \\ a_{21}(i) & a_{22}(i) \\ \vdots & \vdots \\ a_{J1}(i) & a_{J2}(i) \end{bmatrix}, \quad a_{jk}(i) = \frac{\partial T(\mathbf{x}_j, t_i)}{\partial [q_k(1)]} \qquad (7.2.5\mathrm{a},\mathrm{b})$$

where $\mathbf{a}(i)$ is a $J \times p$ matrix with $p=2$ in this case. For the case of p components of $\mathbf{q}$ at each time, $\mathbf{q}(i)$ given by Eq. (7.2.3b) is a vector of p elements, and $\mathbf{a}(i)$ given by Eq. (7.2.5a) is a matrix with p columns. For the special case of one sensor ($J=1$) and one heat flux history ($p=1$), $a_{11}(i)$ is simply $\Delta\phi_{i-1}$ which is used in Chapters 3, 4, and 5.

The components of $\mathbf{X}$ in the most general sense can be considered to be sensitivity coefficients. See Eqs. (6.5.18) and (6.5.19). The $\mathbf{X}$ components can be generated using finite differences, Duhamel's theorem, or other methods.

For the first two times (associated with t_M and t_{M+1}) a partially expanded form of Eq. (7.2.1) is

$$\mathbf{T}(M) = \mathbf{T}(M)|_{\mathbf{q}(M)=0} + \mathbf{a}(1)\mathbf{q}(M) \qquad (7.2.6\mathrm{a})$$

$$\mathbf{T}(M+1) = \mathbf{T}(M+1)_{\mathbf{q}(M)=\mathbf{q}(M+1,=0)} + \mathbf{a}(1)\mathbf{q}(M+1) + \mathbf{a}(2)\mathbf{q}(M) \qquad (7.2.6\mathrm{b})$$

Both these matrix equations represent J scalar equations, some of which are

$$T_1(M) = T_1(M)|_{q_1(M)=\cdots=q_J(M)=0} + a_{11}(1)q_1(M) + a_{12}(1)q_2(M)$$

$$\vdots$$

$$T_J(M) = T_J(M)|_{q_1(M)=\cdots=q_J(M)=0} + a_{J1}(1)q_1(M) + a_{J2}(1)q_2(M) \qquad (7.2.7)$$

$$T_1(M+1) = T_1(M+1)|_{q_1(M)=\cdots=q_J(M)=q_1(M+1)=\cdots=q_J(M+1)=0} + a_{11}(1)q_1(M+1)$$
$$+ a_{12}(1)q_2(M+1) + a_{11}(2)q_1(M) + a_{12}(2)q_2(M)$$

$$\vdots$$

$$T_J(M+1) = T_J(M+1)|_{q_1(M)=\cdots=q_J(M+1)=0} + a_{J1}(1)q_1(M+1) + a_{J2}(1)q_2(M+1)$$
$$+ a_{J1}(2)q_1(M) + a_{J2}(2)q_2(M) \qquad (7.2.8)$$

Notice that in the IHCP there are two unknown q components at the first time (t_M); namely, $q_1(M)$ and $q_2(M)$. There are J measurements at that time and J must be equal to or greater than the number of q components (2 in this case). For the case of $J=2$ it is possible to solve Eq. (7.2.7) simultaneously for $q_1(M)$ and $q_2(M)$ with $T_1(M)$ set equal to $Y_1(M)$, and $T_2(M)$ equal to $Y_2(M)$. Knowing $q_1(M)$ and $q_2(M)$, Eq. (7.2.8) can be solved for $q_1(M+1)$ and $q_2(M+1)$. Such a procedure is analogous to the Stolz method discussed in Sections 4.3 and 5.3.2. This procedure is usually unacceptable due to its extreme sensitivity to measurement errors. Instead the function specification and regularization methods are used in this chapter.

The trial function method of Section 4.6 includes both the function specification and regularization methods. It also allows a smooth transition between the two methods. The trial function criterion is used here in an effort to unify the methods but the two methods of function specification and regularization represent two extreme cases. The trial function criterion given by Eq. (4.6.1) is written as

$$S = (\mathbf{Y} - \mathbf{T})^T \psi^{-1}(\mathbf{Y} - \mathbf{T}) + \alpha[\mathbf{H}(\mathbf{q} - \tilde{\mathbf{q}})]^T \mathbf{W}[\mathbf{H}(\mathbf{q} - \tilde{\mathbf{q}})] \qquad (7.2.9)$$

where ψ is the covariance matrix of the random measurement errors in $\mathbf{Y}$; α is the regularization parameter, a scalar; $\mathbf{H}$ is the matrix for a zeroth or first or other order regularization; and $\tilde{\mathbf{q}}$ is the trial function.

For the function specification method, α is made so large that

$$\mathbf{q} = \tilde{\mathbf{q}} \qquad (7.2.10)$$

and for the regularization method

$$\tilde{\mathbf{q}} = \mathbf{0} \qquad (7.2.11)$$

and α is adjusted to cause (see Sect. 4.5.3.3)

$$(\mathbf{Y} - \hat{\mathbf{T}})^T \psi^{-1}(\mathbf{Y} - \hat{\mathbf{T}}) \cong \text{expected value} \qquad (7.2.12)$$

The $\hat{\mathbf{T}}$ vector in Eq. (7.2.12) is the estimated value of $\mathbf{T}$ as a result of minimizing S. When the first four, sixth, and seventh standard statistical assumptions

SEC. 7.2 TWO INDEPENDENT HEAT FLUXES CASE

(Section 1.4.2) are valid, the expected value of Eq. (7.2.12) is equal to the number of measurements minus the number of **q** components. (See Reference 5, p. 268.)

7.2.1 Sequential Function Specification Method

For the function specification method the **q** function is temporarily set equal to **q̃** as indicated by Eq. (7.2.10). The simplest function to specify is that of **q̃** independent of time. This is a *temporary* assumption that is used to obtain **q**(M). (See Section 4.4.3.) Using this assumption gives

$$\mathbf{q}(M) = \mathbf{q}(M+1) = \cdots = \mathbf{q}(M+r-1) \tag{7.2.13}$$

which is analogous to Eq. (4.4.20). To be more explicit, for **q**(M) with two components, Eq. (7.2.13) gives

$$q_1(M) = q_1(M+1) = \cdots = q_1(M+r-1) \tag{7.2.14a}$$

$$q_2(M) = q_2(M+1) = \cdots = q_2(M+r-1) \tag{7.2.14b}$$

See Figure 7.3. The objective is to estimate $q_1(M)$ and $q_2(M)$ at the Mth time

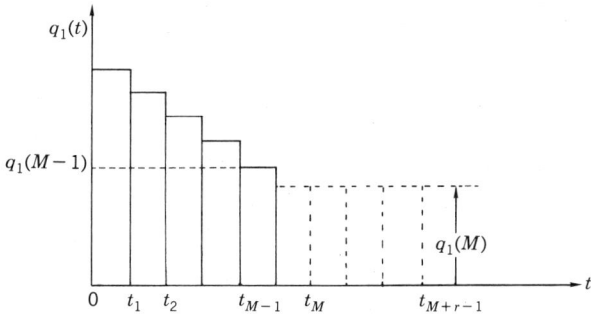

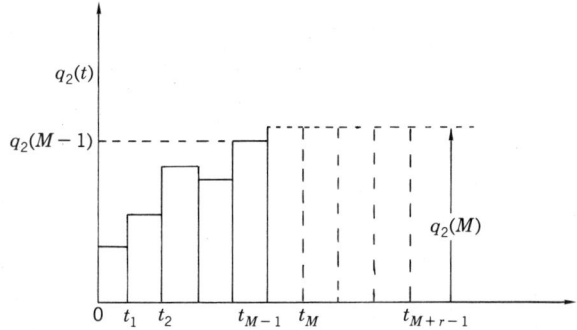

FIGURE 7.3 Two independent heat flux histories showing the temporary constant heat flux assumption.

step. In order to permit greater generality, $\mathbf{q}$ is set equal to

$$\mathbf{q} = \mathbf{A}\boldsymbol{\beta}, \quad \mathbf{A} = \begin{bmatrix} \mathbf{A}(1) \\ \vdots \\ \mathbf{A}(r) \end{bmatrix}, \quad \boldsymbol{\beta} = \begin{bmatrix} q_1(M) \\ q_2(M) \end{bmatrix} \quad (7.2.15)$$

For the constant $\mathbf{q}$ assumption indicated by Eq. (7.2.14a,b), the $\mathbf{A}(i)$ matrix is

$$\mathbf{A}(i) = \begin{bmatrix} 1 & 0 \\ 0 & 1 \end{bmatrix} \quad (7.2.16)$$

Then using

$$\mathbf{Z} = \mathbf{X}\mathbf{A} \quad (7.2.17)$$

the function to minimize with respect to $\boldsymbol{\beta}$ is

$$S = (\mathbf{Y} - \mathbf{T}|_{\boldsymbol{\beta}=0} - \mathbf{Z}\boldsymbol{\beta})^T \boldsymbol{\psi}^{-1} (\mathbf{Y} - \mathbf{T}|_{\boldsymbol{\beta}=0} - \mathbf{Z}\boldsymbol{\beta}) \quad (7.2.18)$$

The matrix derivative of Eq. (7.2.18) with respect to $\boldsymbol{\beta}$ yields the estimator

$$\hat{\boldsymbol{\beta}} = (\mathbf{X}^T \boldsymbol{\psi}^{-1} \mathbf{Z})^{-1} \mathbf{Z}^T \boldsymbol{\psi}^{-1} (\mathbf{Y} - \mathbf{T}|_{\boldsymbol{\beta}=0}) \quad (7.2.19)$$

which gives the $\hat{\mathbf{q}}(M)$ vector as indicated by Eq. (7.2.15). After it is obtained, M is increased by one and Eq. (7.2.19) is used again.

If the first four standard assumptions are valid, $\boldsymbol{\psi}$ simplifies to

$$\boldsymbol{\psi} = \sigma^2 \mathbf{I}, \quad \boldsymbol{\psi}^{-1} = \sigma^{-2} \mathbf{I} \quad (7.2.20)$$

and then $\boldsymbol{\psi}^{-1}$ cancels in Eq. (7.2.19).

EXAMPLE 7.1. For the case of two independent heat flux histories, $q_1(t)$ and $q_2(t)$, use the sequential function estimation procedure with the constant heat flux approximation to give explicit forms of the algorithm. Assume that the standard statistical assumptions apply. Let there be three measurements taken at a time and use two future temperature measurement times.

Solution. The equation to use is Eq. (7.2.19) with $\boldsymbol{\psi}^{-1}$ canceling, $J=3$, and $r=2$. The $\mathbf{Z}$ matrix is

$$\mathbf{Z} = \mathbf{X}\mathbf{A} = \begin{bmatrix} \mathbf{a}(1) & \mathbf{0} \\ \mathbf{a}(2) & \mathbf{a}(1) \end{bmatrix} \begin{bmatrix} \mathbf{A}(1) \\ \mathbf{A}(2) \end{bmatrix}$$

$$= \begin{bmatrix} \mathbf{a}(1)\mathbf{A}(1) \\ \mathbf{a}(2)\mathbf{A}(1) + \mathbf{a}(1)\mathbf{A}(2) \end{bmatrix}$$

$$= \begin{bmatrix} a_{11}(1) & a_{12}(1) \\ a_{21}(1) & a_{22}(1) \\ a_{31}(1) & a_{32}(1) \\ a_{11}(1)+a_{11}(2) & a_{12}(1)+a_{12}(2) \\ a_{21}(1)+a_{21}(2) & a_{22}(1)+a_{22}(2) \\ a_{31}(1)+a_{31}(2) & a_{32}(1)+a_{32}(2) \end{bmatrix} \quad (7.2.21)$$

SEC. 7.2 TWO INDEPENDENT HEAT FLUXES CASE

The $\mathbf{Z}^T\mathbf{Z}$ matrix is

$$\mathbf{Z}^T\mathbf{Z} = \begin{bmatrix} C_{11} & C_{12} \\ C_{12} & C_{22} \end{bmatrix} \quad (7.2.22)$$

where

$$C_{ij} = \sum_{m=1}^{3} a_{mi}(1)a_{mj}(1) + \sum_{m=1}^{3} [a_{mi}(1) + a_{mi}(2)][a_{mj}(1) + a_{mj}(2)] \quad (7.2.23)$$

The $\mathbf{Z}^T(\mathbf{Y} - \mathbf{T}|_{\boldsymbol{\beta}=0})$ matrix is

$$\mathbf{Z}^T(\mathbf{Y} - \mathbf{T}|_{\boldsymbol{\beta}=0}) = \begin{bmatrix} \mathbf{A}(1)\mathbf{a}^T(1)[\mathbf{Y}(M) - \mathbf{T}(M)|_{\boldsymbol{\beta}=0}] + \mathbf{A}(1)\mathbf{a}^T(2) + \mathbf{A}(2)a(1) \\ (\mathbf{Y}(M+1) - \mathbf{T}(M+1)|_{\boldsymbol{\beta}=0}) \end{bmatrix} = \begin{bmatrix} D_1 \\ D_2 \end{bmatrix} \quad (7.2.24)$$

where

$$D_j = \sum_{i=1}^{3} a_{ij}(1)[Y_i(M) - T_i(M)|_{q_1(M) = q_2(M) = 0}]$$

$$+ \sum_{i=1}^{3} [a_{ij}(1) + a_{ij}(2)][Y_i(M+1) - T_i(M+1)|_{q_1(M) = \cdots = q_2(M+1) = 0}] \quad (7.2.25)$$

Finally $\hat{q}_1(M)$ and $\hat{q}_2(M)$ are, in algebraic form,

$$\hat{q}_1(M) = \frac{C_{22}D_1 - C_{12}D_2}{C_{11}C_{22} - C_{12}^2} \quad (7.2.26a)$$

$$\hat{q}_2(M) = \frac{C_{12}D_1 - C_{22}D_2}{C_{11}C_{22} - C_{12}^2} \quad (7.2.26b)$$

□

7.2.2 Sequential Regularization Method

The sequential regularization method can start with the criterion given by Eq. (7.2.9) with $\tilde{\mathbf{q}}$ set equal to the zero vector. Introducing the model given by Eq. (7.2.1) for $\mathbf{T}$, taking the matrix derivative with respect to $\mathbf{q}$, and setting the matrix equation equal to zero, gives the matrix normal equation,

$$[\mathbf{X}^T\boldsymbol{\psi}^{-1}\mathbf{X} + \alpha\mathbf{H}^T\mathbf{W}\mathbf{H}]\hat{\mathbf{q}} = \mathbf{X}^T\boldsymbol{\psi}^{-1}(\mathbf{Y} - \mathbf{T}|_{\mathbf{q}=0}) \quad (7.2.27)$$

This represents a set of pr simultaneous algebraic equations. For the case of two unknown heat flux histories, $p=2$ and the number of unknowns is $2r$.

If both $\mathbf{B}$ and $\tilde{\mathbf{q}}^f$ are zero matrices in Eq. (4.6.7), then Eq. (7.2.27) is obtained because $\mathbf{H}^T\mathbf{W}\mathbf{H}$ can represent any or all of the $\mathbf{H}_i^T\mathbf{W}_i\mathbf{H}_i$ terms in Eq. (4.6.7). The difference is in the interpretation of the various terms to allow for both multiple measurements and also multiple heat flux components at each time. In the present case, there are $J(\geqslant 2)$ measurements at each time $(t_M, t_{M+1}, \ldots, t_{M+r-1})$ and there are $p=2$ components of the heat flux at each time. The $\mathbf{X}$ matrix is given by Eqs. (7.2.4, 5) and $\mathbf{q}$ is given by Eq. (7.2.3) for $p=2$.

The main quantity in Eq. (7.2.27) that has not been discussed is $\mathbf{H}$. As pointed out there are a number of forms that it can take. For a zeroth-order regularization, the simplest form of $\mathbf{H}$ is $\mathbf{I}$. See Eq. (4.5.16a). Such a choice of $\mathbf{H}$ does not distinguish between independent (or unrelated) heat flux histories such as $q_1(t)$ and $q_2(t)$ of Figure 7.3 or the continuous ones of Figure 7.4.

Another $\mathbf{H}$ is for the first-order regularization. It usually involves first differences in time and no differences in space; that is, differences are not taken between $q_1(t)$ and $q_2(t)$. The first differences can be represented by $\mathbf{H}$ given by

$$\mathbf{H} = \begin{bmatrix} -\mathbf{I} & \mathbf{I} & 0 & 0 \\ 0 & -\mathbf{I} & \mathbf{I} & 0 \\ 0 & 0 & -\mathbf{I} & \mathbf{I} \\ 0 & 0 & 0 & 0 \end{bmatrix} \qquad (7.2.28)$$

which is for $r=4$ and $\mathbf{I}$ has the dimensions of $p \times p$ which is 2×2 in the present case. For the case of

$$\mathbf{W} = \mathbf{I} \qquad (7.2.29)$$

$\mathbf{H}^T \mathbf{W} \mathbf{H}$ for $r=4$ becomes

$$\mathbf{H}^T \mathbf{W} \mathbf{H} = \mathbf{H}^T \mathbf{H} = \begin{bmatrix} \mathbf{I} & -\mathbf{I} & 0 & 0 \\ -\mathbf{I} & 2\mathbf{I} & -\mathbf{I} & 0 \\ 0 & -\mathbf{I} & 2\mathbf{I} & -\mathbf{I} \\ 0 & 0 & -\mathbf{I} & \mathbf{I} \end{bmatrix} \qquad (7.2.30)$$

where for $p=2$ [i.e., $q_1(t)$ and $q_2(t)$], $\mathbf{I}$ is

$$\mathbf{I} = \begin{bmatrix} 1 & 0 \\ 0 & 1 \end{bmatrix} \qquad (7.2.31)$$

For this case of two heat flux histories, $p=2$, and there are four future time steps, that is, $r=4$; there are $pr=8$ unknown components of the heat flux. In the sequential procedure only the first components of $q_1(t)$ and $q_2(t)$ are retained, namely $q_1(M)$ and $q_2(M)$. Then the next time step is begun with M increased by one.

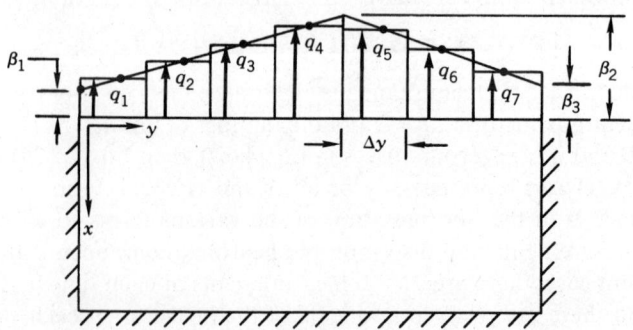

FIGURE 7.4 Multiple heat flux function specification case.

SEC. 7.3 MULTIPLE HEAT FLUX CASE

EXAMPLE 7.2. Construct the partitioned matrices for the matrix in the brackets on the left of Eq. (7.2.27) for X with $r=2$ and $p=2$; $\psi=\sigma^2 I$; and $W = I$. Use first-order time differences.

Solution. For $r=2$, the X matrix from Eq. (7.2.4) is

$$X = \begin{bmatrix} \mathbf{a}(1) & 0 \\ \mathbf{a}(2) & \mathbf{a}(1) \end{bmatrix}$$

and then

$$X^T \psi^{-1} X = X^T X \sigma^{-2} = \begin{bmatrix} \mathbf{a}^T(1)\mathbf{a}(1) + \mathbf{a}^T(2)\mathbf{a}(2) & \mathbf{a}^T(2)\mathbf{a}(1) \\ \mathbf{a}^T(1)\mathbf{a}(2) & \mathbf{a}_1^T(1)\mathbf{a}(1) \end{bmatrix} \sigma^{-2}$$

The $H^T W H$ matrix becomes

$$H^T W H = H^T H = \begin{bmatrix} -I & 0 \\ I & 0 \end{bmatrix} \begin{bmatrix} -I & I \\ 0 & 0 \end{bmatrix} = \begin{bmatrix} I & -I \\ -I & I \end{bmatrix}$$

and thus

$$X^T \psi^{-1} X + \alpha H^T W H = \begin{bmatrix} [\mathbf{a}^T(1)\mathbf{a}(1) + \mathbf{a}^T(2)\mathbf{a}(2)]\sigma^{-2} + \alpha I & \mathbf{a}^T(2)\mathbf{a}(1)\sigma^{-2} - \alpha I \\ \mathbf{a}^T(1)\mathbf{a}(2)\sigma^{-2} - \alpha I & \mathbf{a}^T(1)\mathbf{a}(1)\sigma^{-2} + \alpha I \end{bmatrix}$$

Notice that the effect of the regularization procedure is to increase the diagonal terms and decrease the off-diagonal terms. This results in improved conditioning of the set of equations—it becomes less sensitive to measurement errors. □

7.3 MULTIPLE HEAT FLUX CASE

For the case of a heat flux that is a function of both time and position as shown in Figure 7.2, the variation over the surface is controlled in a manner similar to that for time. There are many ways to accomplish this. In this section examples are given for both the sequential function specification and sequential regularization methods.

In Chapter 1 sensitivity coefficients were plotted for both one-dimensional and two-dimensional cases. To determine the heat flux q_M for the one-dimensional case it was found that a few future time steps were needed. Measurements over the whole time domain were not needed. This point was also discussed in Chapter 5 with reference to the whole domain regularization method. See Section 5.4. This characteristic of needing only a limited number of future time steps is a consequence of the nature of the heat conduction equation which for constant thermal properties and rectangular coordinates is

$$k\left(\frac{\partial^2 T}{\partial x^2} + \frac{\partial^2 T}{\partial y^2}\right) = \rho c \frac{\partial T}{\partial t} \tag{7.3.1}$$

There is a first derivative with time. There is no propagation backward in time. Any change in the surface heating condition at any time t greater than t_M does not influence the temperatures for t less than t_M. This is illustrated by the sensitivity coefficients for various one-dimensional cases in Figures 1.10–1.14. The sensitivity coefficient for the heat flux component q_M (which is constant between t_{M-1} and t_M, and zero otherwise) is zero for time t less than t_M.

The character of Eq. (7.3.1) for a heat flux variation across the surface in the y-direction (see Figure 7.2) is different from that for a time variation. Notice the *second* derivative in Eq. (7.3.1) with respect to y. The temperature at any point $y > 0$ is affected by the heating at both smaller and larger y values. This is illustrated by the sensitivity coefficients shown in Figure 1.17 and implies that all the surface heat flux components at a given time must be found simultaneously. For this reason the estimation procedures for the multiple heat flux case require a solution of a set of simultaneous, full-matrix equations. (A *full-matrix* set of algebraic equations $\mathbf{Ax} = \mathbf{b}$ has a full square matrix $\mathbf{A}$ whereas a *sparse* set of equations may have an $\mathbf{A}$ matrix which is mainly zeros and a relatively small number of nonzero elements. An example is the tridiagonal set of equations.)

Because of the necessity for treating the complete space variation of $q(\mathbf{r}, t)$, the dimensionality of the problem is considerably increased over the case of a single heat flux IHCP. The importance of the sequential-in-time algorithms becomes more apparent also. Suppose that the surface heat flux is divided in space so that there are 10 spatial components. A discrete variation in time could easily have 100 time components. In a whole domain method all $1000 (= 10 \times 100)$ components must be simultaneously estimated. This is computationally much more expensive than the sequential methods where as few as 10 components are found simultaneously. This means that only 10 or so simultaneous equations need be solved rather than 1000. Since the number of the computations in the solution of simultaneous, full-matrix equations varies as the number of the equations cubed, the reduction in computation time using sequential methods is very large indeed.

7.3.1 Sequential Function Specification Method for Multiple Heat Flux Components

The case of multiple connected (i.e., continuous) heat flux components can be treated using the function specification method in a manner similar to that in Section 7.2.1 which is for two unrelated heat flux segments. The method is described with an example.

The spatial variation of the surface heat flux can be described by many different expressions; for example, constant segments, linear-in-position segments, parabolic segments, and splines. Figure 7.4 illustrates the case of linear segments with two segments. Three parameters, β_1, β_2 and β_3, are utilized to describe the two segments; these parameters can change with time. The linear segments shown in Figure 7.4 represent the heat fluxes at time t_M. The linear segments are divided into seven equal increments of Δy and there are seven q components. The heat fluxes to be found are

$$q_1(M), q_2(M), \ldots, q_7(M)$$

SEC. 7.3 MULTIPLE HEAT FLUX CASE

which can be represented by $\mathbf{q}(M)$. The temporary assumption of $\mathbf{q}$ independent in time is used, which is given by Eq. (7.2.13). With this assumption and the relationships implied by Figure 7.4, Eqs. (7.2.15) and (7.2.16) are replaced by

$$\mathbf{q} = \mathbf{A}\boldsymbol{\beta}, \quad \mathbf{A} = \begin{bmatrix} \mathbf{A}(1) \\ \vdots \\ \mathbf{A}(r) \end{bmatrix}, \quad \boldsymbol{\beta} = \begin{bmatrix} \beta_1(M) \\ \beta_2(M) \\ \beta_2(M) \end{bmatrix} \tag{7.3.2}$$

$$\mathbf{A}(i) = \begin{bmatrix} 7/8 & 1/8 & 0 \\ 5/8 & 3/8 & 0 \\ 3/8 & 5/8 & 0 \\ 1/8 & 7/8 & 0 \\ 0 & 5/6 & 1/6 \\ 0 & 1/2 & 1/2 \\ 0 & 1/6 & 5/6 \end{bmatrix} \tag{7.3.3}$$

because the $\mathbf{q}(M)$ components are related, the $\mathbf{A}(i)$ matrix is not diagonal as is Eq. (7.2.16) which is for two *unrelated* heat fluxes.

The solution proceeds in the same manner as for the case discussed in Section 7.2.1. The sequential procedure implied by Eq. (7.2.19) is employed.

7.3.2 Sequential Regularization Method for Multiple Heat Fluxes

In order to describe clearly the sequential regularization multiple heat flux method, the minimization criterion is first expressed in a continuous rather than a discrete form. The continuous form is

$$S = \int_{t=t_M}^{t_{M+r-1}} \int_{y=0}^{L} [Y(x,y,t) - T(x,y,t)]^2 dy\, dt + \alpha' \int_{t_M}^{t_{M+r-1}} \int_0^L \left(\frac{\partial^2 q}{\partial y \partial t}\right)^2 dy\, dt$$

(7.3.4)

Notice that this expression involves a first derivative of the surface heat flux q with respect to t and also to y; it is a first-order regularization criterion in both t and y. Fluctuations in the estimated heat flux in both t and y are reduced by the regularization term. The integral in Eq. (7.3.4) involving the measured temperatures $Y(x, y, t)$ is over y from zero to L. (Experimental temperatures may not be uniformly spaced in y, however.)

A summation form analogous to Eq. (7.3.4) is

$$S = \sum_{i=1}^{r} \sum_{j=1}^{J} [Y_j(M+i-1) - T_j(M+i-1)]^2$$

$$+ \alpha \sum_{i=1}^{r-1} \sum_{j=1}^{J-1} [q_{j+1}(M+i) - q_{j+1}(M+i-1) - q_j(M+i) + q_j(M+i-1)]^2$$

(7.3.5)

The second double summation in Eq. (7.3.5) represents the cross-derivative integral in Eq. (7.3.4). The expression given by Eq. (7.3.5) can be generalized to the matrix expression given by Eq. (7.2.9) with $\tilde{\mathbf{q}}$ set equal to zero. Equation (7.2.9) is more general since it includes the matrices ψ^{-1} and $\mathbf{W}$.

The sequential regularization method for multiple heat fluxes is developed from Eq. (7.2.9) with $\tilde{\mathbf{q}} = 0$. The first differences in space and time are obtained by setting $\mathbf{H}$, a $pr \times pr$ matrix, equal to

$$\mathbf{H} = \mathbf{H}_t \mathbf{H}_y \tag{7.3.6}$$

$$\mathbf{H}_t = \begin{bmatrix} -\mathbf{I} & \mathbf{I} & 0 & 0 & \cdots & 0 \\ 0 & -\mathbf{I} & \mathbf{I} & 0 & \cdots & 0 \\ 0 & 0 & -\mathbf{I} & \mathbf{I} & & \\ \vdots & & \ddots & \ddots & \ddots & \\ 0 & \cdots & & 0 & -\mathbf{I} & \mathbf{I} \\ 0 & \cdots & & 0 & 0 & 0 \end{bmatrix} \tag{7.3.7}$$

$$\mathbf{H}_y = \text{diag}[\mathbf{h} \quad \mathbf{h} \quad \cdots \quad \mathbf{h}] \tag{7.3.8}$$

where $\mathbf{H}_t$ and $\mathbf{H}_y$ are also $pr \times pr$ matrices. The identity matrices in $\mathbf{H}_t$ have dimensions of $p \times p$; the $\mathbf{H}_t$ matrix affects the first time difference. It is the same as Eq. (7.2.28), with $-\mathbf{I}$ along the main diagonal (except the last which is zero) and $\mathbf{I}$ on the diagonal just above the main diagonal. The $\mathbf{H}_y$ matrix is partitioned into elements such that the only nonzeros are the matrices, $\mathbf{h}$, which are along the diagonal of $\mathbf{H}_y$. The $\mathbf{h}$ matrix is $p \times p$ in dimensions and simulates first differences in space,

$$\mathbf{h} = \begin{bmatrix} -1 & 1 & 0 & \cdots & 0 \\ 0 & -1 & 1 & & 0 \\ & & \ddots & \ddots & \\ & & & -1 & 1 \\ 0 & \cdots & & 0 & 0 \end{bmatrix} \tag{7.3.9}$$

When Eqs. (7.3.7) and (7.3.8) are used in Eq. (7.3.6), $\mathbf{H}$ becomes

$$\mathbf{H} = \begin{bmatrix} -\mathbf{h} & \mathbf{h} & 0 & 0 \\ 0 & -\mathbf{h} & \mathbf{h} & 0 \\ \vdots & & \ddots & \ddots \\ 0 & \cdots & -\mathbf{h} & \mathbf{h} \\ 0 & \cdots & & 0 & 0 \end{bmatrix} \tag{7.3.10}$$

The $\mathbf{H}^T\mathbf{W}\mathbf{H}$ matrix portion of Eq. (7.2.27) with $\mathbf{W} = \mathbf{I}$ is

$$\mathbf{H}^T\mathbf{W}\mathbf{H} = \mathbf{H}^T\mathbf{H} = \begin{bmatrix} \mathbf{P} & -\mathbf{P} & & & & \mathbf{0} \\ -\mathbf{P} & 2\mathbf{P} & -\mathbf{P} & & & \\ & -\mathbf{P} & 2\mathbf{P} & -\mathbf{P} & & \\ & & & \ddots & & \\ & & & -\mathbf{P} & 2\mathbf{P} & -\mathbf{P} \\ \mathbf{0} & & & & -\mathbf{P} & \mathbf{P} \end{bmatrix} \qquad (7.3.11)$$

where $\mathbf{P}$ is

$$\mathbf{P} = \mathbf{h}^T\mathbf{h} = \begin{bmatrix} 1 & -1 & & & & 0 \\ -1 & 2 & -1 & & & \\ & -1 & 2 & -1 & & \\ & & & \ddots & & \\ & & & -1 & 2 & -1 \\ 0 & & & & -1 & 1 \end{bmatrix} \qquad (7.3.12)$$

The structures of $\mathbf{H}^T\mathbf{H}$ and $\mathbf{P}$ are similar. See also Eq. (4.5.23). They are tridiagonal with diagonal, positive, and equal elements except the first and last terms are one-half in magnitude. The off-diagonal terms are all negative and equal. Except for the first and last rows, the rows represent the coefficients for second differences.

REFERENCES

1. Bass, B. R., Drake, J. B., and Ott, L., ORMDIM: A Finite Element Formulation for Two-Dimensional Nonlinear Inverse Heat Conduction Analysis, NUREG/CR-1709, ORNL/NUREG/CSD/TM-17, U.S. Nuclear Regulatory Commission, Washington, D.C., December 1980.
2. Imber, M., Two-Dimensional Inverse Conduction Problem—Further Observations, *AIAA J.* **13**, 114–115 (1975).
3. Hsieh, C. K. and Su, K. C., A Methodology of Predicting Cavity Geometry Based on the Scanned Surface Temperature Data—Prescribed Surface Temperature at the Cavity Side, *J. Heat Transfer* **102**, 324–329 (1980).
4. Hsieh, C. K. and Su, K. C. A Methodology of Predicting Cavity Geometry Based on the Scanned Surface Temperature Data—Prescribed Heat Flux at the Cavity side, *J. Heat Transfer* **103**, 42–46 (1981).
5. Beck, J. V. and Arnold, K. J., *Parameter Estimation in Engineering and Science*, Wiley, New York, 1977.

PROBLEMS

7.1. Use the sequential function specification method for two heat fluxes and $r = 3$ to obtain algorithms for $\hat{q}_1(M)$ and $\hat{q}_2(M)$ in the gain function form. Give the gain coefficients K_{1j} and K_{2j} (for $j=1, 2, 3$) in terms of the influence functions ϕ_{ji} where $j=1, 2$ refers to the sensors and $i=1, 2, \ldots, M+r-1$ refers to time.

7.2. Write a computer program for the algorithms of Problem 7.1.

7.3. a. Using exact data from Table 1.1 at $x/L=0$ and 1 with dimensionless time steps of 0.01, find estimates of the heat fluxes at $x/L=0$ and 1 using the computer program for Problem 7.1.
b. Repeat part (a) for the sensors at $x/L=0.25$ and 0.75. Also solve parts for $\Delta t^+ = 1$.

7.4. a. For Problem 7.1 derive digital filter algorithms to obtain estimates of $\hat{q}_1(M)$ and $\hat{q}_2(M)$. Let the filter coefficients be f_{ji}, $j=1, 2$; $i= -2, -1, 0, 1, \ldots$.
b. Obtain numeral values for the filter coefficients, f_{ji}, for the flat plate of Table 1.1. Use $\Delta t^+ =0.05$ and sensors at $x=0$ and L.

7.5. Use the zeroth-order sequential regularization method with $r=3$ to derive algorithms for estimating two heat flux histories, $\hat{q}_1(M)$ and $\hat{q}_2(M)$. Let $\alpha\sigma^2$ be an input.

7.6. Write a computer program for the algorithms of Problem 7.5.

7.7. a. Using the exact data from Table 4.1 at $x/L=0$ and 1 with $\Delta t^+ =0.01$, find estimates of the heat fluxes at $x/L=0$ and 1; use the computer program from Problem 7.5.
b. Repeat part (a) for the sensors at $x/L=0.25$ and 0.75.

CHAPTER 8

HEAT TRANSFER COEFFICIENT ESTIMATION

8.1 INTRODUCTION

The estimation of the heat transfer coefficient, h, from transient temperature measurements has aspects of both the inverse heat conduction problem and parameter estimation.

An example of the treatment as an IHCP is that of a one-dimensional case with a known ambient temperature, $T_\infty(t)$, such as the transient determination of boiling heat transfer coefficients using an initially hot spherical copper solid suddenly immersed in water at its saturation temperature. From transient temperatures measured inside or at the surface of the copper body, the methods of the IHCP can be used to estimate the surface heat flux, $\hat{q}_M$, and the surface temperature, $\hat{T}_{0M}$; the definition of the heat transfer coefficient, h, can be used to obtain the estimate of h given by

$$\hat{h}_M = \frac{\hat{q}_M}{T_{\infty M} - 0.5(\hat{T}_{0M} + \hat{T}_{0,M-1})} \tag{8.1.1}$$

In this expression $\hat{T}_{0M}$ is the estimated surface temperature at time t_M; $\hat{q}_M$, $T_{\infty M}$, and $\hat{h}_M$ are usually most accurately evaluated at $t_{M-1/2}$.

An example of a case that can be treated as a parameter estimation problem is a flat plate over which a fluid is flowing at a constant temperature, T_∞; see Figure 8.1. If the plate is suddenly heated by some electric heaters inside the plate, the plate temperature begins to rise and the heat transfer coefficient is a strong function of position from the leading edge of the plate. In some cases the time variation is small and the basic form of h is a function of x; that is, $h = h(x)$, is known, such as

$$h = \beta x^{-1/2} \tag{8.1.2}$$

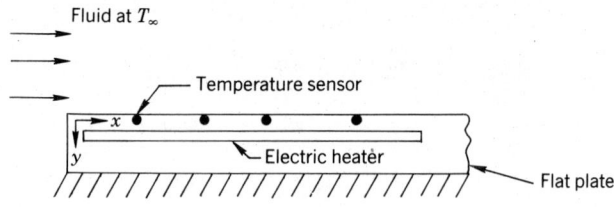

FIGURE 8.1 Electrically heated flat plate.

The determination of β utilizing various time- and space-dependent measurements of T in the solid and T_∞ in the fluid is a parameter estimation problem.

For cases similar to the two just given, basic solution techniques are known. These examples illustrate the large diversity of problems associated with determining the heat transfer coefficient. For this reason a discussion of various types of problems is given next.

The heat transfer coefficient can be:

1. A constant (independent of x and t).
2. A function of t only; that is, $h(t)$.
3. A function of x only; that is, $h(x)$.
4. A function of x and t; that is, $h(x, t)$.

In this list, x is a coordinate parallel to the heated surface; it can also be generalized to two surface coordinates such as $h = h(x, z)$ where both x and z are parallel to the heated surface. In these problems the ambient temperature, T_∞, is assumed to be known but it can also be a function of x and t.

A related problem is the determination of the ambient temperature, T_∞, with h known. The same four categories of (1) $T_\infty = C$, (2) $T_\infty = T_\infty(t)$, (3) $T_\infty = T_\infty(x)$, and (4) $T_\infty = T_\infty(x, t)$ can be listed. These four estimation problems are linear if the heat conduction equation and boundary conditions are linear. An example of a problem wherein the ambient fluid temperature is needed is in the extrusion of plastics. The temperature of the molten plastic is critical and heating must be controlled, but temperature sensors are placed only in the solid mold, not in the flowing plastics.

A more complicated problem is to simultaneously estimate the heat transfer coefficient and the ambient temperature.

There are many concepts and techniques of the IHCP and parameter estimation that can be utilized in the solution of these problems. One concept is that the sensitivity coefficients can be employed to gain much insight into the estimation problems. Another concept is that the use of a sequential procedure can be advantageous in terms of computation speed and for insight. These concepts are briefly explored in this chapter. Due to the large variety of cases in connection with the determination of the heat transfer coefficient, only a few can be treated. The case of $h = h(t)$ is the main one covered. The basic concepts

SEC. 8.2 SENSITIVITY COEFFICIENTS

and examples serve to illustrate procedures which can be modified for different conditions. Section 8.2 covers some sensitivity coefficients. Section 8.3 discusses methods for lumped bodies and Section 8.4 briefly covers methods for bodies with interior temperature gradients.

8.2 SENSITIVITY COEFFICIENTS

In this section the sensitivity coefficients for the heat transfer coefficient, h, are investigated for two cases. The first case is for a lumped body—that is, one in which the temperature is a function of time only. Sensitivity coefficients are given for h constant over the complete time domain and for finite time intervals. Also the sensitivity coefficient of the ambient temperature, T_∞, is found, both for T_∞ constant with time and for T_∞ approximated by a number of constant segments. The second case is for a semi-infinite body that is suddenly exposed to a fluid. The heat transfer coefficient is constant with time.

8.2.1 Lumped Body Case

The differential equation for a lumped body which is suddenly exposed to a fluid at a temperature T_∞ is

$$\rho c V \frac{dT}{dt} = h A (T_\infty - T) \tag{8.2.1}$$

where V is the volume and A is the heated surface area of the lumped body. For convenience in notation, the ratio of V divided by A is denoted L,

$$L = \frac{V}{A} \tag{8.2.2}$$

For an initial temperature of T_0 and with both h and T_∞ independent of time, the solution of Eq. (8.2.1) is

$$T^+ \equiv \frac{T - T_0}{T_\infty - T_0} = 1 - \exp\left(\frac{-ht}{\rho c L}\right) \tag{8.2.3}$$

Note that T^+ can be plotted as a function of the single dimensionless time t^+, where

$$t^+ \equiv \frac{ht}{\rho c L} \tag{8.2.4}$$

For a constant h, the h step sensitivity coefficient, Z_h, is given by

$$Z_h \equiv \frac{\partial T}{\partial h} = (T_\infty - T_0) \frac{t}{\rho c L} \exp(-t^+) \tag{8.2.5}$$

A dimensionless h step sensitivity coefficient is

$$Z_h^+ \equiv \frac{h}{T_\infty - T_0} \frac{\partial T}{\partial h} = t^+ \exp(-t^+) \tag{8.2.6}$$

which is also a function of t^+. The distinction between the pulse and step sensitivity coefficients is discussed in Section 6.2.

The T_∞ step sensitivity coefficient is

$$Z_{T_\infty} = \frac{\partial T}{\partial T_\infty} = 1 - \exp(-t^+) \tag{8.2.7}$$

which is dimensionless and thus rearrangement of terms is not needed. Notice that

$$T^+ = Z_{T_\infty} \tag{8.2.8}$$

and see the upper curve in Figure 8.2. As the temperature rise becomes large for increasing t^+, the T_∞ sensitivity coefficient, Z_{T_∞}, increases in exactly the same manner. As a parameter's sensitivity coefficient becomes large, more information can be gained about the parameter. Consequently T_∞ is relatively easy to measure for a constant T_∞ over the complete time range. Since the T_∞ sensitivity coefficient becomes largest as $t \to \infty$, the most information regarding T_∞ is found by using measured temperatures at large times; that is, $t^+ > 3$.

Also shown in Figure 8.2 is the dimensionless heat transfer sensitivity coefficient, Z_h^+. For the small values of $t^+ < 0.5$, Z_h^+ increases and is nearly equal to T^+; in the range of $0.5 < t^+ < 1.5$, Z_h^+ attains a maximum; and for $t^+ > 1.5$, Z_h^+ decreases toward zero. There are several ramifications of these behaviors of Z_h^+. First, the optimal times for measuring temperature in order to estimate h are near $t^+ = 1$ and measurements at the early and late times contain less

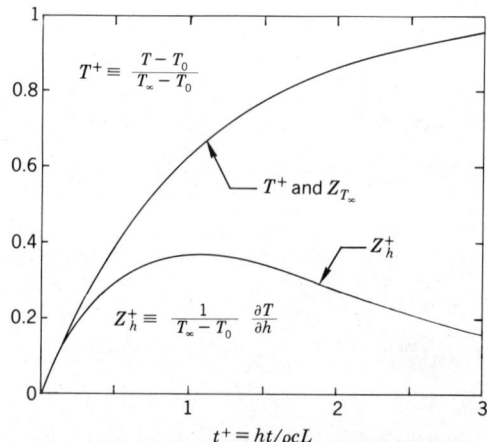

FIGURE 8.2 Temperatures and sensitivity coefficients for convective heat transfer from a lumped body.

SEC. 8.2 SENSITIVITY COEFFICIENTS

information regarding h. Second, the average magnitude of Z_h^+ is considerably less than T^+, hence h is relatively sensitive to measurement errors in temperature. Third, since Z_{T_∞} is so much larger on the average than Z_h^+, T_∞ can be estimated more accurately than h. Finally, it is possible to simultaneously estimate both h and T_∞ but a relatively large dimensionless time range should be covered, say $t^+ = 0$ to 3. When two parameters are estimated, the sensitivity coefficients must not be highly correlated if accurate estimates are to be obtained. This is also discussed in Section 1.6 and in Reference 1.

This case of a lumped body is sufficiently tractable that the sensitivity coefficients for time-variable h and T_∞ can be obtained in algebraic form. The case where $h(t)$ and $T_\infty(t)$ are constant over the time segments,

$$0 < t < t_1, \quad t_1 < t < t_2, \quad t_2 < t < t_3, \ldots$$

is considered as illustrated in Figure 8.3. The heat transfer coefficient components are $h_1, h_2, \ldots$, and the corresponding components for T_∞ are $T_{\infty 1}, T_{\infty 2}, \ldots$. The equations for T for three time intervals are for $0 < t < t_1$,

$$T = T_{\infty 1} + (T_0 - T_{\infty 1}) \exp\left(\frac{-h_1 t}{\rho c L}\right) \qquad (8.2.9a)$$

for $t_1 < t < t_2$,

$$T = T_{\infty 2} + (T_1 - T_{\infty 2}) \exp\left[\frac{-h_2(t - t_1)}{\rho c L}\right] \qquad (8.2.9b)$$

for $t_2 < t < t_3$

$$T = T_{\infty 3} + (T_2 - T_{\infty 3}) \exp\left[\frac{-h_3(t - t_2)}{\rho c L}\right] \qquad (8.2.9c)$$

and the T_1 and T_2 values are those evaluated at t_1 and t_2,

$$T_1 = T_{\infty 1} + (T_0 - T_{\infty 1}) \exp\left(\frac{-h_1 t_1}{\rho c L}\right) \qquad (8.2.10a)$$

$$T_2 = T_{\infty 2} + (T_1 - T_{\infty 2}) \exp\left[\frac{-h_2(t_2 - t_1)}{\rho c L}\right] \qquad (8.2.10b)$$

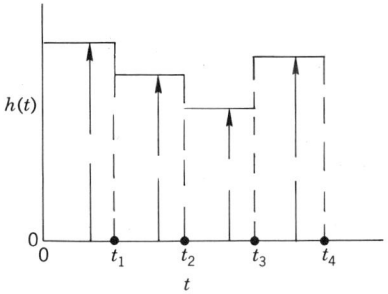

FIGURE 8.3 Heat transfer coefficient history approximated by constant segments.

286 **CHAP. 8 HEAT TRANSFER COEFFICIENT ESTIMATION**

The pulse sensitivity coefficients can be found by differentiation as previously illustrated. The pulse sensitivity coefficient for h_i is

$$X_{h_i} \equiv \frac{\partial T}{\partial h_i} \tag{8.2.11a}$$

and for $T_{\infty i}$,

$$X_{T_{\infty i}} \equiv \frac{\partial T}{\partial T_{\infty i}} \tag{8.2.11b}$$

To reduce the number of dimensionless groups, after the differentiations are completed, the h_i components are made equal,

$$h_1 = h_2 = h_3 = \cdots = h_0 \tag{8.2.12}$$

and the same is done for $T_{\infty i}$,

$$T_{\infty 1} = T_{\infty 2} = T_{\infty 3} = \cdots = T_{\infty} \tag{8.2.13}$$

Figure 8.4 displays the heat transfer coefficient sensitivity coefficients for

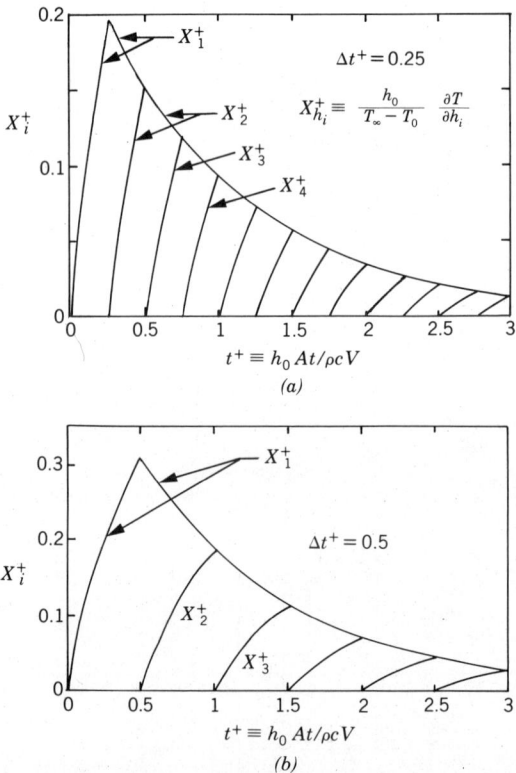

FIGURE 8.4 Heat transfer sensitivity coefficients for h_i constant over segments and $h_1 = h_2 = \cdots = h_0$ and $T_{\infty 1} = T_{\infty 2} = \cdots = T_{\infty}$.

SEC. 8.2 SENSITIVITY COEFFICIENTS

the two cases of dimensionless time steps of 0.25 and 0.5. The $X_{h_i}^+$ values are uncorrelated as can be seen by comparing the values at $t^+ \equiv h_0 At/\rho cV = 0.25$, 0.5, 0.75, and 1 of $X_{h_1}^+$ which are 0.19, 0.15, 0.12, and 0.09 and those for $x_{h_2}^+$ which are 0, 0.15, 0.12, and 0.09. These values are not proportional for all times and hence are uncorrelated. The first values of $X_{h_1}^+$ and $X_{h_2}^+$ are different but the succeeding values are identical. This suggests that a sequential procedure (which emphasizes "recent" values) of estimating $h = h(t)$ would be more effective than a whole domain estimation procedure which uses all the data simultaneously. Another conclusion that can be drawn from Figure 8.4 is that the h_i components become more difficult to estimate as i increases. This is a consequence of the X_{h_i} values becoming smaller in magnitude as i increases.

In Figure 8.5 the sensitivity coefficients for $T_{\infty 1}$ and $T_{\infty 2}$ are displayed. A significant difference between this figure and Figure 8.4b is that the coefficients in the former have the same magnitude for each $X_{T_{\infty i}}$ while the $X_{h_i}^+$ curves have decreasing magnitudes with increasing i. This difference means that all components of $T_{\infty i}$ can be estimated with equal sensitivity to measurement errors while the h_i components become much more sensitive to errors as i increases.

8.2.2 Semi-Infinite Body

For a semi-infinite body at an initial temperature of T_0 and exposed at $t = 0$ to a fluid at T_∞, the temperature at a position x measured from the surface and at time t is

$$T^+ \equiv \frac{T - T_0}{T_\infty - T_0} = \text{erfc}\left[\frac{1}{2(t_x^+)^{1/2}}\right]$$

$$- \exp(B_x + B_x^2 t_x^+) \, \text{erfc}\left[\frac{1}{2(t_x^+)^{1/2}} + B_x(t_x^+)^{1/2}\right] \quad (8.2.14)$$

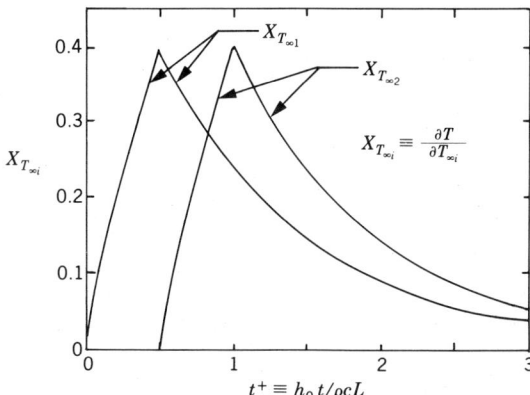

FIGURE 8.5 Ambient temperature sensitivity coefficients for $\Delta t^+ = 0.5$.

where

$$B_x = \frac{hx}{k}, \quad t_x^+ = \frac{\alpha t}{x^2} \qquad (8.2.15a,b)$$

Notice that

$$B_x^2 t_x^+ = \frac{h^2 \alpha t}{k^2} \qquad (8.2.15c)$$

which is independent of x. The dimensionless temperature is plotted in Figure 8.6 versus $B_x^2 t_x^+$ for several values of B_x. For each B_x value, the dimensionless temperature starts at zero, and increases monotonically, and finally approaches unity.

The dimensionless step sensitivity coefficients are given by

$$Z_h^+ \equiv \frac{h}{T_\infty - T_0} \frac{\partial T}{\partial h} = -(B_x + 2B_x^2 t_x^+) \exp(B_x + B_x^2 t_x^+) \operatorname{erfc}\left[\frac{1}{2(t_x^+)^{1/2}} + B_x(t_x^+)^{1/2}\right]$$

$$+ \frac{2}{\pi^{1/2}} B_x (t_x^+)^{1/2} \exp\left(-\frac{1}{4t_x^+}\right) \qquad (8.2.16)$$

and are displayed in Figure 8.7. The notation Z_h^+ is used to emphasize that a single h is being examined. Though the sensitivity coefficients always initially increase, there is a maximum after which the curves decrease and approach

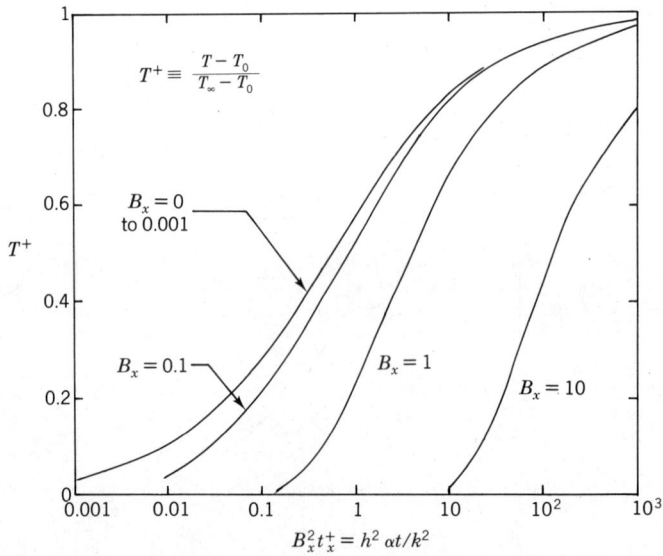

FIGURE 8.6 Dimensionless temperatures for semi-infinite body with a convective boundary condition.

SEC. 8.2 SENSITIVITY COEFFICIENTS

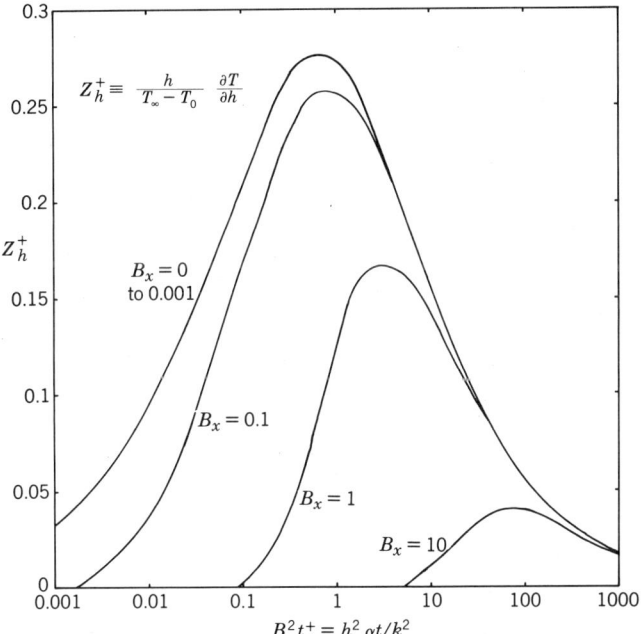

FIGURE 8.7 Dimensionless sensitivity coefficients for h for convectively heated semi-infinite body.

zero. Consequently the moderate times are best for taking temperature measurements. For example, for $B_x = 0$, which corresponds to taking temperature measurements at the heated surface, the sensitivity coefficients are largest for $B_x^2 t_x^+$ between 0.03 and 10. Both the very early and very large times are less effective for determining the heat transfer coefficient. This finding is consistent with that found for the lumped-parameter case depicted in Figure 8.2.

Another important observation regarding Figure 8.7 is that the sensitivity coefficients decrease in magnitude as B_x increases. In this regard there is no correspondence with the lumped-parameter case. Since B_x is defined to be hx/k, large values of B_x result from large h or x values or small k values.

If the ambient temperature, T_∞, is of interest, the sensitivity coefficient for T_∞ must be found. From Eq. (8.2.14) it is readily shown that

$$Z_{T_\infty} \equiv \frac{\partial T}{\partial T_\infty} = T^+ \tag{8.2.17}$$

and thus Figure 8.6 also represents the T_∞ sensitivity coefficients. Again as for the lumped-parameter case, T_∞ can be estimated more accurately than h (if the error statistics are the same in both cases). Furthermore the optimal times to take temperature measurements to estimate T_∞ are the largest.

8.3 LUMPED BODY ANALYSES

In this section the determination of the heat transfer coefficient for the case of a lumped body is discussed. A thermally lumped body is one in which the temperature is uniform but varies with time. Transient calorimeters based on this approach have been used for measuring heat transfer coefficients and heat fluxes for a number of applications including determination of the boiling curve.[2]

This case of a lumped body involves a simple model, Eq. (8.2.1), and thus the details of the estimation procedure can be readily seen. Several procedures are discussed in this section. In Section 8.3.1 exact matching of the model temperature is employed. In Section 8.3.2 some results obtained using regression analysis[1] are discussed. Section 8.3.3 presents a method based on the function specification procedure with the $q=C$ assumption. Finally in Section 8.3.4, function specification with $h=C$ is developed.

8.3.1 Exact Matching of the Measured Temperatures

The exact matching of the model temperatures with the measured temperatures is analogous to the Stolz method for the IHCP. This method uses Eq. (4.3.3) which is

$$\hat{q}_M = \frac{Y_M - \hat{T}_M|_{q_M=0}}{\phi_1} \tag{8.3.1}$$

The heat transfer coefficient equation, Eq. (8.1.1), can be written for the lumped body case as

$$\hat{h}_M = \frac{\hat{q}_M}{T_{\infty M} - 0.5(\hat{T}_{M-1} + \hat{T}_M)} \tag{8.3.2}$$

With exact matching of the calculated and the measured temperatures, the following relations are used:

$$\hat{T}_{M-1} = Y_{M-1}, \quad \hat{T}_M = Y_M, \quad \hat{T}_M|_{q_M=0} = Y_{M-1} \tag{8.3.3a,b,c}$$

The relation given by Eq. (8.3.3c) postulates uniform temperature within the body; thus it is true for lumped bodies. Then the foregoing relations yield

$$\hat{h}_M = \frac{(Y_M - Y_{M-1})}{[T_{\infty M} - 0.5(Y_{M-1} + Y_M)]\phi_1} \tag{8.3.4}$$

which is the desired algorithm.

An expression for ϕ_1 is needed. It represents the temperature rise at time $t=\Delta t$ for a unit step increase in the heat flux. Specifically it is found from the solution of

$$\rho c V \frac{\partial T}{\partial t} = qA \tag{8.3.5}$$

SEC. 8.3 LUMPED BODY ANALYSES

with T replaced by ϕ, q by unity, and the solution evaluated at $t=\Delta t$. The resulting expression is

$$\phi_1 = \frac{A\Delta t}{\rho c V} = \frac{\Delta t}{\rho c L} \tag{8.3.6}$$

EXAMPLE 8.1. For the case of $\rho cL=1$, $T_\infty=1$, h constant, and the measurements equal to the exact values given by

$$Y_M = 1 - \exp(-M\Delta t)$$

the h values are to be estimated using Eq. (8.3.4) for $\Delta t = 0.02$, 0.5, and 1.

Solution. Introducing the given values into Eq. (8.3.4) yields

$$\hat{h}_M = \frac{-\exp(-M\Delta t) + \exp[-(M-1)\Delta t]}{\{1-1+0.5[\exp(-M\Delta t)+\exp[-(M-1)\Delta t]]\}\Delta t}$$

$$= \frac{\exp(\Delta t)-1}{0.5\Delta t[1+\exp(\Delta t)]} = \frac{\tanh(\Delta t/2)}{0.5\Delta t}$$

which is independent of M. For $\Delta t = 0.02$, 0.5, and 1, the $\hat{h}_M$ values are 0.999967, 0.979675, and 0.92423, respectively. Hence, smaller Δt's have smaller errors in $\hat{h}_M$ for exact data; the errors are a result of the numerical approximations. □

EXAMPLE 8.2. For the same cases as in Example 8.1, calculate the errors in $\hat{h}_M$ due to a temperature error of $\varepsilon = 0.001$ at $M = 3$.

Solution. The calculated $\hat{h}_M$ values are:

M	$\Delta t=0.02$	$\Delta t=0.5$	$\Delta t=1$
1	0.999967	0.979675	0.92423
2	0.999967	0.979675	0.92423
3	1.053081	0.988114	0.94011
4	0.946850	0.971226	0.90820
5	0.999967	0.979675	0.92423

The errors at $M=3$ due to ε are about 0.0531, 0.00844, and 0.01588 for $\Delta t = 0.02$, 0.5, and 1, respectively. The smallest Δt results in the largest errors in $\hat{h}_M$. This is consistent with the sensitivity coefficients decreasing as the time steps are reduced as shown in Figure 8.4. Consequently, small Δt's result in smaller numerical approximation errors but greater sensitivity to random measurement errors. □

EXAMPLE 8.3. For the same copper billet mentioned in Example 2.1 find the heat transfer coefficient for the ambient temperature of 81.5°F. The first eleven measured temperatures are given in Table 8.1 and further values are given in Table 2.2.

Solution. The equation for $\hat{h}_M$ is Eq. (8.3.4), $T_{\infty M}$ is 81.5°F, and the Y_M values are taken from Tables 8.1 and 2.2. The reciprocal of ϕ_1 is 31.3 Btu/hr-ft² ($=98.72$ W/m²). The

TABLE 8.1 Results for the Copper Billet Example Using the Function Specification Method with $q=C$ and $r=2$

M	$t_M(s)$	Y_M, °F	$\hat{q}_M$, Btu/hr-ft^2	$\hat{h}_M$, Btu/hr-ft^2-F	$\hat{T}_M$, °F
0		279.59			
1	96	264.87	−443.459	2.3217	265.422
2	192	251.53	−414.012	2.33498	252.195
3	288	239.30	−381.388	2.31702	240.010
4	384	228.18	−359.134	2.35077	228.536
5	480	217.24	−329.576	2.32471	218.006
6	576	207.86	−296.969	2.25382	208.518
7	672	199.36	−268.527	2.18797	199.939
8	768	191.65	−245.945	2.14779	192.082
9	864	184.44	−228.649	2.13832	184.777
10	960	177.64	−212.027	2.12261	178.003

results of the calculations are displayed in Figure 8.8 as points connected by solid lines. There are some fluctuations in the values, particularly near time steps 5 and 12. The method is relatively sensitive to measurement errors, as was noted in Example 8.2. □

8.3.2 Regression Method

Another way of estimating h is to pass a regression line through the measured temperatures. Polynomials of various degrees can be used. In Reference 1 a number of polynomials for T such as

$$T = \beta_1 + \beta_2 \left(\frac{t}{\Delta t}\right) + \beta_3 \left(\frac{t}{\Delta t}\right)^2 + \cdots \quad (8.3.7)$$

were investigated and an F-test was used to determine the "best" curve. After estimating the temperature curve, $\hat{T}$, it can be differentiated and used in Eq. (8.2.1) which can be written in the form

$$\hat{h} = \rho c L \frac{1}{T_\infty - \hat{T}} \frac{d\hat{T}}{dt} \quad (8.3.8)$$

Results using this method are given as crosses in Figure 8.8 for Example 8.3. These results are not as sensitive to errors as those from the exact matching method.

This method can also be made a sequential procedure by considering a limited number of measurements as was done in Example 2.1.

SEC. 8.3 LUMPED BODY ANALYSES

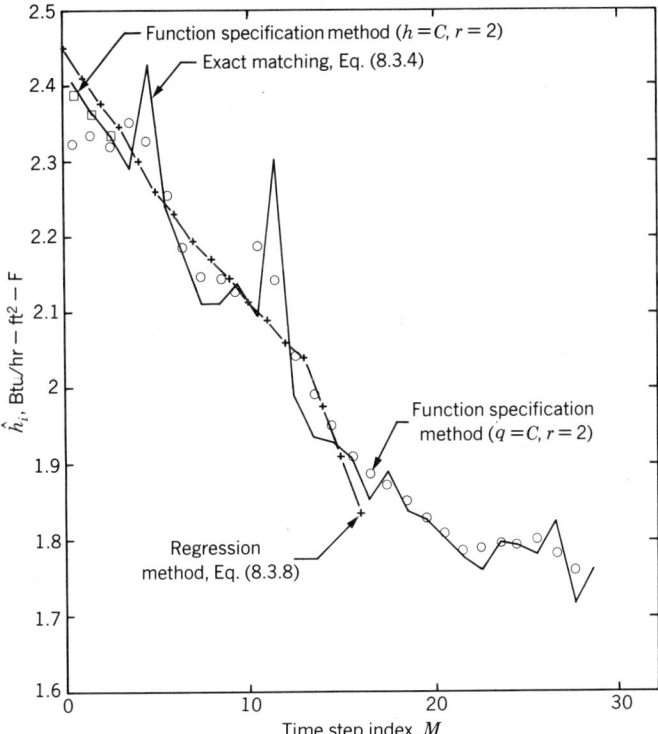

FIGURE 8.8 Heat transfer coefficient for copper billet of Example 8.3.

8.3.3 Function Specification Procedure with q=Constant

The heat transfer coefficient can be estimated using the function specification method. For the temporary assumption of constant q, Eq. (4.4.24) can be utilized. For the lumped body case, a solution of Eq. (8.3.5) for ϕ (with $T \to \phi$, and so on) gives

$$\phi_j = \frac{j\Delta t}{\rho c L}, \quad j = 1, 2, \ldots \tag{8.3.9a}$$

The $\Delta\phi_{j-1}$ values are constant,

$$\Delta\phi_{j-1} = \phi_j - \phi_{j-1} = \frac{\Delta t}{\rho c L} = \phi_1, \quad j = 1, 2, \ldots \tag{8.3.9b}$$

As a consequence Eq. (3.2.29) gives

$$\hat{T}_{M+j-1}\big|_{q_M = \cdots = 0} = T_0 + \phi_1 \sum_{i=1}^{M-1} \hat{q}_i, \quad j = 1, 2, \ldots \tag{8.3.9c}$$

With Eqs. (8.3.9a, b, c) used in Eq. (4.4.24), the equation for $\hat{q}_M$ for the lumped bodies is

$$\hat{q}_M = \frac{\sum_{j=1}^{r}(Y_{M+j-1} - \phi_1 \sum_{i=1}^{M-1}\hat{q}_i - T_0)j}{\sum_{j=1}^{r} j^2} \frac{\rho c L}{\Delta t} \quad (8.3.10)$$

The equation for $\hat{h}_M$ is given by Eq. (8.3.2) with $0.5(\hat{T}_{M-1} + \hat{T}_M)$ given by

$$0.5(\hat{T}_{M-1} + \hat{T}_M) = \phi_1 \left(\sum_{i=1}^{M-1} \hat{q}_i + 0.5\hat{q}_M \right) + T_0 \quad (8.3.11)$$

The algorithm for each time step is Eq. (8.3.10) followed by Eq. (8.3.2) which employs Eq. (8.3.11).

For example 8.3, calculations for $r = 2$ are displayed as circles in Figure 8.8. The results are less sensitive to small irregularities in the measurements than the exact matching procedure and yet follow the exact matching case when the latter moves regularly, as near the time step index of 20 in Figure 8.8. Even less sensitivity to random temperature errors can be achieved using $r = 3$ and 4.

Some numerical values of $\hat{q}_M$, $\hat{h}_M$, and $\hat{T}_M$ are given in Table 8.1. In order to permit verification by the reader, values are given for $M = 1$ to 10.

8.3.4 Function Specification Procedure with h=Constant

To estimate the heat transfer coefficient using the function specification method, a more direct procedure is to estimate h without the intermediate calculations for $q(t)$. Unfortunately the nonlinearity of the problem enters into this approach. In this section a direct sequential estimation procedure for h is investigated for the temporary assumption of constant h_M.

The sum of squares, S,

$$S = \sum_{i=1}^{r} (Y_{M+i-1} - T_{M+i-1})^2 \quad (8.3.12)$$

is minimized with respect to h_M where r is the number of future times over which h_M is temporarily held constant. Taking the partial derivative of S with respect to h_M, replacing h_M by $\hat{h}_M$, and setting the equation equal to zero gives

$$\sum_{i=1}^{r}(Y_{M+i-1} - \hat{T}_{M+i-1})\frac{\partial \hat{T}_{M+i-1}}{\partial h_M} = 0 \quad (8.3.13)$$

Expressions for the calculated temperatures $\hat{T}_{M+i-1}$ can be found analogous to Eq. (8.2.9). To simplify the presentation, however, the ambient temperature, $T_\infty(t)$, is approximated by $T_{\infty M}$ for the r future times; for this case, the temperature is given by

$$\hat{T}_{M+i-1} = T_{\infty M} + (\hat{T}_{M-1} - T_{\infty M})\exp\left[-\frac{h_M(t_{M+i-1} - t_{M-1})}{\rho c L}\right] \quad (8.3.14)$$

SEC. 8.3 LUMPED BODY ANALYSES

for $t > t_{M-1}$. The sensitivity coefficient, Z_{M+i-1},

$$Z_{M+i-1} \equiv \frac{\partial T_{M+i-1}}{\partial h_M} \tag{8.3.15}$$

is found from Eq. (8.3.14) to be

$$Z_{M+i-1} = -(\hat{T}_{M-1} - T_{\infty M}) \exp\left[-\frac{h_M(t_{M+i-1} - t_{M-1})}{\rho c L}\right] \frac{t_{M+i-1} - t_{M-1}}{\rho c L} \tag{8.3.16}$$

which is valid for $t > t_{M-1}$. The $\hat{T}_{M-1}$ value is the converged temperature for the solution for $\hat{h}_{M-1}$. Because this coefficient is a function of h_M, the estimation problem is nonlinear. Due to this nonlinearity, an iterative solution of Eq. (8.3.13) for h_M is needed. One such procedure is the Gauss method.[1]

In the Gauss linearization method, it is first assumed that an estimate of h_M is known for the $(v-1)$st iteration and then an improved value $h_M^{(v)}$ is sought. The sensitivity coefficient given explicitly in Eq. (8.3.13) is evaluated with h_M equal to $h_M^{(v-1)}$ and is denoted $Z_{M+i-1}^{(v-1)}$. Moreover, the calculated temperature in Eq. (8.3.13) is approximated by the two-term Taylor series,

$$\hat{T}_{M+i-1}^{(v)} = \hat{T}_{M+i-1}^{(v-1)} + Z_{M+i-1}^{(v-1)}(h_M^{(v)} - h_M^{(v-1)}) \tag{8.3.17}$$

and Z is given by Eq. (8.3.16) with h_M replaced by $h_M^{(v-1)}$. Using these approximations in Eq. (8.3.13) and solving for $h_M^{(v)}$ gives

$$\hat{h}_M^{(v)} = \hat{h}_M^{(v-1)} + \frac{\sum_{i=1}^{r}(Y_{M+i-1} - \hat{T}_{M+i-1}^{(v-1)})Z_{M+i-1}^{(v-1)}}{\sum_{i=1}^{r}(Z_{M+i-1}^{(v-1)})^2} \tag{8.3.18}$$

This equation is used in an iterative manner until the changes in $\hat{h}_M^{(v)}$ are less than some small amount, such as

$$\left|\frac{\hat{h}_M^{(v)} - \hat{h}_M^{(v-1)}}{\hat{h}_M^{(v)}}\right| < 10^{-4} \tag{8.3.19}$$

After a converged value of $\hat{h}_M^{(v)}$ is found, M is increased by one, $\hat{T}_M$ is calculated, and the procedure is repeated for the new $\hat{h}_M$.

The same copper billet problem previously considered in this section is also investigated using this method. Table 8.2 gives some details of the calculation for the first two time steps for the case of $h_M = C$ and $r = 2$. The sensitivity coefficients are given as functions of M and the iteration index, v. The column labeled Numerator is the numerator of the fraction in Eq. (8.3.18) and Denominator is the denominator of Eq. (8.3.18). Notice that the numerator for fixed M rapidly decreases in magnitude with v, whereas the denominator (sum of squares of Z's) approaches a constant. The initial guess for h_1 was 2.7 Btu/hr-ft²-F and the procedure rapidly converged to 2.3964 Btu/hr-ft²-F. This converged value of $\hat{h}_1$ is used as the initial value for h_2. A few values of $\hat{h}_M$ are plotted in Figure 8.8 as squares. The first few h values are significantly different from the $q = C$, $r = 2$

TABLE 8.2 Details of Copper Billet Calculation for Function Specification Method with $h_M = C$ and $r = 2$

M	v	Z_M	Z_{M+1}	Numerator	Denominator	$\hat{h}_M$, Btu/hr-ft²-F
1	0					2.7
1	1	−5.806	−10.652	−45.05	147.167	2.3939
1	2	−5.863	−10.862	0.3926	152.360	2.3964
1	3	−5.862	−10.860	−1.8E-4	152.315	2.3964
1	4	−5.862	−10.860	−1.4E-5	152.315	2.3964
2	0					2.3964
2	1	−5.430	−10.060	−3.99	171.832	2.3659
2	2	−5.436	−10.080	3.05E-3	133.452	2.3659
2	3	−5.436	−10.080	−1.56E-4	131.153	2.3659

results which are shown as circles. This can be verified by comparing the estimated h_M values for $M = 1$ to 10 in Table 8.3 with those for $q = C$, $r = 2$ given in Table 8.1. The maximum difference is 3%, which is for $M = 1$; for $M = 4$, it is 0.3% and for $M = 10$ it is about 0.05%.

For this example, the $h_M = C$ and $q_M = C$ results are then very close; in general the results are closer than the variability due to small fluctuations in the measured temperature. The $h_M = C$ calculation is more expensive because iteration is needed. The number of iterations for this case is two or three which is not large but even so there is more computation than without the iterations. In addition, the computer program for the $h_M = C$ analysis is more complicated than that for the $q = C$ analysis. For these reasons and at least for this billet example, the $q = C$ sequential analysis is preferred over the $h = C$ method.

TABLE 8.3 Results for the Copper Billet Example Using the Function Specification Method with $h = C$ and $r = 2$

M	$t_M(s)$	$\hat{h}_M$, Btu/hr-ft²-F	Iterations
1	96	2.39643	3
2	192	2.36590	3
3	288	2.32878	3
4	384	2.35714	3
5	480	2.32776	3
6	576	2.25087	3
7	672	2.18316	3
8	768	2.14357	3
9	864	2.13633	2
10	960	2.12164	2

8.4 BODIES WITH INTERNAL TEMPERATURE GRADIENTS

In Section 8.3 it was found that the function specification method for $q_M = C$ gave nearly the same results as for the temporary assumption of $h_M = C$. One case of a lumped body does not prove that the temporary assumptions of constant q and h are equally valid for other cases. More research is needed to determine the relative merits. Until more is known, however, the $q = C$ method is recommended because it is computationally simpler and does not involve iteration. Furthermore, existing programs for the IHCP can be employed. For these reasons, only the function specification method with the constant heat flux temporary assumption is used in this section.

8.4.1 Analysis for r Future Temperatures Using $q = C$ Function Specification Method

A one-dimensional heat conduction model is

$$\frac{1}{r^n}\frac{\partial}{\partial r}\left(r^n \frac{\partial T}{\partial r}\right) = \frac{1}{\alpha}\frac{\partial T}{\partial t} \qquad (8.4.1)$$

where $n=0$ is for a rectangular coordinate, $n=1$ is for a radial cylindrical coordinate, and $n=2$ is for a radial spherical coordinate. The boundary conditions are

$$-k \frac{\partial T}{\partial r}\bigg|_{r=r_0} = h(t)[T_\infty(t) - T(r_0, t)] \qquad (8.4.2)$$

$$\frac{\partial T}{\partial r}\bigg|_{r=r_1} = 0 \qquad (8.4.3)$$

and the initial temperature is

$$T(r, t) = T_0 = \text{constant} \qquad (8.4.4)$$

For simplicity, the thermal conductivity k and the density-specific heat product ρc are assumed to be independent of temperature so that nonlinearity does not enter the problem via the thermal properties. The heat transfer coefficient is assumed to vary only with time. Other "inactive" boundary conditions at $r = r_1$ can be as readily treated as the insulation condition indicated in Eq. (8.4.3). Much more general conditions can be treated but they unduly complicate the presentation.

For a single interior sensor, a function specification algorithm for the $q = C$ temporary assumption is Eq. (4.4.24), repeated here

$$\hat{q}_M = \frac{\sum_{j=1}^{r}(Y_{M+j-1} - \hat{T}_{M+j-1}|_{q_M = \cdots = 0})\phi_j}{\sum_{j=1}^{r}\phi_j^2} \qquad (8.4.5)$$

where r future temperatures are used. This equation can be used to obtain a single $\hat{q}_M$ after which h_M is estimated. Also all of the $\hat{q}_M$ components can be estimated before any of the h_M values is estimated. A heat transfer coefficient is found from

$$\hat{h}_M = \frac{\hat{q}_M}{T_{\infty M} - 0.5(\hat{T}_{0,M} + \hat{T}_{0,M-1})} \tag{8.4.6}$$

The zero subscript is used to denote the surface at which the fluid is located. Unless the sensor is located at the heated surface, the calculated temperature in Eq. (8.4.5) is not the same as that with zero subscripts in Eq. (8.4.6). This is a major difference between this case and that of a lumped body.

Equations (8.4.5) and (8.4.6) are valid for both the Duhamel integral solution and a difference method solution (e.g., FD, FE, FCV) of Eqs. (8.4.1)–(8.4.3). In both cases, ϕ_j in Eq. (8.4.5) represents the temperature rise at time t_j at the location of the sensor for a unit step increase in the surface heat flux [$r=r_0$, for Eq. (8.4.2)]. If accurate solutions of the heat conduction problem are obtained in both approaches, the results for h_M are nearly identical. For simplicity of presentation, Duhamel's theorem approach is selected.

The calculated temperatures in Eq. (8.4.5) are given by Eq. (3.2.29) and can be written as

$$\hat{T}_M|_{q_M=0} = \hat{q}_1 \Delta\phi_{M-1} + \hat{q}_2 \Delta\phi_{M-2} + \cdots + \hat{q}_{M-1}\Delta\phi_1 + T_0$$

$$= \sum_{i=1}^{M-1} \hat{q}_i \Delta\phi_{M-i} + T_0 \tag{8.4.7a}$$

$$\hat{T}_{M+1}|_{q_M = q_{M+1} = 0} = \hat{q}_1 \Delta\phi_M + \cdots + \hat{q}_{M-1}\Delta\phi_2 + T_0 = \sum_{i=1}^{M-1} \hat{q}_i \Delta\phi_{M-i+1} + T_0 \tag{8.4.7b}$$

All the ϕ_j values (recall $\Delta\phi_j = \phi_{j+1} - \phi_j$) are for the location of the sensor.

The surface temperature denoted, $\hat{T}_{0,M}$, is found in a similar manner to the calculation of an internal temperature; the equation is

$$\hat{T}_{0,M} = \sum_{i=1}^{M} \hat{q}_i \Delta\phi_{0,M-i} + T_0 \tag{8.4.8}$$

where the subscript 0 for $\Delta\phi_{0,M-1}$ denotes the heated surface.

For the case of one future temperature, $r=1$, the interior calculated temperature is made to match the measured temperature. For this case, Eq. (8.4.5) simplifies to

$$\hat{q}_M = \frac{Y_M - \hat{T}_M|_{q_M=0}}{\phi_1} \tag{8.4.9}$$

No comparable simplifications occur in Eqs. (8.4.6) and (8.4.8).

SEC. 8.4 BODIES WITH INTERNAL TEMPERATURE GRADIENTS

8.4.2 Examples

The case of a semi-infinite body suddenly exposed to a fluid at T_∞ is considered. The exact solution for the case of constant h is given by Eq. (8.2.14) and is plotted in Figure 8.6. Though h is actually constant, it is estimated as though it is a function of time.

Some results are displayed in Figures 8.9, 8.10, and 8.11 (Reference 5).

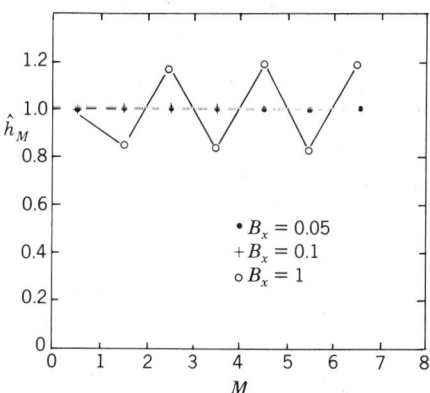

FIGURE 8.9 Semi-infinite body example with exact matching of temperature measurements ($\Delta t_x^+ = 0.3$).

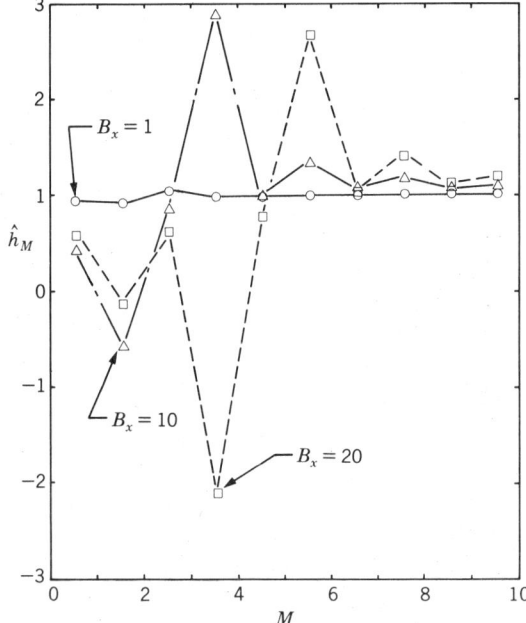

FIGURE 8.10 Semi-infinite body example with exact matching of temperature measurements ($\Delta t_x^+ = 0.5$).

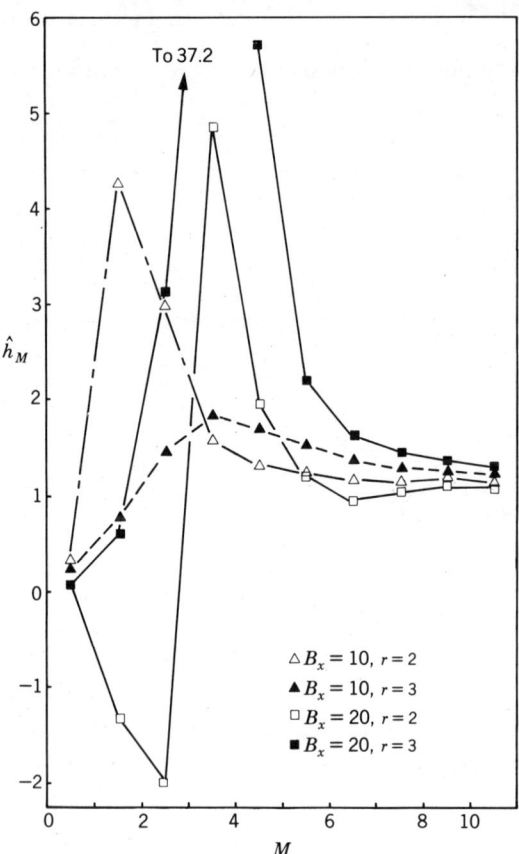

FIGURE 8.11 Semi-infinite body example using function specification algorithm with $q=C$ ($\Delta t_x^+ = 0.25$).

Figures 8.9 and 8.10 are for exact matching of the simulated data. Figure 8.9 is for dimensionless time steps of $\Delta t_x^+ = 0.3$ which are relatively small (i.e., near the stability limit) for exact matching, which is equivalent to the Stolz method. For $B_x = hx/k$ equal to 0.05 and 0.1, the estimated values of h_M are very good for all times, whereas for $B_x = 1$ there are some oscillations on the order of $\pm 20\%$. Evidently larger B_x values cause increased difficulty for a fixed time step with exact data and exact matching of measurements. More results for exact matching are shown in Figure 8.10 with the larger time step of $\Delta t_x^+ = 0.5$. The value of $B_x = 1$ in this figure is more accurate than that shown in Figure 8.9. Consequently increasing the time step can aid in reducing oscillations, but information regarding changes in h can be lost as Δt becomes larger.

Figure 8.10 also reaffirms the finding from Figure 8.9 that increasing B_x makes the estimation process more difficult; this is consistent with the sensitivity coefficients becoming smaller as B_x is increased, as shown in Figure 8.7. (This figure is for the sensitivity coefficient Z which is for h constant over the entire time period. The needed sensitivity coefficients are denoted X and are for individual h_M components. (The Z coefficients are always as large or larger than the X coefficients, however.)

In order to take smaller time steps, future temperatures are needed in the function specification method. Even so, for large values of B_x, such as $B_x \geq 10$, the sensitivity coefficients are very small (see Figure 8.7) and thus accurate estimation is still difficult. This is verified by the results shown in Figure 8.11 which are for $\Delta t_x^+ = 0.25$, a value for which the Stolz method is unstable. Future time steps, such as $r = 2$ or 3, remove the instability but inaccurate results are found, which is not surprising since the sensitivity coefficients are so small (Figure 8.7). A conclusion is that the sensor should be located as near the surface as possible, at least so that $hx/k < 1$, where x is the distance from the heated surface.

8.5 ESTIMATION OF CONTACT CONDUCTANCE

The estimation of the contact conductance from transient measurements is quite similar to the method for determining the heat transfer coefficient. Sensitivity coefficients and optimal considerations are given in References 3 and 4.

REFERENCES

1. Beck, J. V. and Arnold, K. J., *Parameter Estimation in Engineering and Science*, Wiley, New York, 1977.
2. Holman, J. P., *Heat Transfer*, 4th ed., McGraw-Hill, New York, 1976.
3. Beck, J. V., Transient Sensitivity Coefficients for the Thermal Contact Conductance, *Int. J. Heat Mass Transfer* **10**, 1615–1616 (1967).
4. Beck, J. V., Determination of Optimum, Transient Experiments for Thermal Contact Conductance, *Int. J. Heat Mass Transfer* **12**, 621–633 (1969).
5. Osman, A. Personal communication, Aug., 1983.

PROBLEMS

8.1. Starting with Eq. (8.2.14), derive Eq. (8.2.16). Derive also an expression in terms of t_x^+ and B_x for the maximum sensitivity coefficient, Z_h^+, for small values of B_x.

8.2. A thick concrete wall initially at 30 K is suddenly exposed to a fluid at 80 K. Calculate and plot the four temperature histories at 1.2 cm from the heated surface if the heat transfer coefficient has the values of 9, 10, 90, and 100 W/m²-K. Relate the differences between the plots to the sensitivity coefficients. Use $k = 1.2$ W/m-K and $\alpha = 7.5 \times 10^{-7}$ m²/s.

AUTHOR INDEX

Abramowitz, M., 42
Alifanov, O. M., 109, 135, 160, 223, 265
Alkidas, A. L., 42
Alliney, S., 136, 161
Anderson, D. A., 103
Arnold, K. J., 10, 42, 56, 57, 75, 160, 204, 235, 264, 279, 301
Arpaci, V. S., 80, 102
Arsenin, V. Y., 3, 42, 108, 135, 160
Avila, J. H., 41

Backus, G., 42
Baker, A. J., 103
Bass, B. R., 41, 263, 265, 268, 279
Bathe, Klaus-Jurgen, 102
Beck, J. V., 2, 10, 40, 42, 56, 57, 75, 102, 103, 109, 125, 135, 160, 204, 212-213, 235, 237, 262-264, 279, 301
Becker, E. B., 92, 102
Bell, J. B., 109, 135, 160, 265
Beyer, W. H., 212
Blackwell, B. F., 2, 40, 61, 160, 264-265
Bransier, J., 253-254, 265
Burggraf, O. R., 2, 39, 42, 67, 71, 76, 172, 205-206, 266

Cannon, J. R., 42
Carey, G. F., 102
Carlson, R. D., 41
Carnahan, B., 76
Carslaw, H. S., 33, 42, 60, 62, 75, 83, 102
Chiang, T., 41
Chow, L. C., 2, 40
Cialkowski, M. J., 42
Cobble, M. H., 41

Cozdoba, L. A., 109, 135, 160
Crykowsky, P. G., 109, 135, 160
Curry, D. M., 2, 41

Davies, J. M., 109, 160
deBoor, C., 160, 264
de Vries, G., 92, 102
Douglas, J., 42
Drake, J. B., 263, 265, 274
Draper, N. R., 160
D'Souza, N., 249, 252, 264

Farnia, K., 212
France, D. M., 41
Frank, I., 109, 119, 160

Gilbert, F., 42
Graham, N. Y., 140, 161
Grysa, K., 42
Gupta, B. P., 41

Hadamard, J., 42, 102, 108, 159
Haji-Sheikh, A., 42
Hamming, R. W., 161
Hanson, R. J., 42, 135, 139, 161
Heaton, H. S., 92
Hedrick, R. A., 265
Hensel, E. C., 254, 256, 264, 265
Hills, R. G., 41, 254, 256, 264, 265
Hoerl, A. E., 160
Holman, J. P., 301
Howse, T. K. J., 42
Hsieh, C. K., 268, 274

Imber, M., 2, 67, 76

Jaeger, J. C., 33, 42, 60, 62, 75, 102
John F., 42

Kaminski, H., 42
Keltner, N. R., 102
Kennard, R. W., 160
Kent, R., 42
Khan, J., 2, 41, 67, 76
Kover'yanov, V. A., 67, 71, 76

Langford, D., 2, 41, 67, 71, 76
Lawson, C. L., 42, 135, 139, 160, 161
Lemmon, E. C., 92, 102
Lin, D. Y. T., 42
Litkouhi, B., 42, 102, 212, 237, 264
Luikov, A. V., 80, 102
Lundgren, T. S., 42
Luther, H. A., 76

Mandrel, J., 42
Marquardt, D. W., 135, 160
Minkowycz, W. J., 41
Misepassi, T. J., 2, 40
Morton, K. W., 103, 255, 265
Mulholland, G. P., 2, 41
Murio, D., 42, 109, 135, 160
Muzzy, R. J., 41, 263, 265
Myers, G. E., 60, 76, 80, 102, 103

Nolet, G., 42
Norrie, D. H., 92, 102

Oden, J. T., 102
Olmsted, J. M. H., 160
Ott, L. J., 263, 265, 274
Ozisik, M. N., 80, 102

Patankar, S. V., 103
Payne, L. E., 42
Pletcher, R. H., 103
Priemer, R., 41

Randall, J. D., 41
Rawson, H., 42
Raynaud, M., 253, 254, 265
Reinsch, C. H. J., 161
Richtmyer, R. D., 103, 255, 265
Root, R. E., 41, 265

St. Clair, C. R., Jr., 237, 264
San Martin, R. L., 41
Schisler, I. P., 102
Schneider, P. J., 60, 75
Sgallari, F., 136, 161
Shih, T. M., 103
Shumakov, N. V., 2, 40
Smith, G. D., 102
Snider, D. M., 42, 263, 265
Sparrow, E. M., 42
Stegun, I. A., 42
Stolz, G., 2, 37, 40, 109, 116, 127, 130, 132, 158, 160, 176–181, 225
Su, K. C., 279

Tannehill, J. C., 103
Tikhonov, A. N., 3, 42, 108, 109, 135, 140, 160
Twomey, S., 109, 145, 161

Van Nostrand, R. C., 160
von Rosenberg, D. U., 103

Wardlaw, A. B., 109, 135, 160, 265
Weber, C. F., 110, 160, 252, 256, 264, 266
Westwater, J. W., 42
Widder, D. V., 110, 160
Wilkes, J. O., 76
Williams, S. D., 2, 41
Wolf, H., 237, 264
Woo, K. C., 2, 40

Zienkiewicz, O. C., 92, 102

SUBJECT INDEX

Additive errors, 10
Amplitude gain factor, 174
Arithmetic operations, 226

Backus-Gilbert technique, 3
Backward difference, 54, 55, 95
Backward difference method, 249, 250, 257
Backward heat conduction problem, 49, 50
Backward heat equation, 8, 163, 164
Bias, 10
   deterministic, *see* Deterministic error
Biased estimators, 153
Boiling, 7, 290
Boiling transient, 2
Boundary conditions, numerical:
   convective, 93
   heat flux, 90, 92, 97, 98
Bounded input-bounded output, 177, 215

Calorimeters, 54
Calorimetry, 2
Cauchy problem, 8
CDC 7600, 54, 55, 96
Central difference, 54, 55, 96
Collocation method, 92
Composite body, 6
Computer programs, 262-264
Constant heat flux solutions:
   cylinder, 18
   finite plate, 14-16
   semi-infinite body, 17
   sphere, 18-19
Constant variance error, 11
Contact conductance, 6
Convective boundary conditions, 93

Correlation, 11
Covariance matrix of errors, 145
Crank-Nicolson method, 96, 98, 99-102
Cylinder, 18, 71

Damped least squares, 135
Damping, 13
Derivative, matrix, 246
Deterministic error, 110, 153-154, 157-159, 204-205
Difference equations:
   explicit method, 95
   forward difference, 95
   lumped capacitance, 95
Differences:
   backward, 54-57
   central, 54-57
   forward, 54, 56, 57
Digital filter, 109, 148-153, 156, 196-203
Dirac delta function, 92
Direct heat conduction problems, 6, 78-102
Direct problem:
   difference methods, 87-99
   Duhamel's theorem, 80-87
Direct problems, nonlinear, 79
Distribution, normal, 12
Duhamel's theorem, 2, 37, 38, 59, 60, 79-87, 99-102, 109, 115-116, 125, 127, 146, 148, 165-167, 176, 187, 197-200, 259, 298
   derivation, 80-81
   matrix form, 83
   numerical approximation, 82
   temperature form, 59

305

Elliptic partial differential equations, 187
Endothermic or exothermic reactions, 7
Energy generation estimation, 163
Errors, 9, 55, 99, 154-156, 170-172, 182-184, 191-194, 204-212
Estimation, 2
 ambient temperature, 7, 213-214
 contact conductance, 7
 heat transfer coefficient, 7
 sensor location, 9
 volume energy generation, 7-8
Euler method, 93, 94-98
Evaluation criteria of IHCP methods, 38
Exact matching, 37, 109, 115-118, 133-134, 136, 138, 140, 290-292. *See also* IHCP solution methods
Exact solution(s), 13, 54, 59, 67
 direct problem:
  plate:
   q=C, 14-17, 30-31
   step change surface temperature, 61
  semi-infinite body:
   q=C, 17, 25-30
   $q \approx t^n$, 43, 62
   sinusoidal q, 13-14
 of direct problems:
  Laplace transform, 78
  separation of variables, 78
 IHCP, 51-75
  Burggraf method, 67-75
  cylinders, 71
  lumped body, 54-59
  plane wall, 67-71
  semi-infinite body, 59-61
  spheres, 71
  steady state, 52, 53
 solid cylinder, q=C, 18
 solid sphere, q=C, 18-19
Existence, 110-111
Explicit method, 95

Filter, 75
 digital, 55
 moving average, 75
Filter coefficients, 150-152
Finite control volume (FCV), 37, 87-94, 99-102, 115, 119
Finite difference, 2, 37
Finite element:
 Galerkin method, 91
 weighting function, 91-92
Finite elements, 2, 37, 91-93, 98, 99-102, 113-114

Finite plate, constant heat flux, 14, 30
Forward differences, 54, 55, 87, 93, 94-102, 113-114
Fourier number, 14
Fully implicit, 96
Function estimation, 9
Function specification, digital filter, 197-201
Function specification algorithms, 176
Function specification method, 37, 38, 39, 109, 114-134, 142, 148-149, 152, 167, 176-186, 197-201, 208-210, 222-232, 237-243, 259-261, 270-273, 293-301. *See also* IHCP solution methods
Future temperatures, 30, 37, 57, 181-186. *See also* IHCP solution methods

Gain coefficients, 149, 151, 153, 156, 159, 198
Galerkin method, 91-93
Gauss elimination, 98
Gaussian distribution, *see* Distribution, normal, 12
Gaussian quadrature, 61
Gauss-Markov estimator, 154
Gauss method, 295-296
General integrator, 96
Green's functions, 162, 163

Heat flux impulse test case, 134, 190, 194
Heat flux step, 168
Heat flux weighting factors, 134
Heat transfer coefficient estimation, 7, 281-301
 exact matching, 290-291
 function specification, 293-296, 297-301
 lumped body, 290-296
 regression method, 292
Hyperbolic partial differential equation, 187

IHCP, filter form, 148
IHCP classification, 36-37
IHCP criteria:
 mean squared error, 104, 154-159
 minimum variance, 110
 unbiased, 110
IHCP definition, 1-2
IHCP exact solutions, 53-67
IHCP filter form, 148
IHCP solution methods:
 difference methods, 218-264
 exact matching, 109, 115-117, 140, 176-181

SUBJECT INDEX

finite differences and finite elements, 167
function specification, 109, 119-134, 176-186, 222-242
  alternative interpretation, 133-134, 231-232
  future time steps, 128-133, 181-186, 237-243
  linear connected segments, 131-133, 242, 243
  sequential, 125-132, 271-273, 276-277
  whole domain, 119-124, 233-237
generalized sequential function specification, 147-148
multiple sensors, 118-119, 230-233, 241-242
optimal conditions, 206-212
regularization, 134-135, 186-196, 243-247, 260, 261, 277-279
single future time steps, 222-233
space marching, 247-256
Stolz method, 116
trial function, 109, 145-148, 270
IHCP steady state solution, 52
IHCP temperature errors, 55-56, 63-66
IHCP test cases, 167-176
Ill-posed conditions, 108-114
Ill-posed problems, 2, 3, 6, 39
Impulse test case, 173
IMSL Library, 139, 161
Infinite memory, 24, 28
Integral equations, 87-93, 112-113
Integrated complementary error function, 83
Internal combustion engines, 2
Inverse problem, 6

Lagging, 13
Lagrangian interpolation polynomial, 90
Laplace transform, 2
Least square, generalized, 145
Least squares, 30, 45, 46, 47, 52, 53, 55, 118-119, 121, 123, 124, 126, 128, 142
  damped, 135
  nonlinear, 135
  weighted, 53
Linear, 22
Linear dependence, 24, 28
Linear heat flux, connected segments, 242
Linearity, 20, 21
  sensitivity coefficients, 21
Lumped body, 22, 54-59
Lumped capacitance (LC) approximation, 54, 91-93, 95, 98, 290

Matrix:
  full, 276
  normal equation, 139, 236, 247, 273
  sparse, 276
Matrix inverse, 100
Mean squared error, 104, 109, 110, 154-159, 204-212
Measurement errors, 9, 110, 127
Measurements, difficulty at heated surface, 1
Memory, long, 200
Multiple heat flux estimation, 267-279
Multiple sensors, sequential function specification, 127-129
Multiple temperature sensors, 118-119, 141, 241

Nonlinear, 21
Nonlinear IHCP, 5, 218, 294-296
Normal equations, 139, 234, 236, 247, 273
Nuclear reactors, 2
Numerical procedures, 78, 79, 82, 83, 87, 94-102

Optimal conditions, 203-212
  function specification, 206-209
  whole domain regularization, 210-212
Ordinary differential equations:
  backward difference, implicit method, 95, 96
  central difference, Crank-Nicolson, 96, 98
  Euler method, forward difference, 94, 95, 96
  general integrator, 96-98
  solution techniques, 94-96
Orthogonal polynomials, 46

Parameter estimation, 4, 9
Planar geometries, 67-71, 72-75, 87-102, 110-113, 129-130, 132-133
Plastics, extrusion of, 282
Prefiltering measurements, 153, 164
Prior information, 12, 145
Probability density, 12
Pulse sensitivity coefficients, 220

Quasi-linearization, 218

Radiation, 5, 79
Random errors, 170-172, 182-184
Random numbers, 48
Regularization, 3, 37-39, 109, 134-149, 167, 186-196, 243-247, 259-261, 270, 273-279

Regularization (*Continued*)
  1st order, 136, 137, 139, 187, 260-261
  multiple heat fluxes, 277-279
  2nd order, 137, 139, 243-247, 260-261
  sequential, 109, 141-145, 191-193, 243-247
  whole domain, 109, 137-141, 187-191, 201-203
  zeroth order, 136-138, 147
  see also IHCP solution methods
Regularization parameter, 136
  selection, 140-141
Ridge regression, 109, 135, 136

Sample correlation coefficient, 12
Sample mean, 11
Semi-infinite body, 25, 284-287, 299-300
  constant heat flux, 17, 25, 32
  constant surface temperature, 43
  surface temperature measured, 61
  surface temperature varies as t, 43
Sensitivity coefficient(s), 19-36, 53, 219-233, 238-243, 244, 246, 256-257, 259, 264, 275-276, 283-289
  heat transfer coefficient, 282-289
  linear dependence, 23-24
  linearity, 21
  lumped body, 22
  plate, 14-16, 30, 31
  semi-infinite body, 25
  two dimensional, 31-36
  pulse, matrix, 84, 219-222
  step, 214-222, 283-284
Sensitivity coefficient matrix, 236
Sequential, 24, 37
Sequential estimation method, 37, 104, 109, 115, 119, 125-134, 141-147, 156, 191-196, 259-261, 271-274
Sequential function specification, 276. *See also* IHCP solution methods
Sequential procedure, 109
Sequential regularization, 2nd order, 243
Sequential regularization method, 141, 191, 273, 277
Single future time step method, 115-119
Single interior sensor problem, 6
Single-valued decomposition, 3
Sinusoidal heating, 13
Small time step, 13
Space marching techniques, 247-256
Space program, 2
Spheres, 18-19, 71

Stability, 111, 112, 179-181, 187
  bounded input-bounded output, 177, 215
Standard form of temperature equation, 85, 99-102
Statistical assumptions, standard, 10-13, 204
Stolz method, 2, 37, 109, 116-118, 127, 132, 142, 144, 158, 176-181, 193, 200, 270
Superposition, 27, 48, 49, 80
  errors, 56
Surface heat flux, 5

Taylor series, 151, 224, 238, 242, 244, 246, 295
  matrix form, 236
Temperature equation, standard form, 85
Temperature impulse test case, 184, 194
Tent function, 91
Test cases:
  heat flux impulse, 173-174, 178-179, 184, 190, 194-195
  step heat flux, 168, 177, 181
  temperature impulse, 174, 179-181, 184-186, 191, 194-195
  triangular heat flux, 169-170, 177-178, 182-184, 189-190, 193-194
Thermal properties, 62, 174-175
Thomas algorithm, 98, 225
Trial function method, 109, 141, 145-149, 270
Triangular heat flux, 169, 177-178, 182, 189, 193
Tridiagonal algebraic equations, 97, 98
Tridiagonal matrix algorithm, 225
Turbine blades, 5
Two dimensional heat flow, 31

Uncorrelated errors, 11
Uniqueness, 111

Variance, 11, 110, 153-157, 204-205
  definition, 11
Variance of estimated heat flux, 156, 157

Whole domain estimation, 37, 109, 119-124, 135-141, 145-147, 152, 156, 187-191, 210-212, 233
Whole domain regularization, 137, 187
Whole domain regularization filter, 201

Zero mean errors, 10